Food Flavor and Safety

A C S   S Y M P O S I U M   S E R I E S   **528**

Food Flavor and Safety

Molecular Analysis and Design

A. M. Spanier, EDITOR
U.S. Department of Agriculture,
Agricultural Research Service

H. Okai, EDITOR
Hiroshima University

M. Tamura, EDITOR
Kowa Company, Ltd.

Developed from a symposium sponsored
by the Division of Agricultural and Food Chemistry
at the 203rd National Meeting
of the American Chemical Society,
San Francisco, California,
April 5–10, 1992

American Chemical Society, Washington, DC 1993

Library of Congress Cataloging-in-Publication Data

Food flavor and safety: molecular analysis and design / editors, A. M. Spanier, H. Okai, M. Tamura.

p.   cm.—(ACS symposium series, ISSN 0097–6156; 528)

"Developed from a symposium sponsored by the Division of Agricultural and Food Chemistry of the American Chemical Society at the 203rd Meeting of the American Chemical Society, San Francisco, California, April 5–10, 1992."

Includes bibliographical references and index.

ISBN 0–8412–2665–2

1. Food—Biotechnology—Congresses.

I. Spanier, A. M. (Arthur M.), 1948–   . II. Okai, H. (Hideo), 1938– . III. Tamura, M. (Masahiro), 1956–   .   IV. American Chemical Society. Division of Agricultural and Food Chemistry.   V. American Chemical Society.   Meeting   (203rd: 1992:   San   Francisco,   Calif.) VI. Series.

TP248.65.F66F69   1993
664—dc20
                  93–551
                   CIP

The paper used in this publication meets the minimum requirements of American National Standard for Information Sciences—Permanence of Paper for Printed Library Materials, ANSI Z39.48–1984. ∞

PRINTED IN THE UNITED STATES OF AMERICA

Foreword

THE ACS SYMPOSIUM SERIES was first published in 1974 to provide a mechanism for publishing symposia quickly in book form. The purpose of this series is to publish comprehensive books developed from symposia, which are usually "snapshots in time" of the current research being done on a topic, plus some review material on the topic. For this reason, it is necessary that the papers be published as quickly as possible.

Before a symposium-based book is put under contract, the proposed table of contents is reviewed for appropriateness to the topic and for comprehensiveness of the collection. Some papers are excluded at this point, and others are added to round out the scope of the volume. In addition, a draft of each paper is peer-reviewed prior to final acceptance or rejection. This anonymous review process is supervised by the organizer(s) of the symposium, who become the editor(s) of the book. The authors then revise their papers according to the recommendations of both the reviewers and the editors, prepare camera-ready copy, and submit the final papers to the editors, who check that all necessary revisions have been made.

As a rule, only original research papers and original review papers are included in the volumes. Verbatim reproductions of previously published papers are not accepted.

M. Joan Comstock
Series Editor

Contents

The editors would like
to dedicate this book to
their wives,
Kazue, Mitsuko, and Sharon;
and to their children,
Adam, David, Holly, Jiro, Kenichi,
Rebecca, Sae, Taro, and Yuji,
who have been the major focal point
and inspiration behind this effort.

Preface

THE LONG-TERM GOALS OF FOOD TECHNOLOGISTS and biotechnologists are to design foods based on a desired function, such as enriched or enhanced flavor quality. This ambitious yet obtainable goal requires a complete understanding of the factors involved in the perception, development, and deterioration of flavor in foods. This process involves a thorough understanding of the rules governing food structure and function and the basis for any given reaction mechanism in the food. We are still a long way from completely unlocking the puzzling relationships between the structure and flavor of many foods, but some of the technical achievements over the past few decades have provided the basic machinery to study the complex nature of the problem of food quality. Factors such as lipid oxidation, cooking temperature, methods of handling and storage, microbiological contamination, storage time, Maillard product formation and activity, and food proteins and enzymes all play an important role in the generation and deterioration of the flavor of muscle foods. Thus, when we speak of food we are really speaking of a chemical factory and chemical storeroom.

Conversations among the three of us led to the decision to put together a symposium that would bridge the gap among the many disciplines involved in the study of food and food quality. The chapters in this book expose the reader to information of interest to chemists, biochemists, physiologists, microbiologists, statisticians, and food technologists alike. The chapters address such questions as

- What is food quality?

- How is it measured?

- What are the mechanisms and factors contributing to the generation and loss of food quality?

- What are some production and management strategies used to enhance food quality?

- What are some of the nutrition and health concerns?

- What are some of the safety concerns regarding our foods?

The chapters in this volume cover food quality from conception to consumption. We hope that this volume offers something of immediate and lasting value to all involved in food science.

Acknowledgments

We extend our sincere gratitude to Chris Dionigi, Peter Johnsen, Y. Kawasaki, and Allen J. St. Angelo, who willingly and professionally executed their responsibilities as chairmen of the five symposium sessions. Our thanks and gratitude are extended to the Division of Agricultural and Food Chemistry of the American Chemical Society, to Sanshin Chemical Industry Company, Ltd., of Japan, and to Snow Brand Milk Products Company, Ltd., of Japan, whose financial support of the symposium made it possible to have broad international participation. We also extend our sincere gratitude to the many scientists who reviewed and offered their critical comments on the chapters found herein. Lastly, we acknowledge the work and helpfulness of Barbara Tansill, Meg Marshall, Robin Giroux, and Paula Bérard of the ACS Books Department for the dedicated labor and support provided during the development and assembly of the chapters presented in this volume.

ARTHUR M. SPANIER
U.S. Department of Agriculture
Agricultural Research Service
New Orleans, LA 70124

HIDEO OKAI
Hiroshima University
Hiroshima 724, Japan

MASAHIRO TAMURA
Kowa Research Institute
Ibaraki 305, Japan

February 8, 1993

Introduction

As the world agricultural system moves ever closer to a market-based economy, providers of agricultural commodities must become increasingly attentive to specific consumer demands. No longer are yield and production efficiency the principal determinants of successful farming. Today the consumer expects food to meet an ever-increasing standard of quality.

This perception of quality may include foods that have

- visual appeal,

- high levels of desirable flavors,

- freedom from off-flavors,

- desirable nutritional profiles,

- stability in storage and convenience in preparation, and

- freedom from biotoxins.

While demanding all of these positive attributes, the consumer still insists that food be relatively inexpensive and represent high value in the marketplace.

These additional demands of agricultural and food science require that new, innovative technologies be employed to make the desired improvements to the quality of food. No longer are conventional approaches to plant and animal breeding alone adequate to develop desirable traits. No longer are traditional processing methods economical for the preparation of food ingredients. No longer are standard sanitation and inspection approaches sufficient to ensure a safe and nutritious food supply. Advances in each of these areas have been enhanced by the new molecular biology techniques.

Recognizing these developments, the editors of this volume organized a symposium to discuss recent advances in molecular approaches to the improvement of food quality. A particular strength of the symposium, and of this book, is the international participation and the presentation of a variety of new and exciting developments in food science.

The research presented here describes how molecular processes can be measured and manipulated by a variety of new approaches to improve food quality. The development and formation of flavors in both plant and animal systems is detailed. The use of enzyme modification of materials to improve functional properties of foods and molecular strategies to improve the safety of foods receive attention from a number of leading authorities.

The reader will find timely reviews of a number of important research areas as well as very recent research findings. This volume represents a benchmark for the fields of food science and agriculture focused at the molecular level and designed to address quality issues. If agricultural sciences follow the lead demonstrated in this volume, the quality demands of the consumer can be met.

PETER B. JOHNSEN
U.S. Department of Agriculture
Agricultural Research Service
New Orleans, LA 70124

February 8, 1993

Chapter 1

Molecular Approaches to the Study of Food Quality

A. M. Spanier

Southern Regional Research Center, Agricultural Research Service, U.S. Department of Agriculture, 1100 Robert E. Lee Boulevard, New Orleans, LA 70124

Research in the realm of food quality involves many scientific disciplines from cellular to organ-system physiology, from neural impulse transduction to psychomotor and psychological response, from isolated molecules to their combined biochemical interactions, from mathematics to complex statistical correlations, from negatively impacting microbiological issues to positively impacting microbial safety issues, from simple methods of creating food to complex ones, and from human means of assessing food flavor quality to instrumental methods performing the same objective. For continued growth of our global economy, food scientists need to know the answers to questions such as: What is food flavor and food quality? How are they measured? What mechanisms, such as processing, endogenous factors, outside (external) factors, contribute to the production and loss of food quality? What are some food safety, nutritional and health concerns? What are some of the management and production strategies to enhance food quality? This chapter was written to present the most recent information in response to these questions.

For millennia, man has raised animals for food and has utilized the soil for planting crops for food and fiber. Initially via trial-and-error and later through more deliberate methods of breeding or design, man selected high-yielding varieties of animals and plants, fashioned numerous implements for agricultural and husbandry use, and even developed rudimentary methods of pest control to satisfy his needs. With the advent of the Industrial and Scientific revolutions, scientific methods were applied on a large scale to agriculture in order to improve yields of plant and animal food products. The resulting increase in agricultural productivity was essential to the further development and growth of the industrializing community of man; the massive increase in production and distribution of agricultural commodities led to significant increases in the standard of living in most developing nations and in the United

States in particular. This development is even more important today with the maturing global economy, increased global population, and decreased acreage of arable land.

The science of agriculture is dynamic and ever evolving. It requires not only a significant investment in training of research personnel and research but also in the development of skilled individuals who can conduct and translate research into products that contribute to the well-being of mankind. Today's researcher must be knowledgeable of both scientific and non-scientific factors that impinge upon their research. The latter includes consumer desires, commodity interests, socioeconomic factors, political commitments, agribusinesses, government and regulatory agency requirements, and general public concerns. Today's scientist must solve technical agricultural problems to ensure the continued adequate production of high-quality food and other agricultural products to help meet the nutritional needs of consumers, to sustain a viable food and agricultural economy, and to maintain a quality environmental and natural resource base.

Productive and innovative scientists tend to have a detailed working knowledge of their own field and a fairly broad knowledge of science and technology in general. From this knowledge base there often grows over time, an understanding and appreciation of market needs and requirements essential for significant creative innovation. Scientists often start with an idea that is rather vague and general, and that might languish without special effort. The successful inventor/scientist is the one who molds these general ideas into more concrete form and does the work required to transform the initial concept into a finished final result. A scientist may have knowledge, experience, and commitment, but without some outside assistance, knowledge, or insight, all this information may go unexploited.

Agricultural problems are often multidisciplinary, requiring expertise covering a range of disciplines such as the biological, physiological, psychology, physical sciences, engineering, mathematics, and economics. These problems may seem insurmountable when viewed from the perspective of one discipline but may seen amenable to original or even elegant solutions when viewed from another discipline. Communication is thus essential for this cross-disciplinary interaction so that opportunities for innovation can be recognized and exploited. This symposium was created to give those scientists involved in food quality research some basic appreciation of the disciplines involved in the study of quality. This would, hopefully, not only impart important information to the food quality researcher, but would also present a forum or potential avenue for further communication and exchange of ideas between scientists. Since the list of published books, peer reviewed articles, and symposia in various disciplines in food science are so massive that citation of all of them would fill a volume in itself, all literature citations are omitted from this chapter in an effort to not offend any author or group by the listing of selected examples.

FLAVOR PERCEPTION AND COMPOUNDS AFFECTING FLAVOR

The acceptance of a food in the market place is largely dependent upon the flavor quality of that food. Thus, preparation of food products for the consumer requires the food scientist to have knowledge of sensory evaluation, psychology (of sensory panels), biochemistry (at the receptor level), and physiology (nerve impulse recognition and transduction). The perception of a food's flavor quality depends upon a multifaceted series of sensory responses. These include not only taste and olfactory responses, but also the oral sensations of coolness, astringency and irritation, referred to as chemesthesis and mediated through the trigeminal system (*Chapter 2*). Several of these sensations are the result of changes in pH or ionic strength, reactions with salivary constituents or direct interaction with free nerve endings. However, certain oral responses such as sweet, bitter and coolness, and most olfactory responses involve stimulus recognition through a weak, reversible binding to receptor proteins at the surface of a sensory cell. This initial recognition is followed by transduction events within the cell involving either cyclic AMP or the IP_3 cascade as "second messenger", with coupling through G-proteins. In addition, the availability of flavor stimuli at receptors is dependent on their release in the mouth and on nonspecific parameters such as solubility and volatility. Several chapters in this volume discuss issues critical to our understanding of the recognition/reception process. Labows and Cagan discuss these complex interactions at the cellular and subcellular level (*Chapter 2*) presenting a molecular model for taste perception. Nakamura and Okai (*Chapter 3*) also offer a molecular model for sweet taste perception presenting data suggesting that a single receptor might be responsible for the recognition of both sweet and bitter. Tamura and Okai (*Chapter 12*) add additional evidence in support of this hypothesis. Further discussion is presented by Spanier and Miller (*Chapter 6*) who expand upon the Nakamura and Okai hypothesis and suggest that a single receptor (specific for proteins, peptides and amino acids) is stimulated at specific subsites to elicit the individual sensation of sweet, sour, bitter, or savory (*umami*). Distinction is made between olfaction and taste both of which impact the final sensory perception, but to different degrees (*Chapters 2, 3, 6, 12*).

Human perception of flavor occurs from the combined sensory responses elicited by the proteins, lipids, carbohydrates, and Maillard reaction products in the food. Proteins (*Chapters 6, 10, 11, 12*) and their constituents and sugars (*Chapter 12*) are the primary effects of taste, whereas the lipids (*Chapters 5, 9*) and Maillard products (*Chapter 4*) effect primarily the sense of smell (olfaction). Therefore, when studying a particular food or when designing a new food, it is important to understand the structure-activity relationship of all the variables in the food. To this end, several powerful multivariate statistical techniques have been developed such as factor analysis (*Chapter 6*) and partial least squares regression analysis (*Chapter 7*), to relate a set of independent or "causative" variables to a set of dependent or "effect" variables. Statistical results obtained via these methods are valuable, since they will permit the food scientist to design experiments and then use the molecular descriptors as independent variables to predict the molecular outcome of the experiment. These methods may eventually assist in the understanding of natural flavor and the development of synthetic ones (*Chapter 8*).

It is often too expensive to have or maintain an inhouse descriptive sensory panel. Therefore, other ways of measuring flavor need to be developed. Off-flavor in many foods have been measured by using gas chromatography to assess the level of lipid volatiles associated with off-flavor development (*Chapters 5, 6, 9*) such as hexanal or by direct chemical determination of thiobarbituric acid reactive substances (*Chapters 5, 6*) as a marker of the degree of lipid peroxidation. A new method being tested for use in the assessment of food quality is impedance technology. This method is showing promise for use in the seafood industry (*Chapter 20*).

QUALITY ANALYSIS AND RESEARCH APPLICATIONS TOWARDS PRODUCTION OF QUALITY FOODS

One of the main issues confronting today's food scientist is the development of new products for the market. Today's consumer oriented products must address the consumer's desire and demand for nutritionally sound, highly flavorful, and more natural products. Food scientists must also address the ergonomics of the situation and maintain maximal utilization of food crops within that society while at the same time maintaining capital outlays. These are difficult tasks for today's food and agricultural scientist to meet.

Proteins and their amino acid precursors have, for the most part, been overlooked as a significant source of flavor. The fairly recent discovery of amino acid derivatives such as "aspartame" with significant sweetness potential has made a significant impact on consumer product development. New methods for synthesizing and preparing flavor peptides are appearing in the literature (*Chapter 11*). Furthermore, new uses of several classes of protein derived materials have been described (*Chapter 10*). Modes of utilization of the surplus of high yield, high growth foods such as soy (*Chapters 14, 15, 16*), milk protein (*Chapter 17*), and meat by-products (*Chapter 6*) have been developed in recent years. Not only have the constituents of food been directly utilized for the formulation of new food products but also enzymes and other natural products of food have been shown to be useful in the preparation of both food and non-food items. For example, native soy has been shown to contain at least three lipoxygenase isoenzymes which improve the characteristics of dough and thereby enhance the production of white bread (*Chapter 15*).

The lipid components of food are known to be critical in the development of much of a food's flavor. Modifications to lipid modifying enzymes such as lipases have led to new products useful in the rapid preparation of other food components (*Chapter 13, 14*). Better utilization of lipid constituents in food products can be gained from a better understanding of the thermodynamic and physicochemical characteristics of emulsions. Significant advancement in emulsion chemistry and food engineering have recently appeared in the literature and are an important portion of this volume (*Chapter 19*).

While "natural" is the current catch-phrase of today's consumer, research must still be performed for the development of synthetic compounds that can lower the cost of production of food that can be utilized to develop other less costly food items. Amino acids or proteins with O-aminoacyl sugars as part of their residue have been examined for their taste impact (*Chapter 12*). Several of these components have been shown to be potential replacements for salt (NaCl); this would have a significant impact for individuals with high blood pressure or with a propensity to other coronary or renal problems. Some glycosides, represented by some sucrose esters, are approved by the United States Food and Drug Administration for food use. These glycosides have potential use in the preparation of food materials and can lead to more cost effective means of production (*Chapter 18*).

FOOD MICROBIOLOGY AND SAFETY ISSUES

In addition to nutritional value and flavor, health concerns must also be a part of any discussion of food quality. Among the most interesting of food products with a health related function are the CLAs or conjugated dienoic derivatives of linoleic acid. The CLAs were originally found in meat (beef) extracts and have been shown to be a potent inhibitor of carcinogen-induced neoplasia in the epidermis and forestomach in mice and of the mammary in rats (*Chapter 21*).

The flavor enhancer monosodium glutamate (MSG) is currently used in virtually every type of savory prepared-food. Unfortunately, MSG has several deleterious side effects on a large proportion of the population. Fortunately, a naturally occurring peptide isolated from a muscle food (beef) can serve not only as a potential replacement for MSG but also as a nutritional adjuvant. The peptide, called BMP or beefy meaty peptide, acts as a flavor enhancer and is found to occur naturally in beef (*Chapter 6*). Research on BMP suggest that it is not only non-allergenic but, by virtue of its protein composition, is a nutritionally sound replacement for MSG.

A product of microbial origin that has serious negative implications on consumer health are the aflatoxins. Aflatoxins are carcinogens produced by *Aspergillus flavus* and *Aspergillus parasiticus* when these fungi infect crops before and after harvest, thereby contaminating food and feed and threatening both human and animal health (*Chapter 22*). Traditional control methods (such as the use of certain agricultural practices, pesticides and resistant varieties) that effectively reduce populations of many plant pests in the field, have not been effective in controlling aflatoxin producing fungi. Future research must, therefore, consist of acquiring knowledge of the molecular regulation of aflatoxin formation within the fungus, environmental factors and biocompetitive microbes controlling growth of *A. flavus* and aflatoxin synthesis in crops, and enhancement of host plant resistance against aflatoxin through understanding the biochemistry of host plant resistance responses. Such research is currently underway in several laboratories and should lead to the development of novel biocontrol strategies and/or to the development of elite crop lines that are immune to aflatoxin producing fungi.

After aflatoxin contamination, perhaps the next most important factor that has a negative effect on human health and food quality is the presence of food borne bacteria. Several routes for reduction of the risk are currently under extensive investigation. One such means of risk reduction is the utilization of ionizing radiation treatments on meat food products. Ionizing radiation has been demonstrated to be an effective method to reduce or eliminate several species of food borne human pathogens such as *Salmonella, Campylobacter, Listeria, Trichinella*, and *Yersinia* (*Chapter 23*). If proper processing conditions are used, it is possible to produce high quality, shelf-stable, commercially sterile muscle foods.

A second promising means for control of food borne bacteria is use of antimicrobial proteins. Several organisms produce antimicrobial proteins including bacteria (*bacteriocins*), frogs (*magainins*), insect (*cecropins*), and mammals (*defensins*). The common denominator among these agents is their proteinaceous composition and antimicrobial activity. These agents, therefore, add to the quality of food by virtue of their ability to be a nutritional source of amino acids while concurrently contributing to food quality by its ability to enhance the safety of the food (*Chapter 24*). Production of antimicrobial proteins for food use may be fairly simple when one considers the bacteriocins found in the lactic acid bacteria, since these bacteria are found present naturally in many foods.

Certain microbes that are not generally deleterious to the public's health are found to be a problem due both to their presence in the aquaculture ponds of commercially grown catfish and in drinking water supplies. These microbes include some species of bacteria and several of the blue-green algae. These organisms produce metabolites such as geosmin and 2-methyl-isoborneol, that enter the municipal drinking water supplies and deposit in the fat layers of commercially grown catfish, yielding a product with a muddy or musty odor. Therefore, presence of these metabolites decreases the marketability of those catfish containing these undesirable flavors and decreases consumption of and increases complaints by consumers using the contaminated municipal water supply. Significant progress has been achieved in research efforts designed to develop means to control the growth of these organisms (*Chapter 25*).

CONCLUDING REMARKS

In closing, a very real and important problem exists today because of the phenomenal growth in the data base in science and engineering generated during the last quarter century alone. Today's food scientist needs to have not only a strong background in his/her specialty area but also a very strong appreciation for associated areas. Making the most of agricultural materials and foodstuffs in the coming decades, particularly with the overwhelming demographic projections that will significantly burden the world's food system, will require effective knowledge of all of the areas covered in this symposium and knowledge of several other areas such as biotechnology and animal husbandry that were not covered in this volume. A working knowledge of the improvements in the scientific and engineering database of these technologies along with the ability to utilize this information will have far-reaching effects on societies the world over and will lead to enhancement in the quality of life along with improvements in agriculture, aquaculture, manufacturing, food processing, and the economic health of nations.

RECEIVED February 8, 1993

Flavor Perception and Compounds Affecting Flavor

Chapter 2

Complexity of Flavor Recognition and Transduction

John N. Labows and Robert H. Cagan

Technology Center, Colgate-Palmolive Company,
Piscataway, NJ 08855–1343

Perception of a flavor depends upon a multifaceted series of sensory responses. These include tastes - sweet, sour, salty, bitter and umami, olfactory responses, which involve a virtually unlimited number of descriptors, and the additional oral sensory sensations mediating coolness, astringency and pressure, referred to as chemesthesis and mediated through the trigeminal system. The latter sensations are the result of changes in pH or ionic strength, reactions with salivary constituents or direct interactions with free nerve endings. Certain oral responses, such as sweet, bitter and coolness, and most olfactory responses involve stimulus recognition through a weak, reversible binding to receptor proteins at the surface of a sensory cell. This initial recognition is followed by transduction events within the cell involving either cyclic AMP or the IP_3 cascade as "second messengers" with coupling through G-proteins. In addition, the availability of flavor stimuli at receptors is dependent on their release in the mouth and on nonspecific parameters such as solubility and volatility.

Perception of a flavor in a food or oral care product is the result of multiple sensory responses. These include sweet, sour, salty and bitter tastes and olfactory stimulation through the retro-nasal movement of odors, which are defined by a virtually unlimited number of descriptors. Additional oral sensations provide information on temperature, irritation, astringency and pressure. Scientists are learning how these individual sensations interact to provide the complexity of flavor perception. At the molecular level, studies are revealing how the stimulus triggers chemical changes which result in communications between the sensory receptor cells and the brain. The underlying mechanisms involved are described below.

Reviews of taste sensations normally concentrate on four 'basic' tastes - sweet, salty, sour and bitter (1,2); however, other oral sensations can contribute important information to the perceived flavor (3). Examples of stimulants evoking these very different sensory sensations are shown in **TABLE I**. Studies on the mechanisms of perception are usually restricted to sensation-specific stimuli; however, food flavors represent an interaction among the various sensations. This chapter describes recent

0097–6156/93/0528–0010$06.00/0

studies related to this complex of flavor sensations based on the underlying mechanisms of recognition and transduction.

TABLE I. Examples of Stimulants Exhibiting Different Taste/Olfactory Sensations

ORAL SENSATIONS	*STIMULANT*	*SENSORY MECHANISM*
salt	NaCl	ionic differences at membrane
sour	citric acid	pH, anion effects on membrane
bitter	quinine	membrane penetration
	phenylthiourea	ion channel protein binding
sweet	sucrose	specific protein binding
cooling	menthol	ion channel protein binding
astringency	tannic acid	protein complexation
irritation	capsaicin	generalized membrane interaction
NASAL SENSATIONS		
odor quality	various	specific protein binding
burning	ammonia	trigeminal
freshness	menthol	generalized binding
irritation	chlorine	chemical
tingling	carbon dioxide	pressure receptors
stinging	acetic acid	trigeminal

TASTE AND ODOR SENSATIONS BASED ON SPECIFIC BINDING

The sweet taste and olfactory responses to a variety of stimuli are examples of chemical senses that utilize protein receptors for initial detection of the stimulus. Most bitter compounds have a hydrophobic component which enables their direct interaction with the cell membrane; however, some evidence suggests a protein receptor mechanism. The cooling sensation is treated as a chemesthetic sense, where stimulation takes place at the basal membrane. However, compounds that evoke this response have very specific structural limitations, and most are related to menthol. For purposes of discussion, bitter and cooling sensations will be discussed under generalized receptor mechanisms.

Sweet Taste. The mechanism of sweetness perception has been extensively studied because of its commercial importance. Many substances that vary in chemical structure have been discovered which are similar to the taste of sucrose. Commercial sweeteners include sucralose, acesulfame-K, saccharin, aspartame, cyclamate (Canada) and the protein thaumatin (4). Each sweetener is unique in its perceived sensation because of the time to the onset of sweetness and to maximum sweetness, ability to mask other sensations, persistence, aftertaste and intensity relative to sucrose [TABLE II]. For example, the saccharides, sorbitol and

TABLE II. Comparison of Potency of Common Sweeteners

SWEETENER	*RELATIVE SWEETNESS*	*SWEETENER*	*RELATIVE*
sucrose	1	sucralose	600
xylitol	1	hernandulcin	1000
aspartame	140	monellin	1500
acesulfame-K	140	alitame	2000
saccharin	300	thaumatin	2000

aspartame have shorter times for onset of sweetness than do saccharin and cyclamate (5).

The first structural model for sweet compounds included hydrogen bonding donor (A-H) and acceptor (B) sites at a distance of 2.5-4.0Å; this was extended to provide for a hydrophobic site 3.5Å from the A-H site (or as much as 10.8Å) and 5.5Å from the B site. Two amino acids, phenylalanine and aspartic acid, are combined in the sweetener aspartame. According to this model, the acid group acts as a proton acceptor site (A-H), the ammonium ion as a proton donor site (B), and the alkyl side chains provide the hydrophobic interaction (X). In alitame, the hydrophobic site is represented by the alkyl substituted thietane ring, while in sucralose it is a chlorine atom [FIGURE 1]. Other proposed interaction sites [FIGURE 2] involve a cyanide or nitro group (D) when present and an additional electrophilic carbonyl or sulfonyl group (E) (6). Superaspartame (14,000x sucrose) or sucrononic acid (200,000x), with these features present in the correct geometry, exhibit a very high sweetness potency. L-Amino acids are in general sweet while D-amino acids are bitter or at best bitter-sweet. The sweetness potential of amino acids is also reflected in the protein sweeteners, monellin and thaumatin. Antibodies developed against one of the proteins crossreacts with the other protein; the spacing of tyrosine and aspartate amino acids in monellin is similar to that of phenylalanine and aspartate in thaumatin (7,8).

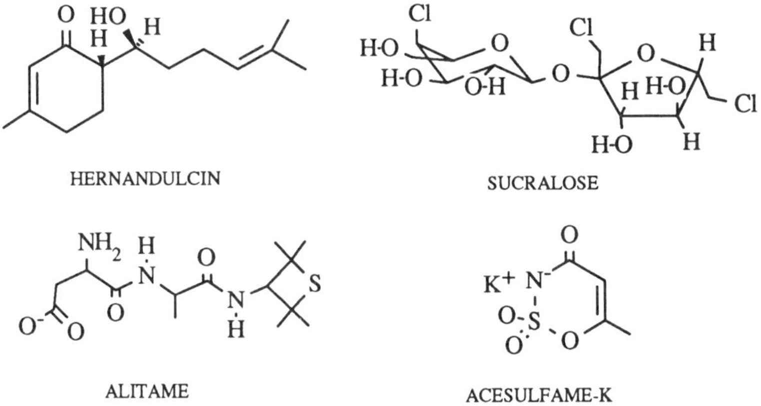

HERNANDULCIN   SUCRALOSE

ALITAME   ACESULFAME-K

FIGURE 1. Chemical structures of potent sweeteners.

Many analogs have been synthesized that incorporate different functional groups arranged to fit the AH, B, X model and they exhibit some degree of sweetness (4). However, sweet compounds have been found that do not fit this concept and bitter compounds that do. The failure of sweeteners to cross-adapt suggests multiple receptors. Multiple receptor sites are also supported by animal studies which show species differences in nerve responses to sweeteners (9). Computerized molecular modeling techniques have been used to develop a more detailed receptor model, which includes electronic und steric properties, for dipeptides and guanidine containing compounds (10,11).

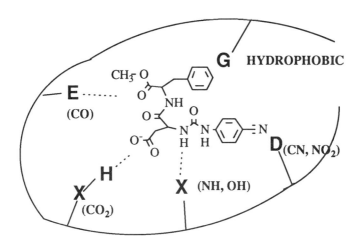

FIGURE 2. Receptor site interactions for maximum sweetness for Superaspartame [Adapted from ref. 6].

Olfaction. Scientists have been very successful in characterizing the chemicals which contribute unique flavors to food materials (*12*). Several impact chemicals have very low thresholds, e.g. geosmin - earthy, 0.5ppt; isobutylmethoxypyrazine - bell pepper, 2ppt; thiaterpineol - grapefruit, 0.1ppt; while others are very characteristic, e.g. phenylhydroxybutanone - raspberry, methylsalicylate - wintergreen.

Recent studies have identified the involvement of specific proteins in the mechanism of olfactory perception. The unique three-dimensional structure of an odorant molecule is translated, via receptor binding and nerve transduction, into a characteristic odor quality. Soluble odorant binding proteins have been characterized in the olfactory mucosa which act to bind chemical stimuli (flavors); it is hypothesized that they transport the molecules to the receptors on the cilia of the olfactory neurons (*13*). The low threshold/high affinity of isobutylmethoxypyrazine was used to study receptor binding. The "pyrazine binding protein" was characterized and later shown to belong to this class of hydrophobic ligand carrier proteins which are not specific for particular odorants. The binding correlates with human detection thresholds for a series of pyrazine analogs, but the protein demonstrates binding with odorants from several chemical classes and odor characters (*14*). Olfactomedin is an olfactory tissue-specific secretory glycoprotein deposited at the chemosensory surface with a possible similar role in olfaction (*15*). Thus the proteins which have been characterized appear to be involved in odor transport to the receptors in the mucosa rather than being the receptors themselves. Proteins, through their role as enzymes, also are involved in inactivating odorants through chemical degradation and/or transformation to either an odor inactive or reduced form.

A very specific relationship exists between chemical structure and odor quality; this has been defined for a number of odorant classes, for example steroids and musk compounds (*16*). The chirality as well as the flexibility of a molecule is also important in defining the odor. A structure that can assume different conformations could interact with more than one receptor type and elicit more than one odor quality. Thus one typically observes several odor descriptors used for compounds that are conformationally mobile, while those having fixed structures, such as polycyclic steroids or aromatics, have fewer descriptors (*17*).

A large number of volatile chemicals, [many from essential oils and present at low concentration], interacting with the olfactory receptors are essential for the full impact and character of a flavor. The appeal of wines and of foods critically depends upon this complexity, which expresses the ability of the flavorist to appeal optimally to the individual's sensory and perceptual abilities.

TASTE SENSATIONS BASED ON GENERALIZED MEMBRANE RESPONSES

Sour and salty. Within the membrane of the taste cell are ion channels which control the movement of ions, such as sodium, potassium and calcium, into and out of the cell. Sour taste sensations are in part due to the effect of hydrogen ions; however, some taste is also a function of the hydrophobicity of the organic acid, such as citric acid (*18*). Acids can produce a decrease in potassium ion conductance (depolarization) in the membrane.

The salty taste is primarily due to sodium ions acting directly on ion channels. Amiloride specifically blocks sodium channels; however, it does not block all responses to salt, indicating more than one mechanism for salty sensation. A different compound, 4-aminopyridine, blocks potassium channels but not sodium. This suggests that receptor proteins and second messengers are not required, and that these stimuli act directly on ion membrane channels. The physiology of the response of cells to salt has been reviewed (*1*).

Bitter. The unpleasant bitter taste may act as a warning system to prevent ingestion of harmful chemicals. Because of the wide variety of chemical structures that are bitter-tasting (*19*), e.g. caffeine, K^+, peptides, denatonium benzoate, the perception of bitter may involve several mechanisms [**FIGURE 3**]. Three mechanisms have been proposed by which these stimulants could affect a potassium ion channel and thus change the membrane potential: a) stimuli may bind to a bitter receptor protein in the membrane and induce the GTP sequence described below; b) stimuli such as small peptides may traverse the membrane and directly induce a change intra-cellularly; and c) the addition of K^+ ions would affect the extracellular concentration and thus block the potassium channel. Additionally, protein complexes, which span the membrane, can create a pore through which ions flow. This pore is normally closed; however an appropriate stimulus, potentially a bitter or cool stimulus, could interact with the protein causing a conformational change and opening of the channel (*20*).

Many bitter compounds contain both hydrophobic and hydrophilic sites which can alter cell membranes through penetration. There is a correlation between bitter intensity and hydrophobicity-solubility indexes such as the octanol/water partition coefficient, logP (*1*). Penetration may directly affect cAMP phosphodiesterase as part of the transduction process (see below). A bitter receptor protein may be involved with certain bitters, such as specific structural requirements with the bitter tasting dipeptides and denatonium salts (*21*). The latter is used in some consumer products to avoid accidental ingestion. A receptor mechanism is also supported by the existence of a genetic "taste blindness" for some bitter materials (see below).

Bitter taste can be masked by sweeteners, by salt or by dipeptides containing aspartic or glutamic acids (*22,23,24,25*). The bitter-masking potential of sugars with quinine was recently assessed, and quinine-equivalent values were derived to predict masking ability of these substances. Attempts to mask bitter taste may be successful only with certain bitter substances.

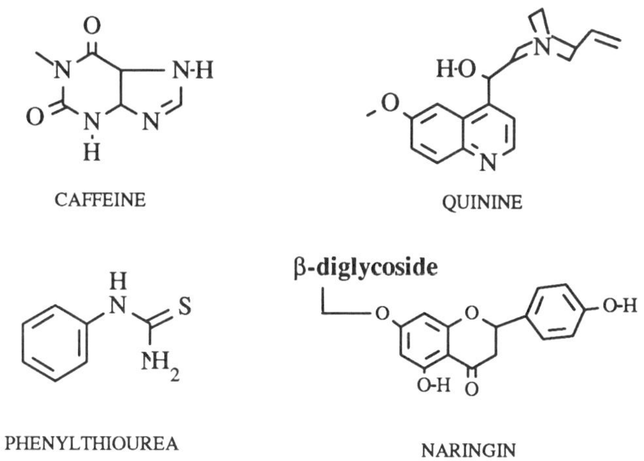

FIGURE 3. Chemical structures of compounds exhibiting bitter taste.

Chemesthesis. The term chemesthesis has been introduced to classify thermal and painful sensations experienced in the mouth (*26*). Chemesthesis refers to a chemical sensibility (mouthfeel) in which certain chemicals directly activate nerve fibers at the level of the basal membrane in the mouth. The sensations are analogous to similar effects at the skin surface where there is a close anatomical and functional relationship. Sensations include the "hot" of capsaicin and piperine, which are active components of chili and pepper, the coolness of menthol and the irritation of chemicals such as salt at high concentrations [**FIGURE 4**]. Some of the descriptive terms used to make qualitative distinctions in food sensations include pungency, freshness, tingling, burning and sharpness.

Preexposure to menthol and capsaicin can desensitize the response to other stimuli, i.e. the thermal response to a warm solution or the irritation upon a second application of capsaicin (*26,27*). After capsaicin desensitization, the nerves no longer respond to 'burning' or 'stinging' but still give sensations of numbness or warmth. This effect is most noticeable upon further stimulation with capsaicin and less for the irritants, cinnamic aldehyde and NaCl. This sensitization and the fact that these sensations build in intensity on repeat application differentiate them from the "basic tastes", which adapt to the sensation on repeated application.

Sensations of irritation, pain and temperature, commonly referred to as trigeminal responses, are also experienced in the nose. Typically strong stimulants include CO_2, menthol, ammonia and acids; however, most odorants do have some chemesthetic component. They have shorter response latencies than in the oral cavity, consistent with a shorter distance required to penetrate to the basal layer containing the nerve endings; the stimulants also increase in intensity with repeated exposures (sniffs) (*28*). Irritants reduce the intensity of odorants; the reverse is also true. The interaction is at the central processing level because an odorant presented in one nostril will be affected by an irritant given in the other nostril.

Cooling. The cooling sensation of menthol-related compounds is different from simple evaporative cooling or cooling from dissolution, as experienced with xylitol, and has a very specific structure-activity relationship. Menthol is the primary cooling agent used in commercial products, for its unique flavor sensation either alone or in combination. It is a major constituent of peppermint (45%; menthone, 20%). Other mint flavorants also provide some cooling, i.e. carvone, menthone. Many compounds that are structural analogs or derivatives of menthol have been synthesized, and these elicit varying degrees of coolness; two carboxamide derivatives have been utilized as nonvolatile cooling agents for skin and oral care products [**FIGURE 5**] (29).

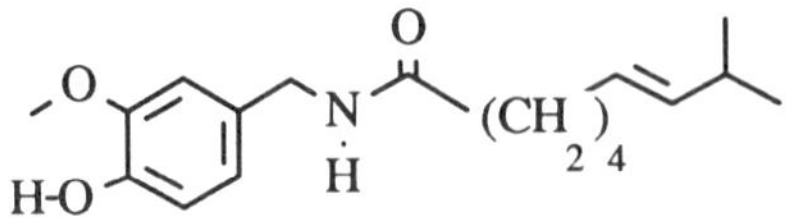

CAPSAICIN 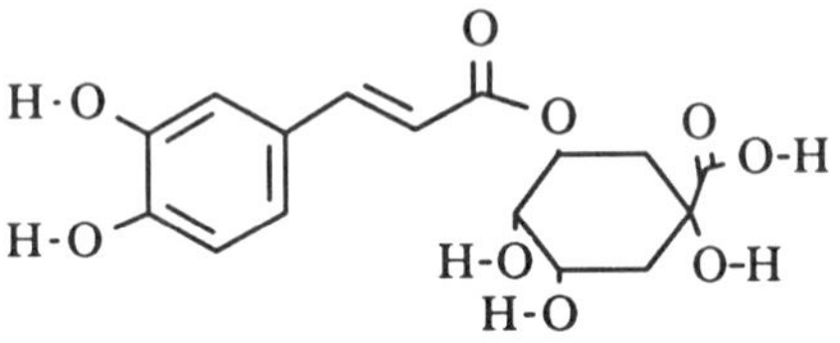   CHLOROGENIC ACID

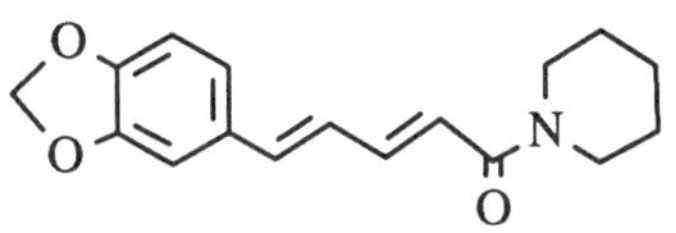

PIPERINE 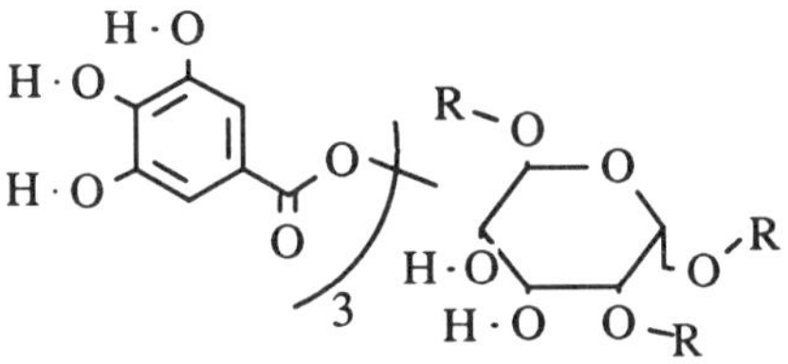   TANNIC ACID

FIGURE 4. Chemicals eliciting irritation and astringency.

Two additional menthyl derivatives, 3-L-menthyloxypropane-1,2-diol (MPD) and menthyl-lactate (Frescolat), and a series of nitroaromatic compounds were developed for this purpose [**FIGURE 5**] (30,31,32). A longer-lasting cooling sensation is claimed for a β-amino-acid/menthol ester, which is initially absorbed in the mouth and slowly hydrolyzes to release menthol (33). The cooling compounds have a hydrogen bonding site, a compact hydrophobic region and a molecular weight between 150 and 350. The fact that the cooling effect can be related to structural parameters suggests that this phenomenon is due to a specific interaction with a receptor protein. Menthol is known to affect Ca^{++} conductance of gut smooth muscle and to sensitize calcium-dependent inactivation of *Helix* neurons (34); in the mouth it may also interact with a membrane protein which directly gates an ion channel.

The N-alkylcarboxamides, originally synthesized for potential cooling activity, were analyzed by molecular modeling (29) (J. Brahms, unpublished studies). The analysis used calculated electronic and steric structural properties, i.e. molecular volume, dipole moment, and logP, to correlate the reported oral threshold values with chemical structure resulting in a predictive model [**FIGURE 6**]. The thresholds of the carboxamide coolants were consistent with this model.

Astringency. The mouthfeel induced by astringents, such as tannic acids, is described as a drying or puckering sensation in the mouth with additional numbing, burning and tingling. Astringent materials include tannins, aldehydes, ethanol or metals, i.e. Zn and Al (35,36). Acidulants, acids which enhance and modify food flavor, evoke strong astringent as well as sour and bitter reponses (37). The tannins, dihydroxy benzene derivatives [**FIGURE 4**], and divalent metal salts can effectively complex with the basic proteins (high content of proline) in saliva and change their physical properties. In addition, there may be direct interaction of astringents with the oral mucosa. This tactile effect, which translates to a drying sensation, may be mediated by mechanoreceptors. Alternatively, studies in rodents suggest that astringency is a taste quality, since a rapid and reversible response was found in the chorda tympani nerve (as opposed to the lingual nerve for mechanoreceptors). This direct response of a taste cell may result from alteration of a membrane protein that is associated with ion transport (*38*). One area of the tongue, the circumvallate papilla, was more sensitive to astringent stimuli. For bitter and astringent compounds found in wine, individuals with low saliva flow rated astringent compounds as more intense and they showed a longer time to reach maximum intensity and a longer persistence of the response.

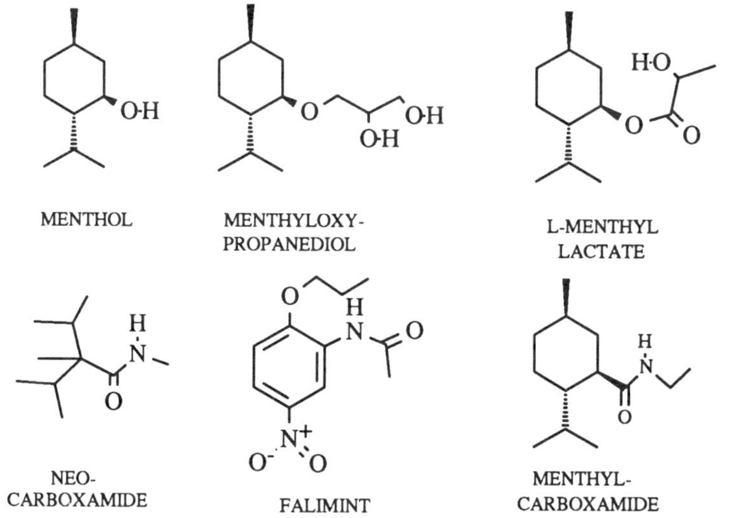

MENTHOL

MENTHYLOXY-
PROPANEDIOL

L-MENTHYL
LACTATE

NEO-
CARBOXAMIDE

FALIMINT

MENTHYL-
CARBOXAMIDE

FIGURE 5. Chemicals eliciting a cooling sensation.

Alum has been suggested as a standard to assess astringency sensitivity in time-intensity studies (*35*) and with dentifrice panels (*39*) because it is primarily an astringent, while tannic acid evokes both astringent and bitter reponses. Bitter and astringent responses can be differentiated by the differences in response during repeated applications of a stimulus. The bitter intensity response plateaus over repeated applications, while the astringency response, like irritation, increases with the number of sequential applications of the astringent, i.e. tannic acid and chlorogenic acids [**FIGURE 7**] (*40,41*). Sweeteners were found to be effective in reducing astringency that was created by repeat applications of tannic acid.

TASTE ENHANCERS

Salt is the best known taste enhancer for a variety of foods. Monosodium glutamate (MSG) and nucleotides, such as inosine monophosphate (IMP) and guanosine monophosphate (GMP), are known to enhance flavor and are recognized as the "umami taste" in Oriental cuisine. They have longer aftertastes than the "basic

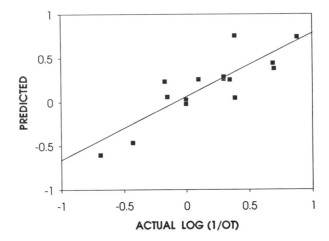

FIGURE 6. Correlation of experimental oral threshold values of alkyl carboxamides with the calculated values based on modeling of electronic and steric properties of the molecules [Adapted from J. Brahms unpublished studies].

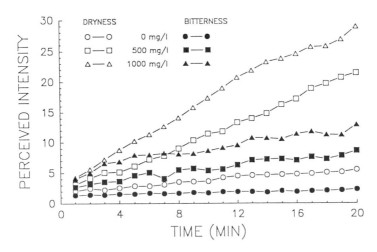

FIGURE 7. Bitter response plateaus while astringency (drying) response continually increases on repeated exposure to tannic acid. [Reproduced from ref. 40 by permission of Oxford University Press]

tastes" *(42)*. In studies published a decade ago, direct binding of tritiated glutamate to bovine taste tissue preparations was demonstrated *(43)*. A marked increase in binding of glutamate was found in the presence of certain 5'-ribonucleotides such as GMP. An allosteric effect was suggested with the nucleotide acting as an effector that enhanced the amount of binding of glutamate *(44)*.

INDIVIDUAL DIFFERENCES IN SENSITIVITY

Taste. Of the fundamental tastes, bitter is unique in showing human genetic differences in sensitivity. Six decades ago, it was reported that phenylthiocarbamide (PTC) tasted extremely bitter to some individuals while being almost tasteless to others *(45)*. The ability to taste PTC was found to be a dominant genetic trait which occurs across gender, age and culture, with 70% of the American population carrying the dominant trait *(46)*. Sensitivity to PTC and propylthiouracil (PROP) are correlated with sensitivities to other bitter tasting compounds, such as caffeine, saccharin (after-taste) and salts of potassium cations and benzoate anions *(47,48,49,50)*. However, in a reexamination of the sensitivity to NaCl and KCl, no differences were found between tasters and nontasters to non-PTC type compounds, and the statistical methods that showed differences were questioned *(51)*. Individuals who do not respond to PTC are not necessarily insensitive to quinine, another intensely bitter compound *(49,50,52)*.

Individuals could be divided into super-, medium- and non-sensitive tasters to PTC/PROP bitter stimuli. Using these classifications on published data allowed a clearer separation of response differences to various bitter stimuli. These now include the casein or hydrolysis products in dry milk powder and Ca^{++} ions in certain cheeses *(53)*. In additional studies, there was a difference in intensity response to all sweeteners except aspartame and dihydrochalcone. With respect to sweetness, nontasters of bitter were "sweet likers", who tasted a purer sweet and liked increasing concentrations of sucrose. Bitter tasters, "sweet dislikers", tended not to like increasing concentrations of sucrose and reported other taste qualities in the solutions *(54)*. Trigeminal stimuli, such as the burning of capsaicin, are more intense in bitter sensitive individuals.

A panel of tasters and nontasters of PTC was formed to evaluate the effectiveness of certain flavors in masking the bitterness of tartar control mouthrinse and dentifrice formulations *(55)*. Panelists used magnitude estimation to scale bitterness and other sensory attributes, particularly astringency and cooling. The results showed that tasters gave higher bitter responses than nontasters to dentifrices and mouthrinses that contained the antitartar ingredient potassium pyrophosphate [FIGURE 8].

None of the studies of cooling or astringent sensations have tested individual sensitivities nor have any past PTC studies examined the perception of cooling and astringency. We have observed differential effects to these responses linked to PTC sensitivity. PTC tasters gave higher responses to the astringency of alcohol in mouthrinses and to the cooling sensation due to menthol in dentifrices [FIGURE 9]. These differences may reflect either a specific deficit or an overall insensitivity of these individuals. Thus, screening for taste acuity for the fundamental tastes, for astringents and for specific flavorants, such as menthol, is necessary for the integrity of oral panels *(39)*.

Odor. Olfaction is an important part of the flavor sensation which allows the individual to distinguish its quality, i.e. raspberry vs lemon. However, not every individual experiences the same olfactory sensations due to specific anosmias, the olfactory equivalent of color-blindness. Specific anosmia occurs when an individual

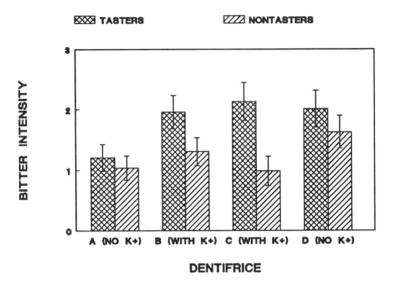

FIGURE 8.   Response differences of PTC tasters and nontasters to bitter taste in dentifrices with and without K^+ ions.   Differences are significant only for dentifrices with K^+ [Adapted from ref. 55].

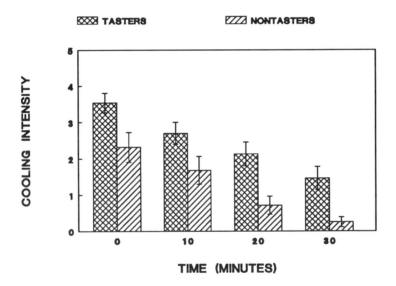

FIGURE 9.   Response differences of PTC tasters and nontasters to cooling response in tartar control mouthrinse over time after use [Adapted from ref. 55].

is unable to detect a particular odorant at a concentration normally detectable by a majority of the population (*56*). Over 12 such anosmias have been characterized and include the odor descriptors sweaty, musky, malty, fishy, minty and camphoraceous; the most notable is the boar taint odor (urinous) of androstenone, which affects ~45% of the population (*57,58*). Within each group are specific chemicals that exhibit the strongest effect, i.e. 1,8-cineole (eucalyptus oil) for camphoraceous, isobutyr-aldehyde for malty and L-carvone (spearmint oil) for minty [**TABLE III**]. Other molecules found within the groups are related on the basis of electronic and steric features. The 'soo-odor' of mutton, 4-ethyl-octanoic acid, has the lowest threshold for an organic acid (1.8ppb), and is not smelled by some individuals (16%) (*59*).

TABLE III. Specific Anosmias to Flavor Chemicals

PRIMARY ODORANT	QUALITY	ANOSMIA %	REFERENCE
isobutyraldehyde	malty	36	56
1,8-cineole	camphor	33	56
L-carvone	spearmint	8	56
ethyloctanoic acid	mutton	16	59
androstenone	boar taint	47	57

For most specific anosmias the individual differences represent a concentration effect; that is anosmic individuals will recognize the odor at some higher level of exposure.   However, in the case of androstenone, individuals find it either unpleasant (urinous, sweaty), pleasant (floral, musky) or odorless, independent of the concentration.   Surprisingly, the anosmia can be reversed in some individuals through constant exposure (*57*). Anosmic subjects exposed to androstenone three times per day (3 minutes) over a 6 week period became osmic to it. A substantial decrease in threshold (i.e. increased sensitivity) was found for 10 of 20 subjects.

It has been suggested that these anosmias refer to a primary odor classification. These olfactory sensitivity differences could potentially contribute significant differences in flavor perception among individuals for a given product.

TRANSDUCTION:

The taste cells are situated in the lingual epithelium with the apical membrane exposed to the mucosal surface of the oral cavity and the basal surface in contact with the nerve [interstitial fluid] [**FIGURE 10**]. Within the basolateral surface are the nerves which respond to the chemesthetic stimulants, i.e. direct nerve stimulation. The microvilli at the apical membrane contain receptor proteins which respond to sweeteners, some bitters and possibly coolants. The olfactory cells are bipolar neurons with dendritic ends containing cilia exposed to the surface and axons linked to the brain, where they synapse in the olfactory bulb. The transfer of information from this initial stimulus-receptor interaction to the brain processing centers involves chemical transduction steps in the membrane and within the receptor cells. The potential chemical interactions at the cell membrane and within the cell are schematically outlined in **FIGURE 10**.

An important goal toward understanding sensation at the molecular level is the characterization of receptor proteins which bind a specific odorant or tastant.   Two approaches have been taken: direct binding studies of stimulant with a receptor preparation, and identification of genes that code for receptor proteins. The weak reversible interactions make it difficult to stabilize such a complex for isolation and

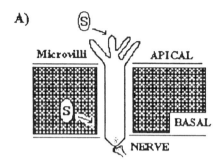

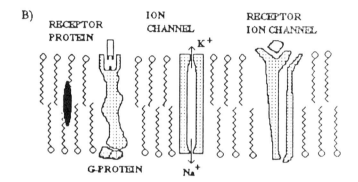

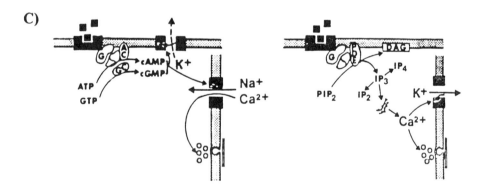

FIGURE 10. A) Schematic of taste cell embedded in epithelium. Shown are the apical membrane which contains the microvilli and the basal membrane which synapses with the nerve; B) Stimulus can interact with a receptor protein embedded in membrane which activates G-protein, with the membrane or directly with an ion channel protein; C) GTP stimulation of phosphatase for formation of cAMP or IP_3 and opening of ion channel (Adapted from ref. 2).

characterization. The first studies on sensory receptors involved the isolation of either the taste membrane or olfactory cilia with functional activity, i.e., the ability to bind relevant stimuli. The channel catfish provides an externalized taste system and a relatively high sensitivity to amino acid stimulants. L-Alanine was shown to bind to a sedimentable fraction from the taste epithelium (*60*). Further binding studies identified several classes of sites (*61*). Electrophysiology and biochemistry suggested different receptor sites for the L- and D- enantiomers of alanine. Studies with mixtures of L- and D-alanine, both of which are taste stimuli to catfish, used both single-label and double-label experiments and electrophysiology to measure responsiveness, cross-adaptation and binding of D- and L-enantiomers to a membrane fraction from catfish taste tissue. The results showed that part of the response occurs through a receptor/transduction process common to both enantiomers, but that an independent process also operates for D-alanine alone (*62*). An L-proline receptor has also been characterized in this taste system (*63*).

This strategy was utilized in the 1970s to isolate olfactory cilia (*64*). Later the technique was utilized to prepare olfactory cilia from frog, where it was demonstrated that GTP was required for stimulation of adenylate cyclase by odorants. Four chemicals, citral, L-carvone, 1,8-cineole and amyl acetate, were used either individually or in a mixture, and adenylate cyclase was measured (*65*). Maximal stimulation by the mixture of 4 odorants was 2.3-fold above the GTP-stimulated value. In further studies a total of 65 odorants were tested individually with frog and rat cilia (*66*). These results indicated that only about a third of the odorants tested stimulated adenylate cyclase by meaningful amounts, with 6 stimulating by 45-65%.

Recently, studies were reported measuring the kinetics of stimulation of both cAMP and IP_3 using a mixing device and rapid quenching, in the millisecond range, in rat olfactory cilia (*67*). The response to a mixture of odorants peaks within 25 to 50 milliseconds, the time frame expected for receptors, with both cAMP and IP_3. Similar measurements of the change in concentration of cAMP or IP_3 were also done in the taste cell. Here mice, which were bred as bitter 'tasters' and 'non-tasters', were used as subjects. The bitter stimuli, sucrose octaacetate, strychnine and denatonium benzoate, were shown to increase IP_3 levels in a membrane preparation from "taster" mice in the presence of GTP-protein and Ca^{++} but not in membranes from "nontaster" mice (*68*).

The genetic approach has recently been successful in identification of a multi-gene family which codes for proteins whose expression is restricted to the olfactory epithelium and which may represent the initial odor receptors. With RNA from rat olfactory epithelium and the polymerase chain reaction (PCR), Buck and Axel amplified homologs of the relevant gene superfamily (*69*). They cloned and characterized 18 different genes of this family. These proteins have seven-transmembranc domains which embed them in the cell membrane. The proteins contain an external sequence for stimulus binding and an internal segment linked to the G-protein. There are three subfamilies with a high degree of amino acid sequence homology which may code for specific odor classes, with the differences matched for small changes in chemical structure and odor quality. The multigene family is extremely large, probably several hundred, in line with the wide variety of known odorants.

The picture that has emerged from these studies is of an initial interaction of a stimulus with a matched portion of a receptor protein embedded in the cell membrane (*13,65*). This initial interaction causes stimulation of the linked G-protein to form cGMP. This is coupled to the reactivity of adenylate cyclase in the cells, leading to increased levels of cAMP, which opens ion channels in the cell membrane. A similar sequence can alternatively activate inositol phosphate as a second messenger. Either odorants, cAMP or cGMP can cause a potential change in the membrane (*13,70,71,72*). As in hormone-sensitive and neurotransmitter-

-sensitive systems, adenylate cyclase in olfactory cells depends upon GTP and G-proteins.

It is apparent that multiple recognition mechanisms are active in taste and odor recognition. There is now emerging evidence of multiple receptors and transduction mechanisms. The complex olfactory component of a flavor requires a mechanism with multiple receptors and several transduction pathways to be sufficiently sensitive and discriminating to distinguish among closely related materials.

FLAVOR RELEASE

In foods, flavors are complexed within the food matrix and partitioned between the oil and water phases. In the mouth the food is broken down through mastication, followed by diffusion and/or partitioning into the saliva. This dilution can lead to differential release of the flavor materials depending on their solubility in the salivary matrix components. Further partitioning occurs into the headspace in the mouth, which conveys the volatile flavorants into the nose. These processes have been recently reviewed (73). Because of these influences, there can be an order of magnitude difference between the threshold of a pure odorant and its threshold in a food sample. The interaction with the food matrix can be an emulsion (oil in water or water in oil), solubilization in the lipids (fats), or binding to proteins and carbohydrates. Headspace analysis techniques have been useful in demonstrating this complexing between flavorants and these macromolecules (74,75). As their concentration increases the flavorants decrease differentially in the headspace. For volatile flavorants this release is crucial to retro-nasal stimulation of the olfactory epithelium thereby affecting flavor perception.

Surfactants can act like lipids or emulsifiers in solubilizing flavor materials in surfactant micelles. Headspace analysis techniques were used to follow the release of several common dentifrice flavorants from a solution containing the surfactant sodium lauryl sulfate. Water/micelle partition coefficients were derived to describe the solubilization of the flavorants in the surfactant micelle (76). Initially, the flavor is solubilized in the surfactant micelle. As both the micelle and flavor concentration decrease on dilution, flavor compounds, which are highly soluble in the micelle, preferentially increase in the headspace [**FIGURE 11**].

SUMMARY

Specific responses - to sweet, some bitters and possibly coolness - and to most olfactory responses involve stimulus recognition through a weak, reversible binding to membrane bound receptor proteins at the surfaces of sensory cells. This initial recognition is followed by transduction events within the cell involving G-proteins coupled with "second messengers", either cyclic adenine monophosphate (cAMP) or inositol triphosphate (IP$_3$). The other sensations are the result of changes in pH or ionic strength, reactions with salivary constituents or direct interaction with the membrane, which initiate changes in cell membrane permeability to ions. In addition, the availability of flavor stimuli at receptors is dependent on their release while in the mouth and on nonspecific parameters such as solubility and volatility.

Individuals may experience a broad array of flavor sensations within one meal or even one food. The interactions among the sensations makes it difficult to separate and quantify individual contributions to the flavor experience. An individual stimulus may elicit more than one sensation, and be processed at the recognition and transduction steps by more than one mechanism. In addition, individual differences in sensory responses may provide different assessments of the same event.

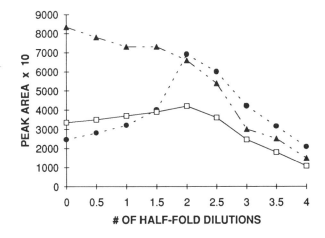

FIGURE 11. Headspace analysis of solutions of menthol [-●-], carvone [-□-] and cineole [-▲-] in SLS (initial concentration, SLS=1.0%, flavorant=0.0075%) showing differences in release of flavor compounds according to their solubility in SLS micelles [Adapted from ref. 77].

LITERATURE CITED

1.   Teeter J., Brand J. In *Neurobiology of Taste and Smell*; Finger T., Silver W., Eds., John Wiley & Sons: New York, 1987, pp. 299-329.
2.   Teeter J.H., Cagan R.H. In *Neural Mechanisms in Taste*; Cagan R.H., Ed., CRC Press: New York, 1989, pp. 1-20.
3.   Kulka K. *J. Agric. Food Chem.* **1967**, *15*, 48-57.
4.   Walters E., Orthoefer F., DuBois G. *Sweeteners: Discovery, Molecular Design and Chemoreception*; ACS Symp. Ser. 450, 1991.
5.   Yamamoto T., Kato T., Matsuo R., Kawamura Y., Yoshida M. *Physiol. Behavior*. **1985**, *35*, 411-415.
6.   Tinti J., Nofre C. In *Sweeteners: Discovery, Molecular Design, and Chemoreception*; Walters D. E., Orthoefer F., DuBois G., Eds., ACS Symp. Ser. 450, 1991, pp. 88-99.
7.   Kim S., Kang C., Cho J. In *Ibid*; pp. 28-40.
8.   Ariyoshi Y., Kohmura M., Hasegawa Y., Ota M., Nio N. In *Ibid*; pp. 41-56.
9.   Jakinovich W., Sugarman D. In *Neural Mechanisms in Taste*; Cagan R.H., Editor, CRC Press: New York, 1989, pp.37-84.
10. Culberson J. , Walters D. E. In *Sweeteners: Discovery, Molecular Design, and Chemoreception*; Walters D. E., Orthoefer F., DuBois G., Eds., ACS Symp. Ser. 450: 1991, pp. 214-223.
11. Douglas A., Goodman M. In *Ibid*; pp. 128-142.
12. Parliment T. *Chemtech*. **1980**, *9*, 284-289.
13. Anholt R. *Chem. Senses* **1991**, *16*, 421-427.
14. Pelosi P., Maida R. *Chem. Senses* **1990**, *15*, 205-215.
15. Pevsner J., Snyder S. *Chem. Senses* **1990**, *15*, 217-222.
16. Ohloff G., Winter B., Fehr C. In *Perfumes: Art, Science and Technology*; Muller P., Lamparsky D., Eds., Elsevier: New York, 1991, pp. 287-327.
17. Labows J., Preti G. In *Fragrance: the Psychology and Biology of Perfume*; Van Toller S., Dodd G., Eds., Elsevier Applied Science: London, 1992.
18. Gardner R. *Chem. Senses* **1980**, *5*, 185-194.
19. Belitz H., Wieser H. *Food Rev. Internat.* **1985**, *1*, 271-354.
20. Margolis F., Getchell T. In *Perfumes: Art, Science and Technology*; Muller P., Lamparsky D., Eds., Elsevier Applied Science: New York, 1991, pp. 481-498.
21. Saroli A. Z. Lebensm. *Unters Forsch.* **1986**, *182*, 118-120.
22. Arai S. *Kagahu to Seibutsu* **1981**, *19*, 22-24.
23. Kroeze J., Bartoshuk L. *Physiol. Behav.* **1985**, *35*, 779-783.
24. Maeyama N., Kusumoto R., Noda K. *Fukuoka Joshi Daig.Ka* **1984**, *15*, 25-27.
25. Mogensen L., Adler-Nissen J. Abs. ACS Ag. & Food Div., International Flavor Conference, **1989**, Greece.
26. Green B. In *Irritation*; Green B., Mason J.R., Kare M.R., Eds., Chemical Senses 2, Marcel Dekker Inc.: New York, 1990, Vol. 2, pp. 171-193.
27. Lawless H. In *Neurobiology of Taste and Smell*; Finger T., Silver W., Eds., John Wiley & Sons: New York, 1987, pp. 401-420.
28. Cain W. In *Irritation*; Green B., Mason J.R., Kare M.R., Eds., Chemical Senses 2, Marcel Dekker Inc.: New York, 1990, Vol. 2, pp. 43-58.
29. Watson H., Hems R., Rowsell D., Spring D. *J. Soc. Cosmet. Chem.* **1978**, *29*, 185-200.
30. Amano A., Moroe M., Yoshida T. Patent 99122707 **1983**; CA 99(*15*) 122707.
31. Technical Report, Haarmann and Reimer Inc., Springfield NJ, **1990**.
32. Fuerst W., Becher M. *Pharmazie* **1990**, *45*, 934-935.
33. Baim M., Koehler D. European Patent 88308801 **1988**.
34. Eccles R. In *Irritation*; Green B., Mason J.R., Kare M.R., Eds., Chemical Senses 2, Marcel Dekker Inc.: New York, 1990, Vol. 2, pp. 275-293.

35. Lee C., Lawless H. *Chem. Senses* **1991**, *16*, 225-238.
36. Ney K., Ernerth M. *Seif. Oele Fette Wachse* **1989**, 115, 183-186.
37. Rubico S., McDaniel M. *Chem. Senses* **1992**, *17*,273-289.
38. Schiffman S., Suggs M., Sostman A., Simon S. *Physiol. Behav.* **1991**, *51*, 55-63.
39. Laraqui A., Audinet C., Ozil P., Verain A. *S.T.P. Pharma* **1989**, *5*, 78-84.
40. Lyman B., Green B. *Chem. Senses* **1990**, *15*, 151-164.
41. Naisch M., Clifford M., Birch G. *Chem. Senses* **1990**, 15, 68.
42. Horio T., Kawamura Y. *Chem Senses* **1990**, *15*, 271-280.
43. Torii K., Cagan R.H. *Biochim. Biophys. Acta* **1980**, *627*, 313-323.
44. Cagan R.H. In *Physiology of Umami Taste*; Kawamura Y., Kare M.R. Eds. Marcel Dekker Inc.: New York, 1987, pp. 155-172.
45. Fox A. *Genetics* **1932**, *18*, 115-120.
46. Whissell-Buechy D. *Chem. Senses* **1990**, *15*, 39-57.
47. Bartoshuk L., Rifkin B., Marks L., Hooper J. *Chem. Senses* **1988**, *13*, 517-528.
48. Gent J., Bartoshuk L. *Chem. Senses* **1983**, 7, 265-272.
49. Hall H., Bartoshuk L., Cain W., Stevens J. *Nature* **1975**, 253, 442-443.
50. Mela D. *Chem. Senses* **1989**, *14*, 131-135.
51. Schifferstein H., Frijters J. *Chem. Senses* **1991**, *16*, 303-317.
52. Fischer R., Griffin F. *Arzneim Forsch.* **1964**, *14*, 675-686.
53. Marino S., Bartoshuk L., Monaco J., Anliker A., Reed D., Desnoyers S. *Chem. Senses* **1991**, *16*, 551.
54. Looy H., Weingarten H. *Chem. Senses* **1991**, *16*, 547.
55. Labows J., Ingersoll D., Cagan R.H., Harrison R., Winter C. *Chem. Senses* **1989**, *14*, 720 abs.
56. Amoore J., Steinle S. In *Genetics of Perception and Communications*; Wysocki C., Kare M., Eds., Chemical Senses 3, Marcel Dekker Inc.: New York, 1991, Vol. 3, pp. 331-351.
57. Wysocki C., Beauchamp G. In *Ibid*, pp. 353-373.
58. Labows J., Wysocki C. *Perf. & Flav.* **1984**, *9*, 21-26.
59. Boelens H., Haring H., de Rijke D. *Perf. & Flav.* **1983**, *8*, 71-74.
60. Krueger J.M., Cagan R.H. *J. Biol. Chem.* **1976**, *251*, 88-97.
61. Cagan R.H. *Comp. Biochem. Physiol.* **1986**, *85A*, 355-358.
62. Brand J.G., Bryant B.P., Cagan R.H., Kalinoski D.L. *Brain Res.* **1987**, *416*, 119-128.
63. Wegert S., Andrews B., Caprio J. *Chem. Senses* **1991**, *16*, 597.
64. Rhein L.D., Cagan R.H. *Proc. Natl. Acad. Sci. USA* **1980**, *77*, 4412-4416.
65. Pace U., Hanski E., Salomon Y., Lancet D. *Nature* **1985**, *316*, 255-258.
66. Sklar P., Anholt R., Snyder S. *J. Biol. Chem.* **1986**, *261*, 15538-15543.
67. Breer H., Boekhoff I. *Chem. Senses* **1991**, *16*, 19-29.
68. Spielman A., Huque T., Brand J., Whitney G. *Chem. Senses* **1991**, *16*, 585.
69. Buck R., Axel L. *Cell* **1991**, *65*, 175-187.
70. Zufall F., Firestein, S., Shepherd G. *J. Neurosci.* **1991**, *11*, 3573-3580.
71. Firestein, S., Zufall F., Shepherd G. *J. Neurosci.* **1991**, *11*, 3565-3572.
72. Shepherd G., Firestein S. *Steroid Biochem. Mol. Biol.* **1991**, *39*, 583-592.
73. Overbosch P., Afterof W., Haring P. *Food Rev. Int.* **1991**, 7, 137-184.
74. Saleeb F., Pickup J. In *Flavor of Foods and Beverages*; Inglett G., Charalambous G., Eds, Academic Press: New York, 1978, pp. 113.
75. Land D. In *Progress in Flavor Research*, Land D., Nursten H., Eds., Applied Science Publishers: London, 1979, pp. 53.
76. Labows J. *J. Am. Oil Chem. Soc.* **1992**, *69*, 34-38.

RECEIVED October 28, 1992

Chapter 3

Do We Recognize Sweetness and Bitterness at the Same Receptor?

Kozo Nakamura and Hideo Okai

Department of Fermentation Technology, Faculty of Engineering, Hiroshima University, Higashihiroshima, Hiroshima 724, Japan

The relationship between taste and chemical structure has been studied. The experimental data suggest that sweetness and bitterness are recognized at the same receptor. Furthermore, the receptor discriminates between bitter and sweet tastes based upon differences in functional unit combination. A new taste receptor model is proposed and presented. The new model explains many taste phenomena that the older model can not.

Current Mechanism for Sweetness and Bitterness Perception.

Numerous sweet substances exist having a wide range of chemical structures. In 1967, Shallenberger and coworkers proposed the AH-B theory showing a possible common molecular feature in sweet substances (1). They reported all sweet substances had a proton donor (AH) and a proton acceptor (B) in the molecule with a distance estimated to be an average of 3Å (2). They concluded that sweet taste was produced by the interaction between AH, B and the corresponding receptor sites. However, the intensity of sweetness could not be explained by the AH-B theory. Many chemists pointed out that a hydrophobic group participate in the intensity of sweetness. Some chemists propagated the theory that a hydrophobic group is one of the sweetness producing units and that sweetness is exhibited by X, AH, and B. This chapter will discuss this newer proposal in light of recent experimental data.

Mechanism for Sweetness Exhibition of Aspartame.

Aspartame, discovered by Mazur in 1969 (3), is 200 times sweeter than sucrose. Aspartame has a large commercial market as an artificial sweetening agent. It is apparent that the sweetness exhibited by aspartame requires amino (AH, electropositive) and carboxyl (B, electronegative) groups of aspartic acid moiety and the hydrophobic side chain (X) of the phenylalanine moiety (4). The sweetness of aspartame is exhibited by the trifunctional units AH, B, and X. It is thought that when the trifunctional units of aspartame, X, AH, and B, fit the corresponding receptor sites, a sweet taste is produced.

An interesting experiment using the model of aspartame (Figure 1) was carried out in 1985. Combinations of two or three different molecules were prepared to examine whether or not sweetness could be produced in the same system as aspartame. The

0097–6156/93/0528–0028$06.00/0

results of the sensory studies show sweetness was produced with several of the test compounds (Table I). It was found that the aspartame molecule is not always necessary for sweetness production, i.e. sweet taste is easily produced by a suitable combination of the components, AH, B, and X (*5*). The strongest sweetness was observed in the compounds with the combination of AH-X and B. The AH-X component produced bitterness while the B component produced sourness.

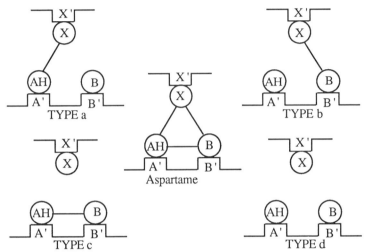

Fig. 1. The Sweetness Production of Aspartame Caused by the Interaction between Three Sweet Units(AH, X, B) and Corresponding Receptor Sites (A', X', B')
(Adopted from ref. 5)

Mechanism for Bitterness Exhibition. About one thousand bitter peptides were systematically synthesized. The relationship between bitterness and chemical structure of the peptides have been studied (*6-12*). These studies have now defined clearly the mechanism for bitterness production by peptides. Bitter peptides were found to possess two active sites, a "biding unit" and "stimulating unit." Hydrophobic groups acted as biding units. Bulky basic groups and hydrophobic groups would acts as stimulating units. Bitterness was produced when these active units attached to a corresponding bitter taste determinant. The distance between the two sites was estimated to be 4.1Å (*11*). A model of the receptor was proposed and is seen in Figure 2. This model is well suited to describe the production of bitterness for all bitter compounds (*13*).

Table I. The Results of Sweetness Production
by the combination of AH,X and B

combination	component**		sweetness
TYPE a	**AH-X**	+ **B**	
	5 mM H-L-Ala-L-Phe-OMe	5 mM acetic acid	yes
	(bitter TV *1.2 mM)	5 mM citric acid	yes
	5 mM H-L-Ala-D-Phe-OMe	5 mM acetic acid	yes
	(bitter TV 1.4 mM)	5 mM citric acid	yes
	5 mM H-L-Gly-L-Phe-OMe	5 mM acetic acid	yes
	(bitter TV 1.2 mM)	5 mM citric acid	yes
	5 mM H-L-Gly-D-Phe-OMe	5 mM acetic acid	yes
	(bitter TV 1.2 mM)	5 mM citric acid	yes
TYPE b	**AH**	+ **X - B**	
	5 mM aq NH_3	5 mM succinyl-L-Phe-OMe	no
	5 mM aq NH_3	5 mM succinyl-D-Phe-OMe	no
	5 mM aq NH_3	5 mM glutaryl-L-Phe-OMe	no
	5 mM aq NH_3	5 mM glutaryl-D-Phe-OMe	no
	5 mM aq NH_3	5 mM Bz-NH-(CH_2)-COOH	no
	5 mM aq Et_3N	5 mM Bz-NH-(CH_2)-COOH	yes
	5 mM $NH_2C_6H_{11}$	5 mM Bz-NH-(CH_2)-COOH	yes
TYPE c	**AH-B**	+ **X**	
	5 mM NH_2-(CH_2)n-COOH	5 mM Bzl-OH	no
	20 mM NH_2-(CH_2)n-COOH	10 mM Bzl-OH	no
TYPE d	**AH** + **X**	+ **B**	
	5 mM aq NH_3 5 mM Bzl-OH	5 mM acetic acid	no
	5 mM aq Et_3N 5 mM Bzl-OH	5 mM citric acid	no
	5 mM aq $NH_2C_6H_{11}$ 5 mM Bzl-OH	5 mM acetic acid	yes
	5 mM aq $NH_2C_6H_{11}$ 5 mM Bzl-OH	5 mM citric acid	yes

SOURCE: Adapted from ref. 5

* Threshold value.

** Individual taste of AH, X, B, and AH-B components: **AH**, 5 mM aq NH_3, 5 mM aq Et_3N, and 5 mM cyclohexylamine, is almost tasteless (undeterminable taste). **X**, 5 mM and 10mM Bzl-OH, is almost tasteless (undeterminable taste). **B**, 5 mM acetic acid, 5 mM citric acid is sour. **AH-B**, all of them (5 mM, 20 mM), is tasteless.

Similarity of Sweetness and Bitterness Perception. Figure 3 shows the model for both sweetness and bitterness production. The X component of a sweet compound corresponds to the binding unit of a bitter compound. The AH component of a sweet compound corresponds to the stimulating unit of a bitter compound. The AH component of a sweet compound corresponds to the stimulating unit of a bitter compound. The distances between AH and X, and between the biding units and stimulating units are essentially the same. The lack of the B component is the only factor that distinguishes between bitterness and sweetness perception. Although the taste phenomenon that is changing bitterness to sweetness by adding sourness appeared strange at first, it is explained readily by this new receptor model.

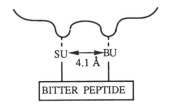

BU: binding unit (hydrophobic group)
SU: stimulating unit (hydrophobic or basic group)

Fig. 2. Bitterness Receptor Model (Reproduced with permission from ref. 10. Copyright 1988 Japan Society for Bioscience, Biotechnology, and Agrochemistry)

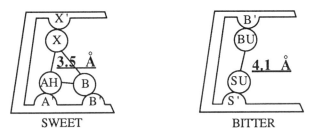

SWEET   BITTER

Fig. 3.   The Model of Sweetness and Bitterness Production at the Taste Receptor

Sweetness Production by the Combination of Bitter and Sweet Tastes. Sensory tests using typically bitter compounds such as brucine, strychnine, phenylthiourea, caffeine and bitter peptides were performed. Sensory tests using typically bitter compounds such as brucine, strychnine, phenylthiourea, caffeine and bitter peptides were performed. Sensory taste impression were also measured for combinations of acetic acid (sour) and typical bitter compounds (5). The data from these studies indicated that the tastes of these bitter/sour mixtures changed to a sweet taste regardless of their chemical structure of the bitter component (Table II).
    From the experimental results, it could be concluded that sweetness and bitterness are recognized in the same taste receptor and the receptor discriminates between bitter and sweet tastes based upon the difference in the functional unit combination. The currently accepted theory is that sweetness and bitterness are recognized by individual specific taste receptors. Data from this laboratory propose an alternative mechanism. Data is presented below to demonstrate which theory is most plausible.

Table II. Tasting Behavior of Combination of Typical
Bitter Compounds and Acetic Acid

bitter compound		B	taste****
0.002 mM brucine	(TV *0.0008 mM)	+ 5 mM acetic acid	bitter → sweet
0.005 mM strychnine	(TV 0.003 mM)	+ 5 mM acetic acid	bitter → sweet
0.1 mM phenylthiourea	(TV 0.025mM)	+ 5 mM acetic acid	bitter → sweet
0.2 mM cyclo(-Leu-Trp-)**	(TV 0.06 mM)	+ 5 mM acetic acid	bitter → sweet
0.2 mM Arg-Gly-Pro-Pro-Phe-Ile-Val***			
	(TV 0.05 mM)	+ 5 mM acetic acid	bitter → sweet

SOURCE: Adapted from ref. 5
*   Threshold value.
**   A bitter peptide by Shiba and Numami, 1974.
***   A bitter peptide by Fukui et al. , 1973.
****   Bitterness + Sourness → Sweetness. The taste were determined according
          to the method which has been described in detail in ref. 5.

Taste Behavior in Mixed Sweet and Bitter Solutions.

The current accepted theory suggests that a bitter compound and a sweet compound bind independently at specific receptors. This situation will be referred to as "independent" in this report. The data to follow will demonstrate that a bitter compound and a sweet compound bind at the same receptor in a competitive manner. Therefore, this situation will be referred to as "competitive" in this report. Which theory was the functioning mechanism of taste reception should be determinable when one measured the taste intensities of mixed solutions of bitter and sweet tasting compounds. In this experiment the mechanism could be predicted to elicit a considerable difference in taste intensity and response that was varying based on the final concentration of each component. The "independent" receptor mechanism would be expected to yield data in which the intensities of bitter and sweet would be unaffected by mixing the two tastes, no matter what the concentration. On the other hand, with the "competitive" receptor mechanism one would expect both flavors to become altered, i.e., one stronger and the other weaker, as component concentrations varied; the latter would occur because of competition of the substances for the same site.

Two kinds of mixed solutions were provided to test the hypothesis. One was a mixed solution of D-phenylalanine and L-phenylalanine. The second was a mixed solution of sucrose and L-phenylalanine. While the chemical properties of D- and L-phenylalanine are the same, for the most part, they do possess different tastes. Sucrose has a structure and taste that is significantly different from L-phenylalanine. The sensory data obtained from a five member sensory panel is shown in Table III.

Table III. The score of sweetness and bitterness

SCORE*		0	1	2	3	4	5
	L-Phe	0 ~	20 ~	30 ~	40 ~	50 ~	60 ~
CONC**	D-Phe	0 ~	2 ~	4 ~	8 ~	14 ~	20 ~
(mM)	Sucrose	0 ~	5 ~	10 ~	20 ~	30 ~	40 ~

* The scores represent the intensity of sweetness or bitterness. Score 1
  indicates bitterness of 1mM caffeine solution or sweetness of 5mM
  sucrose solution. The taste-intensities become progressively stronger as the
  scores increase.
**The concentration of the aqueous solution of each compound.

Theoretical sensory sweetness and bitterness scores for the two theories are described below for mixed solutions of sweet and sour. This exercise considers the case of a mixture of 20 mM D-phenylalanine (D-PA) and 60 mM L-phenylalanine (L-PA). Individually each of these solutions yield a sensory score of 5.0, i.e. the sweetness of 30 mM D-PA is 5 and the bitterness score of 60 mM L-PA is 5. Theoretically, if both compounds bind independently at different taste receptors, the sweetness and bitterness should each be 5. On the other hand, if both compound bind at the same taste receptor in a competitive manner, the sweetness and bitterness scores should decrease. In regards to the latter, the sweet and bitter compound were mixed in a ration of 1:3 (20 mM to 60 mM) which yields a probability of fitting a taste receptor of 1-in-4 for sweet and 3-in-4 for bitter. If these probabilities are multiplied by the original concentrations, i.e. 20 mM and 60 mM, the products would be in the expected concentration of 5 mM (1/4 of 20mM) and 45 mM (3/4 of 60mM). The sweetness of a 5 mM solution and the bitterness of a 45 mM solution corresponds to a sensory score of 2 and 3, respectively, and represent the expected sensory score for the competitive hypothesis.

Results from sensory evaluation of mixed solution are seen in Table IV. The data list the theoretical response for both the independent and competitive receptor hypothesis as well as the actual sensory score. The actual sensory scores were found to agree fairly well with the competitive model. The minor dissimilarity between the actual and theoretical is due to the inability of individual to taste bitterness in solutions that are extremely sweet, i.e., there is some masking of overall sensory perception which is concentration dependent. The data, therefore, clearly indicate that sweetness and bitterness act in a competitive manner and should be considered to compete for the binding sites at the same receptor.

Table IV. The Results of Sensory Analysis of   Mixed Solution

| combination | | original score | expected value | | RESULTS* |
sweet	bitter		INDEPENDENTLY	COMPETITIVELY	
D-Phe + L-Phe					
20mM	- 60mM	5 - 5	5 - 5	2 - 3	1 - 3
20	- 40	5 - 3	5 - 3	2 - 2	2 - 2
20	- 20	5 - 1	5 - 1	3 - 0	3 - 1
40	- 20	5 - 1	5 - 1	5 - 0	5 - 0*
60	- 20	5 - 1	5 - 1	5 - 0	5 - 0*
sucrose + L-Phe					
20mM	-60mM	3 - 5	3 - 5	0 - 3	0 - 3
20	- 40	3 - 3	3 - 3	1 - 2	0 - 3
20	- 20	3 - 1	3 - 1	2 - 0	2 - 1
40	- 20	5 - 1	5 - 1	3 - 0	3 - 0
60	- 20	5 - 1	5 - 1	5 - 0	5 - 0

* D-Phe possess some bitterness in the high concentration solutions.
The bitterness of 40mM D-Phe solution equals to score 1.
The bitterness of 60mM D-Phe solution equals to score 2.
The results is obtained by considering the bitterness.

Other experiments were conducted to verify the conclusions drawn and described above. For example, benzoyl-ε-aminoaproic acid (BACA), possesses a sour taste despite having a hydrophobic group in the molecule, i.e., it has an X-B component. If there is single receptor for sweet and bitter (and perhaps sour), then prebinding or preloading of the bitter receptor site of the receptor with BACA should change the final taste perception of subsequently added chemical since the newly added component cannot bind additional bitter molecules. In this experiment, two bitter tasting solutions caffeine (weak bitterness) and strychnine (strong bitterness) were given after BACA. Data in Table V show that the weak bitterness of caffeine is eliminated and the strong bitterness of strychnine is weakened if they are added after BACA administration thereby supporting the competitive receptor hypothesis.

Table V. Tasting Behavior of Combination of Typical Bitter Compounds and Benzoyl-ε-aminocoproic acid

bitter compound	X-B	taste**
0.005 mM strychnine	+ 5mM Bz-NH-(CH2)5-COOH	bitter → no
0.05mM strychnine(TV*0.003mM)	+ 5mM Bz-NH-(CH2)5-COOH	bitter → weak
2 mM caffeine	+ 5mM Bz-NH-(CH2)5-COOH	bitter → no
5 mM caffeine(TV1.0 mM)	+ 5mM Bz-NH-(CH2)5-COOH	bitter → weak

SOURCE: Adapted from ref. 5
* Threshold value.
** Decrease of bitterness by X-B component

Hexamethylenetetramine and cyclohexylamine are both tasteless solutions. Hexamethylenetetramine, a highly water soluble compound, is tasteless even though it is composed of a AH-X component (Figure 1). Its molecular size is too small to exhibit the bitter taste proposed by the theory presented in Figure 1. Cyclohexamine, similarly, is tasteless. When both solutions are mixed together, a bitter flavor is exhibited. Interestingly, the bitter flavor of the mixed components is changed to sweet when acetic acid is added (5), a fact consistent with the competitive theory.

A New Receptor Model

The data described above led to the prediction of a single taste receptor that recognizes both sweetness and bitterness and to the development of a new molecular model of a taste receptor (Figure 4). According to this model, bitterness is produced when a hydrophobic group labeled "X" and electropositive or a hydrophobic group labeled "AX" attach at the corresponding receptor sites. Sweetness is produced when an electronegative group labeled "B" attaches at the remaining receptor site (Figure 4). This model readily explains all of the experimental data obtained thus far. For example, benzoyl-ε-aminocaproic acid (BACA), describe above, produced only a sour taste in spite of having a hydrophobic group in the molecule. According to the model (Figure 4) BACA could not produce a sweet or bitter taste since it lacks an AX component.

The new model describes the bifunctional unit of bitter reception as the "hydrophobic group" and the "electropositive of hydrophobic group" whereas the 1985 model of bitter (6-12) described the bifunctional units of bitterness production as the "binding unit" and the "stimulating unit." The new model in Figure 4 describes the perception of sweetness as a AH-B-X response rather than an AH-B response as the proposed by the theory of Shallenberger (1). The new model offers a molecular approach more consistent with the experimental data. Further investigations to develop the theory and widely apply it to various flavor compounds are currently in progress and should prove to be both exciting to the scientist and useful to producers of foods.

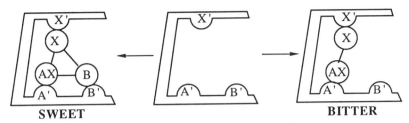

SWEET   BITTER

X ; hydrophobic group B ; electronegative group
AX ; hydrophobic group or electropositive group

Fig. 4. A New Model of Taste Receptor

Literature Cited
(1) Shallenberger, R.S.; Acree, T. E. *Nature (London)* **1967**, 216, 480.
(2) Shallenberger, R.S.; Acree, T. E.; Lee, C. Y. *Nature (London)* **1969**, 221, 555.
(3) Mazur, R.; Schlatter. J. M.; Goldkamp. A. H. *J. Am. Chem.* **1969**, 91, 2684.
(4) Fujino, M.; Wakimasu, M.; Mano, M.; Tanaka, K.; Nakajima, N.; Aoki, H. *Chem. Pharm. Bull.* **1976**, 24, 2112.
(5) Shinoda, I.; Okai, H. *J. Agric. Food Chem.* **1985**, 33, 792-795.
(6) Ishibashi, N.; Arita, Y.; Kanehisa, H.; Kouge, K.; Okai, H.; Fukui, S. *Agric. Biol. Chem.* **1987**, 51, 2389.
(7) Ishibashi, N.; Sadamori, K.; Yamamoto, O.; Kanehisa, H.; Kouge, K.; Kikuchi, E.; Okai, H.; Fukui, S. *Agric. Biol. Chem.* **1987**, 51, 3309.
(8) Ishibashi, N.; Ono, I.; Kato, K.; Shigenaga, T.; Shinoda, I.; Okai, H.; Fukui, S. *Agric. Biol. Chem.* **1988**, 52, 91.
(9) Ishibashi, N.; Kubo, T.; Chino, M.; Fukui, S.; Shinoda, I.; Kikuchi, E.; Okai, H. *Agric. Biol. Chem.* **1988**, 52, 95.
(10) Ishibashi, N.; Kouge, K.; Shinoda, I.; Kanehisa, H.; Okai, H. *Agric. Biol. Chem.* **1988**, 52, 819.
(11) Oyama, S.; Ishibashi, N.; Tamura, M.; Nishizaki, H.; Okai, H. *Agric. Biol. Chem.* **1988**, 52, 871.
(12) Shinoda, I.; Nosho, Y.; Kouge, K.; Ishibashi, N.; Okai, H.; Tatumi, K.; Kikuchi, E. *Agric. Biol. Chem.* **1987**, 51, 2103.
(13) Ishibashi, N. In The dissertation *Pepuchido ni okeru nigami hatugen kikou.* Hirosima University (1988).

RECEIVED February 16, 1993

Chapter 4

Heterocyclic Compounds in Model Systems for the Maillard Reaction

Reactions of Aldehydes with Ammonium Sulfide in the Presence or Absence of Acetoin

Gaston Vernin[1], Jacques Metzger[1], Adel A. M. Sultan[2], Ahmed K. El-Shafei[2], and Cyril Párkányi[3]

[1]Chimie des Arômes-Oenologie, Laboratoire de Chimie Moléculaire, URA 1411, Faculté des Sciences et Techniques de Saint-Jérôme, F–13397 Marseille Cédex 13, France
[2]Department of Chemistry, Assiut University, Sohag, Egypt
[3]Department of Chemistry, Florida Atlantic University, Boca Raton, FL 33431–0991

Aldehydes formed by Strecker degradation of amino acids in the presence of α-diketones react with ammonium sulfide in the presence or in the absence of acetoin (3-hydroxy-2-butanone) giving a variety of oxygen, sulfur, and/or nitrogen-containing heterocyclic compounds. The analytical data for selected groups of these compounds are reported and typical examples of their mechanism of formation are suggested. The compounds formed can be identified by GC-MS and the use of a SPECMA computer data bank containing information about their mass spectra and Kováts indices. Examples of computer-predicted (hitherto not found experimentally) heterocyclic compounds are given. Also, the sensory characteristics of several groups of heterocyclic compounds formed in these reactions are presented.

Numerous review articles *(1-20)* and a number of books *(21-28)* have been devoted to the Maillard reaction, *viz.*, the non-enzymatic browning reactions involving amino acids and reducing sugars. These reactions play a very important role in the formation of the various heterocyclic food flavors and thus are directly related to food quality. The various steps in the Maillard reaction, such as the formation of Amadori and Heyns intermediates, the retro-aldolization of rearranged sugars, and the Strecker degradation of α-amino acids lead to highly reactive, low-molecular weight compounds, such as aldehydes, α-hydroxyketones (or α-diketones), ammonia, and hydrogen sulfide (formed from sulfur-containing amino acids). These key compounds subsequently undergo a ring closure affording a variety of heterocyclic compounds with very pronounced and characteristic sensory properties. Both aromatic and partially or fully saturated heterocycles are formed. The heterocycles discussed in this contribution include, but are not limited to, thiophenes, dithiolanes, trithiolanes, thiopyrans, dithiins, trioxanes, dioxathianes, oxadithianes, trithianes, dithiazines, pyridines, and pyrazines.

0097–6156/93/0528–0036$06.00/0

The goal of this contribution is to review the formation of at least some of these compounds in model systems (consisting of aldehydes, acetoin, and ammonium sulfide), their identification, and analytical characterization (mass spectra, Kováts indices) accomplished by using the GC-MS-SPECMA data bank.

Experimental

Many of the experimental data and results referred to in this text are those obtained by Sultan in his doctoral research (29). The general procedures used in the various reactions of the model systems were are follows.

Reaction of Aldehydes with Ammonium Sulfide. An aldehyde (0.1 mol, e.g, acetaldehyde, propionaldehyde, butyraldehyde, capro-aldehyde) was added to a 20% aqueous solution of ammonium sulfide (50 ml) and the mixture was stirred for 3 h at room temperature. Then it was extracted with ether, the extract was washed with water, dried over anhydrous sodium sulfate, and evaporated under reduced pressure. The remaining liquid residue was analyzed by GC and GC-MS.

Reaction of Acetoin (3-Hydroxy-2-butanone) with Ammonia. Aqueous solution of ammonium hydroxyde (20%, 100 ml) was added to acetoin (17.6 g, 0.2 mol) and the reaction mixture was stirred for 30 min at 50°C and then for 6 h at room temperature. The precipitated product was filtered off, the filtrate was neutralized with 10% hydrochloric acid, and extracted with ether (continuous overnight extraction). The extract was washed with water, dried over anhydrous sodium sulfate, and concentrated on a spinning-band distillation apparatus. The residual solution was then analyzed by GC and GC-MS.

Reaction of Aldehydes with Acetoin and Ammonium Sulfide. To an aqueous solution of acetoin (17.6 g, 0.2 mol) and ammonium sul-fide (20%, 100 ml), an aldehyde (0.2 mol) was added dropwise and the mixture was stirred for 6 h at room temperature. Then it was neutralized with 10% hydrochloric acid and the precipitate that formed was filtered off. The filtrate was extracted with ether, the ether extract was washed with water, dried over anhydrous sodium sulfate, and evaporated under reduced pressure. The residual solution was fractionally distilled under reduced pres-sure and each fraction was analyzed by GC-MS. In the case of acetaldehyde, the filtered reaction mixture was divided into two parts and one of them (pH 11.0) was neutralized with 10% hydro-chloric acid. Each portion was extracted with ether and subse-quently analyzed by GC and GC-MS.

GC and GC-MS Analyses. The concentrated extracts were analyzed on an Intersmat IGC 120F gas chromatograph with an ICR 1B inte-grator using an SGE FFAP quartz capillary column (free fatty acid phase, 50 m x 0.22 mm i.d.); injection temperature 250°C, oven temperature 60°C for 8 min, operating temperature 230°C (tempera-ture gradient 2°C/min).

GC-MS analyses were carried out on a Sigma 3B Perkin-Elmer gas chromatograph coupled with a computerized VG ZAB-HF mass spectrometer (VG Analytical, Manchester, England). The source temperature was 190°C, the ionizing current 100 μA, and the ioniz-ing potential (electron ionization) 70 eV. The SPECMA data bank containing mass spectral data and Kováts indices for 1,100 het-erocyclic aroma compounds was used to identify the products. The use of the SPECMA data bank is based on the mass spectral data available for the compound in question (molecular weight, frag-mentation pattern) and its Kováts index (a gas-chromatographic index). Detailed information about the procedure can be found in our previous publications (29-31).

Reactions with Aldehydes, Hydrogen Sulfide, and Ammonia (or Ammonium Sulfide)

These reactions were studied in detail by several groups of authors (29,32-37) and reviewed by Kawai and co-workers (38). The resulting heterocyclic systems contain five-and six-membered rings with one or more heteroatoms (N,O,S).

Five- and Six-Membered Ring Heterocycles with One, Two, or Three Sulfur Atoms. A summary of the various sulfur heterocycles formed from aldehydes in the presence of hydrogen sulfide and the corresponding analytical data are presented in Table I. Examples of aldehydes used in this reaction include propionaldehyde, butyraldehyde and caproaldehyde.

Reaction of various aldehydes with hydrogen sulfide leads to substituted thiophenes, dihydrothiophenes, dithiolanes and trithiolane, as well as to six-membered ring thiopyran derivatives and dithiins. Ledl (33) obtained 2,4-dimethylthiophene (1, R = Me) as a product of the reaction of propionaldehyde with hydrogen sulfide in the presence of ammonia. Sultan (29) reported the formation of 2,4-diethylthiophene (1, R = Et), 2,4-dibutyl-thiophene (1, R = Bu), and their dehydro derivatives from the reaction of ammonium sulfide with butyraldehyde and caproaldehyde (hexanal), respectively. The mechanism suggested for their formation is depicted in Scheme 1. Space limitations do not allow us to discuss the mechanism here in detail (for additional information, see ref. 29).

The simple 1,2-dithiolenes, viz., 1,2-dithiolene (3, R = H), 3-methyl-1,2-dithiolene (3, R = Me) and the saturated derivative of the latter were detected by Takken and co-workers (2) with crotonaldehyde and butanedione as the starting materials. Ledl (33) identified 2-ethyl-4-methyl-1,3-dithiolene (4) in the reaction mixture containing propionaldehyde, hydrogen sulfide and ammonia, and the isomeric 2,4,5-trimethyl-1,3-dithiolane (5) was obtained by Sultan (29) from the reaction of acetaldehyde, acetoin, and ammonium sulfide.

1,2- and 1,3-dithiolenes and their monomethyl and dimethyl derivatives were patented by Wilson and co-workers (39) because of their roasted and onion-like notes (40) (the threshold values are from 2 to 20 ppb).

The most important sulfur heterocycles with a five-membered ring and three sulfur atoms are the diastereomeric 3,5-dialkyl-1,2,4-trithiolanes 6. The dimethyl derivatives (6, R = Me) have been found in various cooked foods, such as mushrooms (41), boiled beef (42) commercial beef extract (43), boiled antarctic krills (44), red algae (45), and several model systems containing a source of sulfur (2,19,29). The diethyl derivative (6, R = Et) was identified by Ledl (33) and by Sultan, with propionaldehyde as the starting material. The dipropyl and diisopropyl derivatives (6, R = Pr, i-Pr) were obtained from the corresponding aldehydes (butyraldehydes, isobutyraldehyde) and ammonium sulfide, in the presence or in the absence of acetoin (29), while no such compounds were identified with methional and ammonium sulfide as the starting materials. Similarly, the higher aldehydes did not lead to the desired heterocycles, except for isovaleraldehyde (35). The proposed mechanism of their formation is shown in Scheme 2. No discussion of the mechanism is possible here (cf. ref. 29). Cleavage of the 2-3 and 5-6 bonds upon electron ionization

Table I. Sulfur-Containing Five- and Six-Membered Ring Heterocyclic Compounds

Compound[a]	R	Mol. formula	Mol. wt.	KIP[b]	m/z[c]	Ref.
1	Me	C_6H_8S	112	1160	111, 97, 112	33
	Et	$C_8H_{12}S$	140	1355	125, 140	29
	Pr	$C_{10}H_{16}S$[d]	168	(1540)	$-$[d]	—
	Bu	$C_{12}H_{20}S$	196	(1725)	155, 141, 196[e]	29
2	Et	$C_8H_{14}S$[f]	142	1377	113, 85, 142	29
	Et	$C_8H_{14}S$[g]	142	1404	113, 85, 142	29
	Bu	$C_{12}H_{22}S$	198	1573	141	29
3	H	$C_3H_4S_2$	104	—	$-$[d]	2
	Me	$C_4H_6S_2$	118	—	$-$[d]	2
4	—	$C_6H_{10}S_2$	146	(1560)	117, 45, 146, 41[e]	33

Continued on next page

Table I. Continued.

Compound[a]	R	Mol. formula	Mol. wt.	KIP[b]	m/z[c]	Ref.
5	—	$C_6H_{10}S_2$	146	1586	85, 86, 146, 45	29
6	H	$C_2H_4S_3$	124	1400	124, 78, 45	29
	Me	$C_4H_8S_3$[f]	152	1555	152, 59, 92, 88	29
	Me	$C_4H_8S_3$[g]	152	1560	152, 92, 59, 88	29
	Et	$C_6H_{12}S_3$[f]	180	1684	180, 41, 74, 106	29, 33
	Et	$C_6H_{12}S_3$[g]	180	1695	180, 41, 74, 106	29, 33
	i-Pr	$C_8H_{16}S_3$[f]	208	1775	55, 208, 143, 87, 88	29
	i-Pr	$C_8H_{16}S_3$[g]	208	1784	55, 208, 143, 87, 88	29
	Pr	$C_8H_{16}S_3$[f]	208	1810	55, 88, 208	29
	Pr	$C_8H_{16}S_3$[g]	208	1820	55, 88, 208	29
	i-Bu	$C_{10}H_{20}S_3$	236	(~1900)	69, 41, 43, 45, 60[e]	35
7	—	$C_7H_{14}S_3$	194	(1860)	69, 194, 101, 129, 92[e]	29

8	—	$C_9H_{14}S$ [h]	154	—	125, 154, 91[e]	33
9	—	$C_6H_{10}S_2$ [h]	146	—	85, 86, 45, 146, 59, 60, 41, 39[e]	33
10	—	$C_9H_{16}S_2$ [h]	188	—	41, 99, 45, 114, 159[e]	33

[a] Examples of aldehydes used to obtain the respective heterocycles are mentioned directly in the text. [b] This work. KIP values are Kováts indices on a polar column (FFAP); the values in parentheses are estimates based on additivity of index increments. [c] Base peak and main fragments. Unless indicated otherwise, the mass spectra were obtained in this work and are in agreement with the literature data (see the ref.). [d] Not reported. [e] Literature data. [f] First isomer. [g] Second isomer. [h] Found by Ledl (*33*) after the reaction of acetaldehyde with hydrogen sulfide.

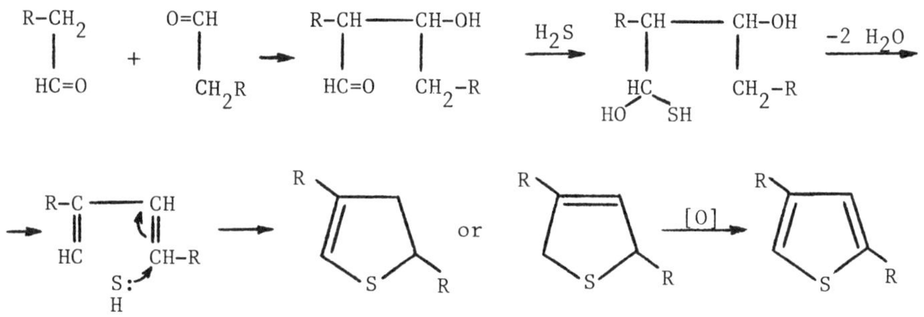

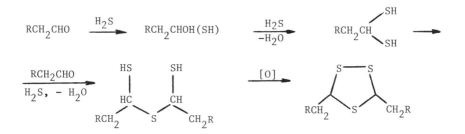

Scheme 1.   Suggested formation of 2,4-dialkylthiophenes from
aldehydes and hydrogen sulfide.

Scheme 2.   Suggested formation of 3,5-dialkyltrithiolanes.

generates a cation radical, $RCH_2.CHS^{\pm}$, which loses a hydrogen, gives a cyclic cation, and finally a fragment with m/z 45.

Reaction of acetaldehyde and valeraldehyde gave two isomeric 3-methyl-5-n-butyl-1,2,4-trithiolanes (7). 3,5-Dimethyl-2-ethyl-2H-thiin (8) 2,6-dimethyl-1,3-dithiin (9), and 2,6-diethyl-5-methyl-1,3-dithiin (10) were identified by Ledl (33) in the reaction of propionaldehyde with hydrogen sulfide, alone or in the presence of acetaldehyde. Finally, Badings and co-workers obtained 2,6-dimethyl-5,6-dihydro-2H-thiopyran-3-carboxaldehyde from crotonaldehyde an hydrogen sulfide (Badings, H. T.; Maarse, H.; Kleipool, T. J. C.; Tas, A. C.; Naeter, R.; Ten Noever de Brauw, M. C., presented at the CIVO Aroma Symposium, Zeist, The Netherlands, May, 1975).

The various heterocyclic systems, their gas-chromatographic data (Kováts indices), and the mass spectra are summarized in Table I.

Trioxanes, Dioxathianes, Oxadithianes, and Trithianes. The information on these compounds is compiled in Table II.

Trimerization of aldehydes affords trialkyl-1,3,5-thioxanes 11. Compounds of this type were obtained from acetaldehyde, propionaldehyde (29,33), isobutyraldehyde (47), and isovaleraldehyde (35). The two isomers normally formed can be separated by GC.

Takken and co-workers (2) determined that, in the presence of hydrogen sulfide at atmospheric pressure, aldehydes give cyclic sulfur-containing products, such as dioxathianes 12, oxadithianes 13, and trithianes 14. Mass spectra of the corresponding triethyl derivatives were reported by Ledl (31).

In our study we were able to show (47) that electron ionization fragmentation of 2,4,6-triisopropyl-1,3,5-trithiane gives a base peak at m/z 88 corresponding to the cation radical $C_4H_8S^{\pm}$. Because of their low volatility, the higher homologs have not been detected, the single exception being the 2,4,6-triisobutyl derivative (35). The above compounds are most likely to be formed by condensation of three molecules of an aldehyde with one, two or three molecules of hydrogen sulfide and the subsequent nucleophilic closure of a heterocyclic ring accompanied by elimination of one, two, or three molecules of water.

1,3,5-Dithiazines. The analytical data for 2,4-6-trialkyl-5,6-dihydro-1,3,5-dithiazines are presented in Table III.

The symmetrical compounds (17, $R^1 = R^2 = R^3$) are formed from an aldehyde and ammonium sulfide (2,29,35) or hydrogen sulfide and ammonia (33). Thialdine (17, $R^1 = R^2 = R^3 = Me$) is an important aroma compound found in the volatiles of beef broth (48), pressure-cooked meat (49), and fried chicken (50). It is also obtained from acetaldehyde (2,29,37) or from ß-mercaptoacetaldehyde and ammonium sulfide (37). In our experiments, it was synthesized as a white powder from a reaction of acetaldehyde with ammonium sulfide in 60% yield.

Similarly, with acetaldehyde and propionaldehyde as the starting materials, one gets the 2,4-dimethyl-6-ethyl- and 2-ethyl-4,6-dimethyl derivatives (33,37). In the presence of acetaldehyde, the higher aldehydes give various substituted

Table II. 2,4,6-Trialkyl-1,3,5-trioxanes (11), 2,4,6-Trialkyl-1,3,5-dioxathianes (12), 2,4,6-Trialkyl-1,3,5-oxadithianes (13), 2,4,6-Trialkyl-1,3,5-trithianes (14), 3,6-Dialkyl-1,2,4,5-tetrathianes (15), and 3,5,7-Trialkyl-1,2,4,6-tetrathiepanes (16)

Compound	R	Mol. formula	Mol. wt.	KIP[a]	m/z[b]	Ref.
11	Me	$C_6H_{12}O_3$[c]	132	1070	45, 43, 89, 97	46
	Et	$C_9H_{18}O_3$[d]	174	1254/1285[e]	59, 57, 117	29, 33
	i-Pr	$C_{12}H_{24}O_3$[f]	216	1160/1230[e]	73	29
	i-Bu	$C_{15}H_{30}O_3$[g]	258	1410/1450[e]	87, 69, 173, 43	35
12	Me	$C_6H_{12}O_2S$[h]	148	-	-[h]	-
	Et	$C_9H_{18}O_2S$[d]	190	1550/1585[e]	74, 41, 57, 132	33
13	Et	$C_9H_{18}OS$[d]	206	(1810)	74, 41, 148[i]	33

Structure	R	Formula		KIP	Base peak and main fragments[b]	Ref.
14	Me	$C_6H_{12}S_3$ [c]	180	1810	60, 59, 180, 55	2, 29
	Et	$C_9H_{18}S_3$	222	2040	73, 41, 74, 64	33
	i-Pr	$C_{12}H_{24}S_3$ [f]	264	2070	88	2, 29
	Pr	$C_{12}H_{24}S_3$ [j]	264	(2300)	88, 55, 87 [i]	2
15	Me	$C_4H_8S_4$ [c]	184	1800	59, 60, 184, 124	29
	Et	$C_6H_{12}S_4$ [d]	212	1970	73, 41, 74, 45	29, 33
	i-Pr	$C_8H_{14}S_4$ [f]	240	2130	87	47
	Pr	$C_8H_{14}S_4$ [h]	240	(2160)	-[h]	-
16	*i*-Pr	$C_{12}H_{24}S_4$ [f]	296	2275	143, 55	47

[a] This work. KIP values are Kováts indices on a polar column (FFAP); the values in parentheses are estimates based on additivity of index increments. [b] Base peak and main fragments. Unless indicated otherwise, the mass spectra were obtained in this work and are in agreement with the literature data (see the ref.). [c] From propionaldehyde. [d] Two isomers. [e] Two isomers. [f] From isobutyraldehyde. [g] From isovaleraldehyde. [h] Not reported. [i] Literature data. [j] From butyraldehyde.

Table III.   5,6-Dihydro-2,4,6-4H-1,3,5-dithiazines Identified in Aldehyde Ammonium Sulfide Model Systems

R^1	R^2	R^3	Mol. formula	Mol. wt.	KIP^a	m/z^b	Ref.
Me	Me	Me	$C_6H_{13}NS_2$	163	1670	44, 163, 71, 70, 60	29, 37
Me	Et	Et	$C_7H_{15}NS_2$	177	(1730)	44, 58, 177[c]	33, 37
Me	Et	Me	$C_7H_{15}NS_2$	177	1750	44, 71, 70, 77	33
Et	Me	Et	$C_8H_{17}NS_2$	191	(1790)	58, 41, 98, 191[c]	33
Me	Et	Et	$C_8H_{17}NS_2$	191	(1810)	44, 58, 41, 84[c]	33
Et	Et	Et	$C_9H_{19}NS_2$	205	1835	58, 41, 98, 74	29, 33
Me	Bu	Me	$C_9H_{19}NS_2$	205	(1930)	86, 112, 205, 70[c]	37
Bu	Me	Me	$C_9H_{19}NS_2$	205	(1910)	71, 70, 44, 103[c]	37
i-Pr	i-Pr	i-Pr	$C_{12}H_{25}NS_2$	247	1930	72	29
Pr	Pr	Pr	$C_{12}H_{25}NS_2$	247	(2100)	-[d]	29
Me	Bu	Bu	$C_{12}H_{25}NS_2$	247	(2200)	154, 86, 98, 247[c]	37
i-Bu	i-Bu	i-Bu	$C_{15}H_{31}NS_2$	289	(2300)	43, 41, 102, 69, 96[c]	35
Bu	Bu	Bu	$C_{15}H_{31}NS_2$	289	(2360)	154, 86, 98, 155[c]	37

[a]This work. KIP values are Kováts indices on a polar column (FFAP); the values in parentheses are estimates based on additivity of index increments. [b]Base peak and main fragments. Unless indicated otherwise, the mass spectra were obtained in this work and are in agreement with the literature data (see the ref.). [c]Literature data. [d]Not reported.

dithiazines. Takken and co-workers (2) have indicated that
trialkyldihydrodithiazenes to be quite unstable under reaction
conditions used and especially at higher temperatures (150°C).
Most of these compounds possess green and vegetable-like odors.

We have suggested a mechanism of formation of these compounds that involves a dithiol intermediate which interacts with
ammonia giving the corresponding amino derivative, followed by a
nucleophilic attack of the amino group upon another molecule of
an aldehyde, the formation of a precursor intermediate, and
heterocyclization with an elimination of a molecule of water.

Mass fragmentation of 5,6-dihydro-4H-13,5-dithiazines has
been discussed by Hwang and co-workers (37).

Pyridines. Another series of heterocycles formed in the
aldehyde-ammonium model systems are the pyridines, **18**, and dihyd-
ropyridines, **19**.

Two alkylpyridines were reported by Ledl (**18**, R^1 = Et, R^2 =
R^4 = Me, R^3 = H; **18**, R^1 = H, R^2 = R^4 = Me, R^3 = Et) from propionaldehyde and ammonia (33). Four additional similar compounds were
obtained from propionaldehyde, acetaldehyde, and ammonia (33).

Hwang and co-workers (37) confirmed 2-butyl-3,5-
dipropylpyridine and 2-pentyl-3,5-dihydropyridine and their
dihydro derivatives after the reaction of valeraldehyde and
caproaldehyde in the presence of ammonium sulfide, respectively.
These compounds were also obtained in the presence of
acetaldehyde.

Shu and co-workers (35) identified 2-isobutyl-3,5-
diisopropylpyridine, 2-pentyl-3,5-dimethylpyridine, and its
dihydro derivative obtained under similar conditions. Sultan (29)
confirmed the presence of 3,5-diethyl-2-propylpyridine in a model
system consisting of butyraldehyde and ammonium sulfide. Our
proposed mechanism of their formation (20) consists of three
steps: 1) aldol condensation of the starting aldehydes to 2,4-
alkadienals, 2) imine formation with ammonia, and 3) subsequent
cyclization and oxidation to corresponding pyridines. An
alternate mechanism, suggested by Shu and co-workers (33), takes
into consideration the isolated dihydro derivatives. Hwang and
co-workers described another dihydro derivative (**19**, R^1 = Bu, R^2 =
R^4 = Pr, R^3 = H) (37).

Under the conditions corresponding to the roasting of
coffee, serine, threonine, and sucrose yield various substituted
pyridines (51), furans, and furanones (52). Thirty-three pyridine
derivatives were identified by Baltes and co-workers (51).
Recently, 3-methylthiomethylpyridine was identified as one of the
products of thermal degradation of the glucose-methionine Amadori
intermediates (53).

Other Compounds. Other types of compounds formed under the
proper conditions include, but are not limited to, N-
alkylidenealkenylamines (35,37) and various aliphatic sulfur
compounds (e.g., sulfides, disulfides, etc.) (33,35).

Reaction of Acetoin with Ammonia. Space limitations do not allow us to discuss the remaining reactions in detail. Among other products, the reaction of acetoin with ammonia gives pyrazines (this work, cf. also ref. *54)*.

Reactions of Acetoin and Aldehydes with Ammonium Sulfide. This reaction leads to the various substituted oxazolines, thiazolines, and imidazolines. Additional heterocyclic systems identified in this reaction include thiophenes and pyrazines. All these products are important aroma compounds in the food industry.

New Heterocyclic Ring Systems Predicted by Computer-Assisted Organic Synthesis (CAOS). A computer program written for this purpose *(55,56)* can be used to predict the formation of additional heterocyclic systems from the reactants mentioned in the above sections. These are the systems which, of course, have not been experimentally detected as reaction products so far.

This approach utilizes various computer programs which, when provided with an appropriate input (starting materials, reaction conditions), can generate structures of chemical compounds which can be potentially formed from their respective precursors. The information can be of help to food chemists in their search for various new, hitherto unidentified heterocyclic compounds in food flavors *(55,56)*.

As a typical example, Scheme 3 shows the possible formation of five-membered ring systems with two heteroatoms in the acetoin-ammonium sulfide model system (Barone, R., Université d'Aix-Marseille III, unpublished results, 1986). Similar predictions can be made for the formation of substituted pyrrolidones and for additional five- and six-membered rings.

Sensory Characteristics

As mentioned before, the sensory properties of the various heterocyclic compounds discussed in this contribution are one of the important factors determining food quality. The data on sensory characteristics of the various numerous compounds formed through the reaction of aldehydes with ammonia or ammonium sulfide, in the presence or in the absence of acetoin, are scattered in the literature *(57)* and thus are not easy to find. At the same time, information on sensory characteristics of compounds of this type is of primary importance to food chemists. Sultan *(29)* has compiled much of this information which is presented here in Table IV where also the appropriate references to the original literature are given.

Conclusion

The goal of this contribution was to show the diversity of the various products obtained from reactions involving acetoin, aldehydes, a source of hydrogen sulfide, and ammonia. It seems that many of these compounds are formed as a result of the degradation of reducing sugars and amino acids, or their condensation products (Amadori intermediates). These compounds play an important role as flavoring substances in various foods (meats, nuts, vegetables, etc.). It is clear that computer-assisted synthesis will make it possible to add a number of new substances to the currently existing inventory of heterocyclic food flavors and aromas.

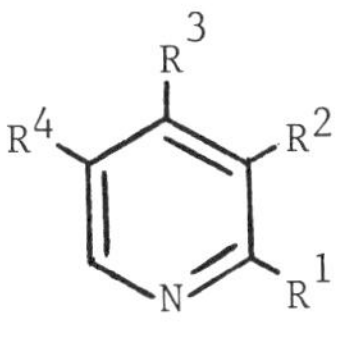

18

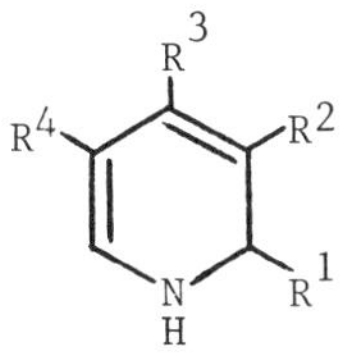

19

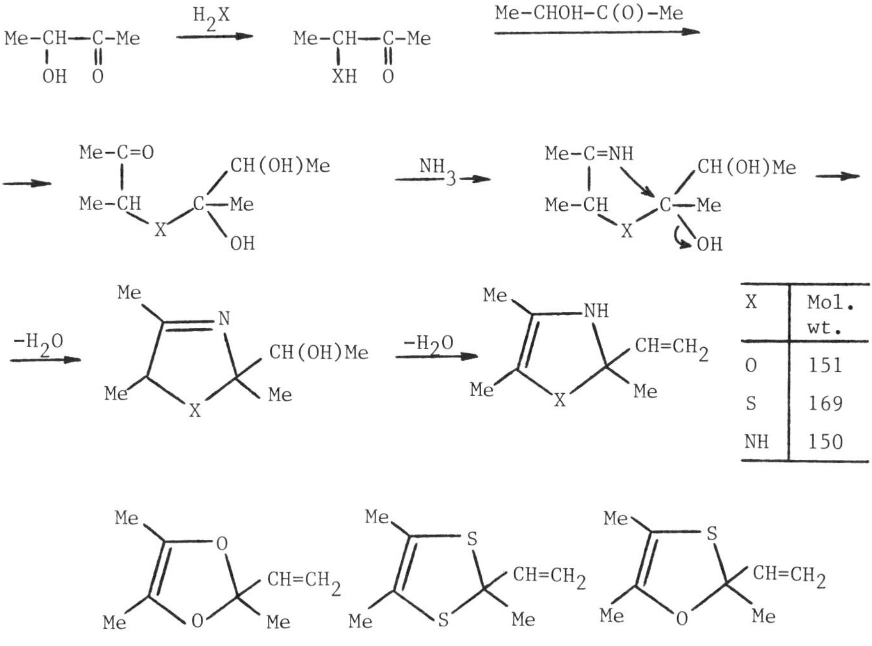

X	Mol. wt.
O	151
S	169
NH	150

Mol. wt. 140   Mol. wt. 172   Mol. wt. 156

Scheme 3. New five-membered ring heterocyclic compounds predicted from the acetoin-ammonium sulfide model system.

Table IV. Sensory Characteristics of Selected Heterocyclic Compounds

Compound	R^1	R^2	R^3	R^4	Description[a]	Threshold value[b]	Occurrence[c]	Ref.
Pyrazines								
	Me	Me	Me	Me	O: Fermented soybeans		F, S	5
					O: Lard			16
					O: Pungent; in dilution chocolate type	O 1000 ppb/water		58
						O 10 ppm/water		59
						38 ppm/oil		59
Thiophenes								
	Me	H	Me	H	O: Fried onion		F, M	60
	Me	H	H	Me	O: Fried onion	O 1.3 ppb/water	F, M	61
					O: Greenish			60
Oxazoles								
	Me	Me	Me	—	F: Boiled beef		F, S	62
					F: Nutty, sweet, green	F 5 ppb		63
Δ^3-Oxazolines								
	Me	Me	Me	—	F: Woody, musty, green	F 1.0 ppm/water	F, S	63
					O: Boiled beef			42

Thiazoles

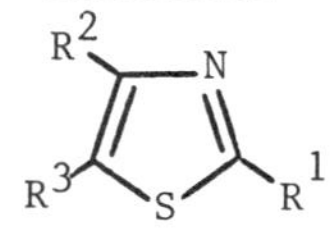

$$R^2-N=,\ R^1,\ R^3-S$$

Compound					Description	Threshold	Methods	Ref.
Thiazoles	H	Me	Me	–	F: Roasted, nutty, green	O 0.5 ppm	F, M, S	*64*
					F: Braised meat, hazelnut			*16*
					F: Meaty, broiled poultry	O 470 ppb/water		*65*
	Me	Me	Me	–	F: Cocoa, nutty		F, M, S	*64*
					O: Cocoa, hazelnut, dark chocolate, green vegetables (green beans)	O 0.05 ppm/water		*16, 17*
Δ³-Thiazolines	Me	Me	Me	–	F: Meaty, nutty, onion-like at 0.2–1 ppm: Freshly chopped, meat-like with light sour effect	F 0.5 ppm	F, M, S	*63*
					F:			*66*
					F: at 2 ppm: Chemical			
1,3,5-Dithiazines	Me	Me	Me	–	O: Roasted beef-like		F, M, S	*44*
					O: Typical note of heated meat			*62*
1,2,4-Trithiolanes	H	H	–	–	O: Roast beef		M	*67*
					O: Sulfurous			*60*
	Me	Me	–	–	O: Boiled beef	F 10 ppb/water	F, M, S	*44, 65*
					O: Onion-like			*1*
	Et	Et	–	–	O: Garlicky		F, S	*44*

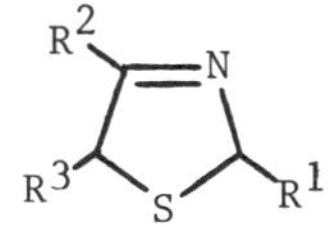

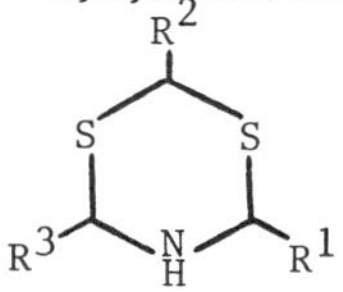

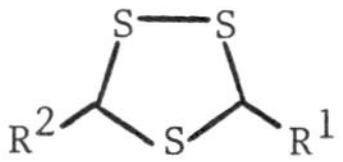

Δ³-Thiazolines

$$R^2-N,\ R^1,\ R^3-S$$

1,3,5-Dithiazines

$$R^2-S,\ R^1,\ R^3-S-N-H$$

1,2,4-Trithiolanes

$$R^2-S-S,\ R^1,\ S$$

Continued on next page

Table IV. Continued.

Compound	R¹	R²	R³	R⁴	Description[a]	Threshold value[b]	Occurrence[c]	Ref.
1,3,5-Trithianes	H	H	H	–	O: Sulfurous	F 0.04 ppb/water F 60 ppb/skim milk F 0.08 ppb/protein hydrolysate (2% solution)	F, M	67 68
	Me	Me	Me	–	F: Dissolved in water: dusty, earthy, nutty		F, M	69
	Et	Et	Et	–	F: Dissolved in water: green, allium (onion garden cress)		M	69

[a] Flavor (=odor + taste), O odor only. [b] O odor threshold (medium given), T taste threshold (medium given).
[c] Occurrence: F foods, M model systems, S synthetic.

Literature Cited

1. Schutte, L. *CRC Cri. Rev. Food Technol.* 1974, *4*, 457.
2. Takken, H. J.; Van der Linde, L. M.; De Valois, P. J.; Van Dort, H. M.; Boelens, M. In *Phenolic, Sulfur, and Nitrogen Compounds in Food Flavors*; Charalambous, G.; Katz, I., Eds.; ACS Symposium Series No. 26: American Chemical Society: Washington, DC, 1976; p. 114.
3. Hurrell, R. R. In *Food Flavors, Part A: Introduction;* Morton I. D.; MacCleod, A. J., Eds.; Elsevier: Amsterdam, 1982; p. 399.
4. Kawamura, S. In *The Maillard Reaction in Foods and Nutrition*: Waller, G. R.; Feather, M. S., Eds.; ACS Symposium Series No.215; American Chemical Society: Washington, DC, 1983; p. 3.
5. Maga, J. A.; Sizer, C. E. *CRC Crit. Rev. Food Technol.* 1973, *4*, 39.
6. Maga, J. A.; Sizer, C. E. *J. Agric. Food Chem.* 1973, *21*, 22.
7. Maga, J. A. *CRC Crit. Rev. Food Sci. Nutr.* 1975, *6*, 241.
8. Maga, J. A. *CRC Crit. Rev. Food Sci. Nutr.* 1976, *8*, 1.
9. Maga, J. A. *J. Agric. Food Chem.* 1978, *26*, 1049.
10. Maga, J. A.; Sizer, C. E. *Lebensm.-Wiss. Technol.* 1978, *11*, 181.
11. Maga, J. A. *CRC Crit. Rev. Food Sci. Nutr.* 1979, *11*, 355.
12. Maga, J. A. *J. Agric. Food Chem.* 1981, *29*, 691.
13. Maga, J. A. *CRC Crit. Rev. Food Sci. Nutr.* 1982, *16*, 1.
14. Ohloff, G. *Fortsch. Chem. Org. Naturst.* 1978, *35*, 431.
15. Ohloff, G.; Flament, I.; Pickenhagen, W. In *Flavor Chemistry; Food Reviews International:* Teranishi, R.; Hornstein, I., Eds.; M. Dekker: New York, NY, 1985; Vol. 1, p. 99.
16. Vernin, G. *Parf. Cosm. Arômes* 1979, *27*, 77.
17. Vernin, G. *Ind. Aliment. Agric.* 1980, 433.
18. Vernin, G. *Riv. Ital. EPPOS* 1981, *63*, 2.
19. Vernin, G.; Metzger, J. *Bull. Soc. Chim. Belg.* 1981, *90*, 553.
20. Vernin, G. *Perfum. Flavor.* 1982, *7*, 23.
21. *Fenaroli's Handbook of Flavor Ingredients*; 2nd Ed.; Furia, T. E.; Bellanca, N., Eds.; CRC Press: Cleveland, OH, 1975.
22. *Phenolic, Sulfur, and Nitrogen Compounds in Food Flavors;* Charalambous, G.; Katz, I., Eds.; ACS Symposium Series No. 26; American Chemical Society: Washington, DC, 1976.
23. *Progress in Flavor Research*; Land, D. G.; Nursten, H. E., Eds.; Applied Science: London, 1979.
24. *Food Flavors;* Morton, I. D.; MacLeod, J. A., Eds.; Elsevier: Amsterdam, 1982.
25. *The Chemistry of Heterocyclic Flavouring and Aroma Compounds;* Vernin, G.; Ed.; Ellis Horwood: Chichester, England, 1982.
26. *The Maillard Reaction in Foods and Nutrition:* Waller, G. R.; Feather, M. S., Eds.; ACS Symposium Series No. 215; American Chemical Society: Washington, DC, 1983.
27. *Maillard Reaction in Food: Chemical, Physiological and Technological Aspects;* Erickson, C., Ed.; Pergamon Press: Oxford, 1985.
28. Nursten, H. E. In *Development of Food Flavors;* Birch, G. G.; Lindley, M. G., Eds.; Elsevier Appl. Sci.: London, 1986; pp. 173-190.
29. Sultan, A. M. Ph.D. Thesis, Assiut University, Sohag, Egypt, 1986.
30. Vernin, G.; Petitjean, M.; Poite, J. C.; Metzger, J.; Fraisse, D.; Suon, K.-N. In *Computer Aids to Chemistry*; Vernin, G.; Chanon, M., Eds.; Ellis Horwood: Chichester, England, 1986, pp. 294-333.

31. Vernin, G.; Petitjean, M.; Metzger, J.; Fraisse, D.; Suon, K.-N.; Scharff, C. In *Capillary Gas Chromatography in Essential Oil Analysis*; Bicchi, C.; Sandra, P., Eds.; Huethig Verlag: Heidelberg, Germany, 1987, pp. 287-328.

32. Wiener, C. (Polak's Frutal Works, Inc.) *U.S. Pat.* 3,863,013, 1975.

33. Ledl, F. *Z. Lebensm.-Unters. Forsch.* 1975, *157*, 28.

34. Shibamoto, T.; Russell, G. F. *J. Agric. Food Chem.* 1977, **25**, 109.

35. Shu, C. K.; Mookherjee, B. D.; Bondarovich, H. A.; Hagedorn, M. L. *J. Agric. Food Chem.* 1985, *33*, 130.

36. Bruening, J.; Emberger, R.; Hopp, R.; Koepsel, M.; Sand, T.; Werkhoff, P. (Haarmann und Reimer GmbH) *Ger. Öffen.* DE 3,447,209, 1986; *Chem. Abstr.,* 1986, *105*, 172518c.

37. Hwang, S. S.; Carlin, J. T.; Bao, Y.; Hartman, G. J.; Ho, C. T. *J. Agric. Food Chem.* 1986, *34*, 538.

38. Kawai, T.; Irie, M.; Sakaguchi, M. *Koryo* 1986, *151*, 53.

39. Wilson, R. A.; Mussinan, C. J.; Katz, I.; Giacino, C.; Sanderson, A.; Shuster, E. J. (International Flavors and Fragrances, Inc.) *U. S. Pat.* 3,863,013, 1975.

40. Ledl, F. *Z. Lebensm.-Unters, Forsch.* 1975, *157*, 229.

41. Chen, C. C.; Chen, S. D.; Chen, J. J.; Wu, C. M. *J. Agric. Food Chem.* 1984, *32, 999.*

42. Chang, S. S.; Hirai, C.; Reddy, B. R.; Herz, K. O.; Kato, A.; Sipma, G.*Chem. Ind.* (London) 1968, 1639.

43. Flament, I.; Wilhalm, B.; Ohloff, G. In *Flavor of Foods and Beverages: Chemistry and Technology;* Charalambous, G.; Inglett, G. E., Eds.; Academic Press: New York, NY 1978; p. 15.

44. Kubota, K.; Kobayashi, A.; Yamanishi, T. *Agric. Biol. Chem.* 1980, *44,* 2677.

45. Ohloff, G.; Flament, I. *Fortschr. Chem. Org. Naturst.* 1978, *36*, 231.

46. Jennings, W. G.; Shibamoto, T. *Qualitative Analysis of Flavor and Fragrance Volatiles by Glass Capillary Gas Chromatography;* Academic Press; New York, NY, 1980.

47. Vernin, G.; Boniface, C.; Metzger, J.; Obretenov, T.; Kantasubrata, J.; Siouffi, A. M.; Larice, J. L.; Fraisse, D. *Bull. Soc. Chim. Fr.* 1987, 681.

48. Brinkman, H. W.; Copier, H.; De Leeuw, J. J. M.; Tjan, S. B. *J. Agric. Food Chem.* 1972, *20*, 177.

49. Wilson, R. A.; Mussinan, C. J.; Katz, I.; Sanderson, A. *J. Agric. Food Chem.* 1973, *21*, 873.

50. Tang, J.; Jin, Q. Z.; Shen, G. H.; Ho, C. T.; Chang, S. S. *J. Agric. Food Chem.* 1983, *31*, 1287.

51. Baltes, W.; Bochmann, G. *Z. Lebensm.-Unters. Forsch.* 1987, *184*, 179.

52. Baltes, W.; Bochmann, G. *Z. Lebensm.-Unters. Forsch.* 1987, *185,* 5.

53. Murello, M. H. Thesis, Université d'Aix-Marseille III (Saint-Jérôme), Marseille, 1986.

54. Rizzi, G. P. *J. Agric. Food Chem.* 1988, *36,* 349.

55. Vernin, G.; Párkányi, C.; Barone, R.; Chanon, M.; Metzger, J. *J. Agric. Food Chem.* 1987, *35,* 761.

56. Barone, R.; Chanon, M. In *Computer Aids to Chemistry;* Vernin, G.; Chanon, M., Eds.; Ellis Horwood: Chichester, England, 1986, pp. 19-102.

57. Fors, S. In *Maillard Reaction in Foods and Nutrition;* Waller, G. R.; Feather, M. S., Eds.; ACS Symposium Series No. 215; American Chemical Society: Washington, DC, 1983; p. 185.

58. Calabretta, P. J. *Perfum. Flavor.* 1978, *3,* 33.

59. Koehler, P. E.; Mason, M. E.; Odell, G. V. *J. Food Sci.* 1971, *36*, 816.
60. Sakaguchi, M.; Shibamoto, T. *J. Agric. Food. Chem.* 1978, *26*, 1260.
61. Boelens, M.; De Valois, P. J. Wobben, H. J.; Van der Gen, A. *J. Agric. Food Chem.* 1971, *19*, 984.
62. Ohloff, G.; Flament, I. *Heterocycles* 1978, *11*, 663.
63. Mussinan, C. J.; Wilson, R. A.; Katz,; I. Hruza, A.; Vock, M. H. In *Phenolic, Sulfur, and Nitrogen Compounds in Food Flavors;* Charalambous, G.; Katz, I. Eds.; ACS Symposium Series No. 215; American Chemical Society: Washington, DC, 1976, p. 133.
64. Pittet, A. O.; Hruza, D. E. *J. Agric. Food. Chem.* 1974, *22*, 264.
65. Buttery, R. G.; Guadagni, D. G.; Lundin, R. E. *J. Agric. Food Chem.* 1976, *24*, 1.
66. Vock, M. H.; Giacino, C.; Hruza, A.; Withycombe, D. A.; Mookherjee, B. J. Mussinan, C. J. (International Flavors and Fragrances, Inc.) *U.S. Pat.* 4,243,688, 1981.
67. Nishimura, O.; Mihara, S.; Shibamoto, T. *J. Agric. Food Chem.* 1980, *28*, 39.
68. Golovnja, R. V.; Rothe, M. *Nahrung* 1980, *24*, 141.
69. Boelens, M.; Van der Linde, L. M.: De Valois, P. J.; Van Dort, H. M.; Takken, H. J. *J. Agric. Food Chem.* 1974, *22*, 1071.

RECEIVED October 28, 1992

Chapter 5

Natural and Synthetic Compounds for Flavor Quality Maintenance

Allen J. St. Angelo

Southern Regional Research Center, Agricultural Research Service, U.S. Department of Agriculture, 1100 Robert E. Lee Boulevard, New Orleans, LA 70124

Flavor compounds are formed in meat when lipid constituents are heated during cooking. The resulting products play a major role in determining the distinctive species character of the meat, i. e., in beef, lamb, and pork. When cooked meats are stored at 4°C, lipids oxidize and produce secondary reaction products that contribute to off-flavor. During this process, called warmed-over flavor, or more recently, meat flavor deterioration (MFD), a loss of desirable meaty flavors and an increase in off-flavors occur. The primary catalysts of lipid oxidation in cooked meat appear to be metal ions or other free radical generating systems. Antioxidants that function primarily as free radical scavengers and those that are chelators were used as probes to investigate the mechanisms involved in MFD. This paper presents results of these studies and compares several classes of antioxidants in retarding lipid oxidation and maintaining flavor.

Meat flavor deterioration (MFD), formerly referred to as warmed-over flavor, is described as the loss of desirable meaty flavor with an increase in off-flavors (1-3). During this process, the increase in off-flavors is primarily contributed by lipid oxidation reactions. As lipids oxidize, they produce mixtures of aldehydes, ketones and alcohols that contribute to the off-flavors observed. Many of these compounds have been identified and the increase in their intensities during storage have been well documented (1, 4-6).

Over the years, scientists have used many types of antioxidants in a variety of foods to retard or inhibit lipid oxidation and, thus, increase shelf-life and preserve quality. These antioxidants include free radical scavengers, chelators, and oxygen absorbers. While there are numerous antioxidants available to food scientists, the objective of this report was to discuss several of these antioxidants as they relate to meat flavor quality research and to show how they were used to retard lipid oxidation and prevent meat flavor deterioration in ground beef patties.

Antioxidants

According to Chipault (*7*), an antioxidant may be described as a substance that in a small quantity is able to prevent or greatly retard the oxidation of materials such as fats. Later, Sherwin noted that food antioxidants are substances that function as free radical inhibitors by interfering with the free radical mechanism fundamental to autoxidation (*8*). In its simplest terms and according to Hamilton (*9*), the classical free radical route to the oxidation of lipids depends on the production of free radicals, R•, from unsaturated lipid molecules, RH, by their interaction with oxygen in the presence of a catalyst. This reaction represents the initiation phase in which oxygen reacts with a carbon-carbon double bond. Consequently, the lipid molecule becomes a free radical, which is then oxidized. These free radicals are continually being generated during the propagation phase that continues until some event takes place that terminates the reaction. The antioxidant reacts with the free radicals being generated in the propagation phase to terminate the reaction. Once reacted with a substrate molecule, the antioxidant free radical formed cannot initiate or propagate the oxidation reaction (*8*). It is important to be aware that substances that contain a high degree of polyunsaturated fatty acids are more susceptible to oxidation than those that contain less. It is also equally important to realize that in the practical world, antioxidants do not prevent oxidation *in toto*, but they do delay the oxidative process and thus, increase shelf-life.

Antioxidants: Synthetic and Natural

Synthetic Antioxidants. Some of the more popular synthetic antioxidants used by the food industry are phenolic compounds such as butylated hydroxyanisole (BHA), butylated hydroxytoluene (BHT), tertiarybutylhydroxyquinone, (TBHQ) and propyl gallate (PG). These compounds, referred to as "primary antioxidants" are very effective in terminating the oxidation of lipids. For example, BHA and BHT are fairly stable to heat and are often used in foods that are processed by heating (*10*). Antioxidants that are stable to heat have the property referred to as "carry through" or "carry over". On the other hand, PG decomposes at 148°C, whereas TBHQ has poor carry through properties in baking applications, but excellent carry through properties in frying applications (*11*). These antioxidants function by interfering with the free radical mechanism. Some antioxidants, particularly BHA and BHT, are also used in various combinations with the resulting effects being synergistic. Synergism can be defined as that instance when the combined effect from two antioxidants is greater than the sum of the individual effects obtained when they are used alone (*8*). BHA is also synergistic with PG; however, the combination of BHT and PG can also result in negative synergism, in which the shelf life of a food is less than that anticipated from the sum of the effects of each when used alone (*12*).

Chelators. The type of positive synergistic effects described above are not to be confused with synergism as referred to when an antioxidant is added with an acid chelator, such as citric acid. Acid chelators, or sequestrants, are referred to as secondary antioxidants (13). These include such compounds as citric, tartaric, oxalic, malic and succinic acids and sodium triphosphate and pyrophosphate, all of which are used as antioxidants in food systems. Phytic acid (inositol hexaphosphate) was recently reported to inhibit lipid oxidation (2, 14-15). However, chelators are not antioxidants in the sense that they do not function as do the phenolic-type antioxidants.

When acid chelators are combined with one of the phenolic-type antioxidants, they can act synergistically. This type of synergism has been referred to as acid synergism (8). In this case, the primary role of a chelator is to bind metals, or metalloproteins that promote oxidation and thus, allow the antioxidant to perform its function and capture free radicals. In this manner, acid synergism is different than the BHA/BHT system.

Natural Antioxidants. Because of the growing concern over the safety of synthetic antioxidants, such as BHA and BHT, in regards to health, there is an increase in the use of naturally derived antioxidants. One that is widely used is ascorbic acid, whose chemistry is multifarious. Ascorbic acid and its esterified derivatives function as antioxidants by protecting double bonds and scavenging oxygen. Ascorbic acid can act as an oxygen scavenger by preferentially being oxidized over other oxidizable components in the food system owing to the 2- and 3-positions being unsubstituted. In this case oxygen is removed and converted into water at the expense of ascorbic acid, which is oxidized to form dehydroascorbic acid. When a metal is present, such as iron, the ability of ascorbic acid to scavenge oxygen is faster due to a chelated intermediate. In this system, a reducing agent, such as reduced glutathione, is needed to convert the dehydroascorbic acid back to ascorbic acid (16).

Important chemical characteristics of ascorbic acid include its redox properties, which enables ascorbic acid to function as a reducing agent and a free radical scavenger. As such, ascorbic acid can donate a hydrogen atom for free radical chain inhibition. In this way, it can act as a synergist with tocopherol by converting oxidized tocopherol back to the reduced form. In these reactions, ascorbic acid can be regenerated by its action with reduced glutathione or NADH+ (16).

Another function of ascorbic acid is to generate Fe(II) from Fe (III), which is part of the Haber-Weiss-Fenton Reaction (17). In that reaction, peroxide is reduced to the hydroxyl radical, HO•, an extremely powerful oxidizing agent. It should also be noted that ascorbic acid can chelate metals (18), and will promote the carbohydrate-amine browning reaction (19-20).

Interest has been shown in using derivatives of ascorbic acid as antioxidants. One such compound is ascorbyl palmitate. In the structure of ascorbyl palmitate, the 2- and 3-positions are occupied by hydroxyl groups; the 6-position contains the substituted fatty acid. Other derivatives, synthesized by Scib and associates were ascorbate-2 phosphate and ascorbate 2-triphosphate (21). Both of these compounds were reported to inhibit lipid oxidation in ground meat as measured by chemical means (2). Ascorbate-2-phosphate was also found to inhibit MFD in beef as measured by sensory means (2). The ascorbate-2-triphosphate was not tested as an inhibitor of MFD in this study, but one assumes that it too should prevent MFD. Ascorbate-2-phosphate was later used to inhibit off-flavor development in cooked frozen turkeys (*22*). For comprehensive reviews on ascorbic acid, the reader is referred to Bauernfeind and Pinkert (*23*) and Seib and Tolbert (*24*).

Spices and Vegetable Oils. There is renewed interest in the use of naturally occurring compounds isolated from spices and vegetable oils. Chipault et al. (*25-26*) found many spices and herbs that contained compounds with antioxidant activity in food systems. Rosemary and sage were two that had the highest activity. Houlihan and Ho (*27*) reported that oleoresin rosemary extracts contained several compounds, including rosmanol, carnosol, rosmaridiphenol, and rosmariquinone that possesses antioxidant activity. All compounds exhibited antioxidant activity similar to or greater than BHA. Vegetable oils and fruits also contain compounds that exhibit antioxidant activity as reported in several reviews (*27-30*). However, the use of extracts from vegetable oils and fruits as antioxidants is beyond the scope of this report.

Effect of Antioxidants on Stored Ground Beef Patties

As ground beef is stored at 4°C, complex chemical reactions begin to take place. One of the first reactions that occurs is the oxidation of polyunsaturated fatty acids, which leads to the production of secondary reaction products (*1*). Within the first few hours after storage, hexanal, a decomposition product of linoleic acid (*31*), can be observed by direct gas chromatography (*4*). Soon afterwards, other volatile compounds are also observed. As these chemical reactions take place, a trained sensory panel can measure the onset of off-flavors and a loss of desirable flavors (*2, 32*). These markers are indicators of the MFD process (*5*). One of the ways to prevent the MFD process from developing is by using antioxidants. Therefore, experiments with ground beef patties, treated with primary antioxidants, chelators and compounds extracted from natural sources, were begun. In some cases, vacuum was used to obtain a synergistic effect. The experimental samples (and controls) were examined by chemical, instrumental and sensory methods to assess their quality in regards to MFD. Results of these experiments are presented below.

Analytical Methodologies. Chemical attributes, such as thiobarbituric acid reactive substances (TBARS), were measured by the distillation method of Tarladgis et al. (*33*), and are reported as mg malondialdehyde per kg sample. A standard curve was made with 1,1,3,3-tetraethoxypropane. Other chemical attributes, such as hexanal, pentanal, 2,3-octanedione and nonanal, were determined by obtaining the volatile profile of beef samples utilizing the direct gas chromatographic (GC) method of Dupuy et al. (*4*). In the direct GC method, ground beef is placed into a glass liner, which is inserted into a heated (110°C) external closed inlet device (ECID) connected to a GC. The sample is purged for 10 minutes, whereby the volatiles are deposited onto the column where they are eluted by normal GC procedures. Sensory profiles of grilled beef patties, fresh and stored, were performed by the method of Meilgaard et al. (*34*). Evaluations were accomplished by a trained panel of 12 using the Spectrum™ universal intensity scale, 0-15, as described by Meilgaard et al (*34*). In this method, 0 is not detectable whereas 15 is the most intense. Sensory attributes included the following descriptors as originally described by Johnsen and Civille (*35*) and modified by Love (*32*): cooked beef/brothy, browned/caramel and sweet (the most significant desirable flavor notes) and painty, cardboardy, sour and bitter (the most significant off-flavor notes). In the present study, cooked beef/brothy was further broken down into two descriptors, namely beefy/meaty and brothy. All results were statistically analyzed using SAS for analysis of variance. Chemical and sensory values were presented in the tables as means with standard error (±s.e.m.).

Correlation of Chemical, Instrumental and Flavor Attributes. As MFD progressed during storage of precooked ground beef at 4°C, several volatile compounds increased in their intensities. Hexanal is the most abundant single volatile component and is an excellent marker for lipid oxidation and MFD. Other volatile compounds that can be used as markers are pentane, pentanal, 2-heptanone, heptanal, 2,3-octanedione, 2-pentylfuran, nonanal, and the *trans, cis-* and *trans, trans*-2,4-decadienals (*1-6*). Concurrently with the increase in volatile compounds, there is also a significant increase in TBARS numbers. TBARS are known to be highly correlated to hexanal and 2,3-octanedione (*1, 3, 5, 36*). Hexanal is the only volatile compound that increased with storage and is highly correlated to one of the off-flavor notes, painty. As painty increased with storage, so did hexanal. On the other hand, the desirable flavor note, cooked beef/brothy, decreased (*5*).

Effects of Primary Antioxidants on Stored Ground Beef

Standards and Controls. In all experiments, the 85 g standard patties were made from freshly ground top round steaks (excess fat trimmed) and immediately frozen in covered glass petri plates until the day of the assay. The fat content was routinely from 4-5%, determined by the method of Koniecko (*37*). The standards generally had relatively low values for hexanal, total volatiles (TV) and TBARS, and low intensity values for painty (PTY), cardboardy (CBD), sour (SUR) and bitter (BTR). These results indicated the absence of lipid oxidation and no formation of off-flavors. As expected, the desirable flavor notes, cooked beef/brothy (CBB), beefy/meaty (BM), brothy (BRO), browned/caramel (BRC) and sweet (SWT) had high intensity values.

Control samples containing added water, 1.18%, the carrier for the antioxidants, were allowed to sit and marinate overnight (generally 18 hours) at 4°C. The next morning, 85 g patties were made and placed in covered glass petri plates. Half of the patties were cooked by grilling (7 min/side) and stored for 2 days at 4°C to develop MFD. The remaining patties were frozen for 2 days. On the morning of the assay, all frozen samples were thawed and cooked. The MFD experimental samples were rewarmed in an oven prior to assaying. Analytical assessment of the raw "0-day" experimental patties, see Cntrl-0 in Tables 1-4, indicated that the initial stage of MFD had already begun. For

Table 1. Effect of propyl gallate on stored raw and cooked beef patties (N=3)

	HX	TV	TBA	CBB[a]	BM[b]	BRO[b]	BRC	SWT	PTY	CBD	SUR	BTR
Raw patties, marinated (stored) 17 hrs, 4°C.												
Standard[c]	14	158	4.03	6.17	3.82	2.35	2.64	1.21	0.26	0.33	0.71	0.41
(±s.e.m.)	10	6	2.61	n.a	0.04	0.08	0.23	0.07	0.04	0.06	0.12	0.05
Cntrl-0[d]	23	176	6.10	5.91	3.77	2.37	2.95	1.14	0.52	0.72	0.86	0.65
(±s.e.m.)	17	28	4.61	n.a	0.03	0.04	0.30	0.07	0.14	0.05	0.11	0.03
PG-25[e]	9	169	2.17	5.64	3.58	2.24	2.65	1.05	0.42	0.68	0.86	0.56
(±s.e.m.)	2	28	1.23	n.a	0.08	0.06	0.13	0.04	0.06	0.10	0.02	0.09
PG-50	16	173	1.87	6.30	3.65	2.26	2.75	1.11	0.32	0.44	0.81	0.57
(±s.e.m.)	10	19	0.91	n.a.	0.03	0.05	0.17	0.11	0.12	0.09	0.11	0.06
Cooked patties stored 2-days, 4°C.												
Cntrl-0[d]	23	176	6.10	5.91	3.77	2.37	2.95	1.14	0.52	0.72	0.86	0.65
(±s.e.m.)	17	28	4.61	n.a	0.03	0.04	0.30	0.07	0.14	0.05	0.11	0.03
Cntrl-2[f]	61	248	8.97	5.07	3.24	1.96	2.38	0.86	1.21	1.14	1.07	0.84
(±s.e.m.)	33	51	2.87	n.a.	0.07	0.04	0.09	0.07	0.28	0.29	0.02	0.12
PG-25	27	202	3.87	5.09	3.26	2.01	2.39	0.98	0.87	0.85	0.96	0.83
(±s.e.m.)	17	5	1.62	n.a.	0.06	0.06	0.20	0.04	0.23	0.11	0.00	0.08
PG-50	14	196	2.20	5.45	3.55	2.23	2.47	1.08	0.55	0.78	0.85	0.61
(±s.e.m.)	5	22	1.03	n.a.	0.12	0.01	0.05	0.02	0.09	0.27	0.09	0.10

[a] N=1

[b] N=2

[c] Raw, stored frozen 2-days, thawed, cooked, assayed.

[d] Control sample (0-day), stored frozen 2-days, thawed, cooked, assayed.

[e] Concentrations of additives in parts per million.

[f] Control sample (MFD), cooked, stored 2 days, rewarmed, assayed.

Abbreviations: HX, hexanal, peak area x 1000; TV, total volatiles, peak area x 1000; TBA, 2-thiobarbituric acid reactive substances, mg MDA/kg sample; CBB, cooked beef/brothy, all sensory values are mean intensity scores; BM, beefy/meaty; BRO, brothy; BRC, browned/caramel; SWT, sweet; PTY, painty; CBD, cardboardy; SUR, sour; BTR, bitter.

Table 2. Effect of TBHQ on stored raw and cooked beef patties (N=3)

	HX	TV	TBA	CBB[a]	BM[b]	BRO[b]	BRC	SWT	PTY	CBD	SUR	BTR
Raw patties, marinated (stored) 17 hrs, 4°C.												
Standard[c]	31	177	3.80	6.31	3.58	2.13	2.63	1.09	0.20	0.31	0.60	0.40
(±s.e.m.)	16	24	1.15	n.a.	0.03	0.04	0.18	0.14	0.03	0.09	0.07	0.03
Cntrl-0[d]	10	133	5.13	5.20	3.48	2.44	2.51	0.91	0.42	0.58	0.91	0.59
(±s.e.m.)	4	30	2.18	n.a.	0.04	0.11	0.10	0.10	0.20	0.09	0.07	0.10
TBHQ-25[e]	6	151	1.10	5.92	3.47	2.39	2.60	0.99	0.24	0.42	0.79	0.46
(±s.e.m.)	3	13	0.15	n.a.	0.02	0.12	0.12	0.19	0.06	0.16	0.12	0.03
TBHQ-50	3	85	1.33	5.63	3.56	2.40	2.62	1.12	0.22	0.44	0.72	0.45
(±s.e.m.)	1	13	0.22	n.a.	0.03	0.10	0.18	0.07	0.02	0.03	0.07	0.11
Cooked patties stored 2-days, 4°C.												
Cntrl-0[d]	10	133	5.13	5.20	3.48	2.44	2.51	0.91	0.42	0.58	0.91	0.59
(±s.e.m.)	4	30	2.18	n.a.	0.04	0.11	0.10	0.10	0.20	0.09	0.07	0.10
Cntrl-2[f]	63	241	9.80	4.59	3.15	2.15	2.30	0.89	1.22	1.27	0.91	0.76
(±s.e.m.)	21	15	2.44	n.a.	0.03	0.02	0.17	0.09	0.33	0.26	0.08	0.11
TBHQ-25	8	144	1.23	5.34	3.27	2.26	2.28	1.03	0.23	0.43	0.76	0.49
(±s.e.m.)	2	6	0.26	n.a.	0.04	0.23	0.20	0.15	0.01	0.04	0.09	0.10
TBHQ-50	10	122	0.90	5.52	3.43	2.36	2.50	1.02	0.23	0.39	0.72	0.61
(±s.e.m.)	2	6	0.90	n.a.	0.05	0.03	0.24	0.13	0.04	0.12	0.19	0.04

[a] N=1
[b] N=2
[c] Raw, stored frozen 2-days, thawed, cooked, assayed.
[d] Control sample (0-day), stored frozen 2-days, thawed, cooked, assayed.
[e] Concentrations of additives in parts per million.
[f] Control sample (MFD), cooked, stored 2 days, rewarmed, assayed.
  Abbreviations: HX, hexanal, peak area x 1000; TV, total volatiles, peak area x 1000;
  TBA, 2-thiobarbituric acid reactive substances, mg MDA/kg sample; CBB, cooked beef/brothy,
  all sensory values are mean intensity scores; BM, beefy/meaty; BRO, brothy; BRC,
  browned/caramel; SWT, sweet; PTY, painty; CBD, cardboardy; SUR, sour; BTR, bitter.

Table 3. Effect of TENOX 20 (TBHQ/Citric Acid) on stored raw and cooked beef patties (N=3)

	HX	TV	TBA	CBB[a]	BM[b]	BRO[b]	BRC	SWT	PTY	CBD	SUR	BTR
Raw patties, marinated (stored) 17 hrs, 4°C.												
Standard[c]	14	124	2.14	6.18	3.25	2.27	2.41	1.08	0.18	0.20	0.61	0.33
(±s.e.m.)	4	24	0.99	n.a.	0.15	0.07	0.31	0.15	0.02	0.05	0.08	0.06
Cntrl-0[d]	24	167	3.15	5.47	2.94	1.91	2.48	0.93	0.38	0.46	0.74	0.54
(±s.e.m.)	8	35	0.99	n.a.	0.06	0.02	0.31	0.17	0.02	0.09	0.05	0.18
TNX20-25	6	150	1.09	5.62	3.08	1.94	2.50	1.01	0.24	0.34	0.66	0.41
(±s.e.m.)	2	21	0.62	n.a.	0.09	0.05	0.26	0.15	0.05	0.10	0.06	0.09
TNX20-50	6	123	0.92	5.57	3.09	1.99	2.36	1.01	0.29	0.44	0.67	0.56
(±s.e.m.)	2	10	0.53	n.a.	0.14	0.01	0.33	0.18	0.05	0.02	0.10	0.11
Cooked patties stored 2-days, 4°C.												
Cntrl-0[d]	24	167	3.15	5.47	2.94	1.91	2.48	0.93	0.38	0.46	0.74	0.54
(±s.e.m.)	8	35	0.99	n.a.	0.06	0.02	0.31	0.17	0.02	0.09	0.05	0.18
Cntrl-2[f]	62	223	8.16	4.66	2.80	1.90	2.11	0.87	1.09	1.05	0.84	0.62
(±s.e.m.)	26	65	1.41	n.a.	0.29	0.07	0.21	0.12	0.34	0.30	0.19	0.14
TNX20-25	22	144	3.09	5.49	2.92	2.01	2.28	0.85	0.43	0.57	0.69	0.59
(±s.e.m.)	6	11	1.07	n.a.	0.13	0.05	0.23	0.16	0.06	0.09	0.07	0.16
TNX20-50	11	154	1.37	5.62	3.12	1.97	2.50	0.99	0.33	0.34	0.63	0.50
(±s.e.m.)	5	8	1.00	n.a.	0.08	0.09	0.38	0.18	0.05	0.01	0.04	0.18

[a] N=1
[b] N=2
[c] Raw, stored frozen 2-days, thawed, cooked, assayed.
[d] Control sample (0-day), stored frozen 2-days, thawed, cooked, assayed.
[e] Concentrations of additives in parts per million.
[f] Control sample (MFD), cooked, stored 2 days, rewarmed, assayed.
Abbreviations: HX, hexanal, peak area x 1000; TV, total volatiles, peak area x 1000; TBA, 2-thiobarbituric acid reactive substances, mg MDA/kg sample; CBB, cooked beef/brothy, all sensory values are mean intensity scores; BM, beefy/meaty; BRO, brothy; BRC, browned/caramel; SWT, sweet; PTY, painty; CBD, cardboardy; SUR, sour; BTR, bitter.

Table 4. Effect of TENOX 4A (BHA/BHT) on stored raw and cooked beef patties (N=3)

	HX	TV	TBA	CBB[a]	BM[b]	BRO[b]	BRC	SWT	PTY	CBD	SUR	BTR
Raw patties, marinated (stored) 17 hrs, 4°C.												
Standard[c]	12	179	3.31	6.24	3.49	2.22	2.69	1.15	0.25	0.29	0.64	0.43
(±s.e.m.)	3	42	0.28	n.a.	0.18	0.07	0.31	0.11	0.03	0.03	0.05	0.06
Cntrl-0[d]	10	133	5.13	5.53	3.10	2.06	2.50	1.21	0.30	0.49	0.73	0.49
(±s.e.m.)	4	30	2.18	n.a.	0.11	0.03	0.18	0.10	0.03	0.02	0.04	0.05
TNX4A-25[e]	5	162	1.74	5.59	3.28	2.09	2.44	1.12	0.32	0.43	0.65	0.56
(±s.e.m.)	2	32	0.16	n.a.	0.11	0.05	0.13	0.07	0.06	0.04	0.08	0.02
TNX4A-50	4	148	1.64	5.76	3.31	2.09	2.56	1.15	0.28	0.39	0.58	0.59
(±s.e.m.)	0	19	0.20	n.a.	0.24	0.10	0.37	0.11	0.04	0.03	0.04	0.08
Cooked patties stored 2-days, 4°C.												
Cntrl-0[d]	10	139	3.80	5.53	3.10	2.06	2.50	1.21	0.30	0.49	0.73	0.49
(±s.e.m.)	4	30	2.18	n.a.	0.11	0.03	0.18	0.10	0.03	0.02	0.04	0.05
Cntrl-2[f]	48	240	9.66	4.29	2.84	1.77	2.12	0.96	0.99	1.09	0.92	0.84
(±s.e.m.)	4	50	0.56	n.a.	0.09	0.03	0.08	0.08	0.08	0.14	0.10	0.13
TNX4A-25	13	191	2.41	5.56	3.02	1.93	2.51	1.15	0.42	0.57	0.70	0.58
(±s.e.m.)	2	42	0.36	n.a.	0.10	0.07	0.25	0.12	0.10	0.06	0.04	0.08
TBX4A-50	6	150	2.46	5.23	3.28	2.16	2.40	1.03	0.54	0.41	0.67	0.51
(±s.e.m.)	1	17	0.90	n.a.	0.17	0.13	0.10	0.03	0.20	0.04	0.02	0.14

[a] N=1
[b] N=2
[c] Raw, stored frozen 2-days, thawed, cooked, assayed.
[d] Control sample (0-day), stored frozen 2-days, thawed, cooked, assayed.
[e] Concentrations of additives in parts per million.
[f] Control sample (MFD), cooked, stored 2 days, rewarmed, assayed.
Abbreviations: HX, hexanal, peak area x 1000; TV, total volatiles, peak area x 1000;
TBA, 2-thiobarbituric acid reactive substances, mg MDA/kg sample; CBB, cooked beef/brothy,
all sensory values are mean intensity scores; BM, beefy/meaty; BRO, brothy; BRC,
browned/caramel; SWT, sweet; PTY, painty; CBD, cardboardy; SUR, sour; BTR, bitter.

example, the MFD and lipid oxidation indicators were present, i.e., the intensities of the desirable flavor notes were slightly decreased, the undesirable flavor notes and the chemical markers were slightly increased when compared to those of the standards. However, assay values for the 2-day samples, Cntrl-2, indicated that MFD and lipid oxidation were progressing as expected. These phenomena were confirmed by increased intensities for hexanal, TV, TBARS, PTY, CBD, SUR and BTR. Simultaneously, the intensities of CBB, BM, BRO and SWT were decreased. Clearly, MFD had progressed in the sample stored for 2 days at 4°C.

Antioxidant-Treated Beef Patties. The effect of several primary antioxidants, PG, TBHQ, Tenox 20 (which contains TBHQ and citric acid) and Tenox 4A (which contains BHA and BHT) were also evaluated by instrumental, chemical and sensory methods for their effectiveness in raw/stored and cooked/stored beef, see Tables 1-4. These particular antioxidants were chosen based on their antioxidant effects on MFD as determined previously (*14*) and because of their GRAS status. Experimental samples were prepared similarly to those of the 0- and 2-day controls, except the antioxidants, either dissolved or suspended in water, were mixed into the raw ground meat.

PG, TBHQ and the two Tenox formulations at the 25 or 50 ppm levels were found to be very effective in preventing MFD in the raw (stored overnight) and the 2-day cooked/stored samples. In the raw antioxidant-treated samples, the 0-day controls, the MFD and lipid oxidation markers were generally decreased and the on-flavor notes increased in comparison to those of the 0-day raw controls. These trends were particularly observed in the samples treated with 50 ppm. Of the four antioxidants examined, TBHQ was the most effective.

When the antioxidants were used in the cooked/stored samples, data indicated that they were very effective in inhibiting lipid oxidation and MFD. The chemical and off-flavor indicators were reduced and the on-flavor notes were increased. Thus, phenolic-type primary antioxidants that function as free radical scavengers are very effective tools for preventing lipid oxidation and MFD in ground beef. It should also be noted that the intensity of the desirable flavor notes remained at very high levels, which meant that the patties retained their beefy tastes. Therefore, for an antioxidant to be highly effective, it should not only prevent lipid oxidation, but it should also retain the desirable flavor properties of the food commodity.

Effects of Secondary Antioxidants on Stored Ground Beef

Secondary antioxidants, i.e., sequestrants or chelators, are important compounds in the prevention of lipid oxidation. The effect of chelators tested varied with the different compounds. Of those chelators tested, ethylenediaminetetraacetic acid (EDTA, tetrasodium salt) and sodium phytate were the most effective inhibitors of lipid oxidation (so indicated by low hexanal and TBARS values) and MFD (as seen by high CBB and low PTY and CBD intensity values), see Table 5. Sodium phytate was previously shown to chelate iron and thus, was proposed as a food antioxidant(*15*). Sodium citrate at a concentration of 500

Table 5. Effect of secondary antioxidants on chemical and sensory properties of cooked beef patties stored 2 days at 4°C

	Hexanal[a]	TBARS[b]	CBB[c]	PTY[c]	CBD[c]
Standard[d]	16	1.99	6.4	0.1	0.1
(±s.e.m.)	4	0.20	0	0	0
Control[e]	12	4.51	5.7	0.3	0.5
(±s.e.m.)	2	0.07	0.1	0	0
Control[f]	82	6.97	4.9	1.7	1.7
(±s.e.m.)	10	0.57	0.1	0.1	0.1
SC[c] (500)[g]	81	5.42	5.3	1.5	1.6
EDTA (125)[g]	5	1.34	5.8	0.2	0.4
EDTA (500)[g]	0.2	0.04	5.6	0.2	0.4
STP (500)[g]	112	5.43	4.7	1.8	1.9
STP (1000)[g]	111	6.60	5.2	1.2	1.0
SPP (125)[g]	127	4.91	5.2	1.2	2.0
SPP (500)[g]	33	3.35	4.7	1.4	1.4
SP (500)[g]	19	1.00	5.6	0.9	0.9

[a] Peak area x 1000
[b] mg MDA/kg ground meat
[c] Abbreviations: CBB, cooked beef/brothy; PTY, painty; CBD, cardboardy; SC, sodium citrate; STP, sodium triphosphate; SPP, sodium pyrophosphate; SP, sodium phytate.
[d] Fresh cooked ground beef frozen immediately after cooking, N=16
[e] Marinated overnight with water, cooked and assayed immediately; zero-day sample, N=13
[f] Marinated overnight with water, cooked, stored 4°C for 2 days, assayed, N=13
[g] Parts per million
SOURCE: Adapted from ref. 14.

ppm was totally ineffective in both inhibition of lipid oxidation and prevention of MFD. Sodium triphosphate (STP) and sodium pyrophosphate (SPP) were slightly active inhibitors; however, SPP at 500 ppm was more effective in prevention of lipid oxidation than STP at the same concentration. This table includes data from preliminary experiments and thus, represents trends of secondary antioxidants as inhibitors of MFD. More research is needed in this area to determine which chelators are the most effective inhibitors of lipid oxidation while maintaining desirable flavor.

Synergism

Another contributor of MFD is the presence of molecular oxygen. An initial study determined the optimum level of free radical scavenger and chelator required to prevent MFD when added to a meat sample under vacuum (*38*). Based on those results, subsequent experiments were designed to investigate the synergistic effects of the two compounds, propyl gallate and EDTA (tetrasodium salt), on MFD both in the presence or absence of oxygen.

Experimental Design. Fresh raw ground beef patties (prepared from top round as described above) were divided into 10 separate packages of 850 grams each. Each portion was mixed with either 10 ml of water containing either additives or water only. The additives were 50 ppm of the primary antioxidant, PG, and 100 ppm of the secondary antioxidant, EDTA. The control samples having no vacuum and no additive, were stored frozen. A second control, ground beef plus water, was marinated (stored) overnight and then cooked (grilled), and either stored frozen as 0-day samples or stored at 4°C for 3 days. The experimental samples were mixed with the additive or marinated overnight, cooked and then stored similarly to those of the controls. All samples were made into 85 gram patties and stored in covered glass petri plates.

Results from Experiments with PG and EDTA, with and without Vacuum. The data, summarized in Table 6 and described in more detail in a previous report (*5*), showed that the 3-day control sample, ENO, developed MFD based on the observed increases in hexanal, TBARS, PTY and CBD and a decrease in CBB. When no vacuum was used but PG was added to the system, MFD was greatly decreased (ENO *vs* ENA). However, EDTA alone was not as effective as PG. Moreover, EDTA had no effect on CBD (ENO [1.35] *vs* ENB [1.39]). By combining the two, PG plus EDTA, sample ENX, there was a synergistic effect observed as indicated by the TBARS, PTY and CBD values.

When ground meat that contained no additives was stored with vacuum, EVO, MFD was minimized (ENO *vs* EVO). By using vacuum on samples treated with PG, there was a reduction in MFD as judged by hexanal, PTY and CBD values (EVA *vs* ENA). Samples containing EDTA that were stored under vacuum also had reduced MFD (EVB *vs* ENB). More noticeably, CBD also decreased, from 1.39 in ENB to 0.68 in EVB. By storing samples containing PG and EDTA with vacuum, a definite synergistic effect was observed (EVX *vs* EVA and EVB). Hexanal, TBARS, PTY and CBD had the lowest intensities and CBB had a high intensity score in the EVX sample.

Table 6. Effect of propyl gallate (50 ppm) and EDTA (100 ppm) with and without vacuum on chemical and sensory properties of cooked beef patties stored 3 days at 4°C

	Hexanal[a]	TBARS[b]	CBB[c]	PTY[c]	CBD[c]
BNO[c]	45	3.02	5.88	0.32	0.40
(±s.e.m.)	6	0.48	0.39	0.24	0.03
ENO	232	10.16	4.82	1.36	1.35
(±s.e.m.)	33	0.62	0.69	0.88	0.11
ENA	63	1.42	5.82	0.61	0.90
(±s.e.m.)	4	0.13	0.63	0.51	0.08
ENB	164	7.15	4.78	1.21	1.39
(±s.e.m.)	16	1.41	0.80	0.88	0.26
ENX	82	1.14	5.57	0.42	0.51
(±s.e.m.)	12	0.25	0.59	0.40	0.03
EVO	136	5.79	5.22	0.89	0.99
(±s.e.m.)	14	1.18	0.52	0.67	0.11
EVA	37	1.77	5.43	0.47	0.59
(±s.e.m.)	6	0.25	0.61	0.38	0.03
EVB	43	1.86	5.53	0.47	0.68
(±s.e.m.)	12	0.54	1.07	0.49	0.16
EVX	30	1.47	5.67	0.36	0.39
(±s.e.m.)	9	0.22	0.50	0.39	0.02

[a] Peak area x 1000

[b] mg MDA/kg ground meat

[c] Abbreviations: CBB, cooked beef/brothy; PTY, painty; CBD, cardboardy; BNO, blind standard control with no vacuum (N) and no additives (O); ENO, the 3-day MFD experimental sample; ENA, ENB, or ENX represented the MFD samples plus PG (A), EDTA (B), and PG plus EDTA (X) with no vacuum; EVO, the 3-day MFD sample plus vacuum (V) with no additives; EVA, EVB, or EVX, the 3-day MFD samples plus PG and EDTA plus vacuum.

Additional particulars should also be noted. First, vacuum when used in combination with EDTA, showed a very large decrease in lipid oxidation when compared to those EDTA-treated samples without vacuum (EVB *vs* ENB). This was not the case when the primary antioxidant, PG, was used with or without vacuum (EVA *vs* ENA). In this latter case, PG was effective with or without vacuum, although there were some intensity differences observed and by combining with vacuum, MFD and lipid oxidation was decreased. Next, vacuum alone lowered lipid oxidation (EVO *vs* ENO). Finally, by combining vacuum with EDTA a strong synergistic effect was observed.

Application of Natural Antioxidants in Foods

Effect of Sodium Ascorbate on Stored Ground Beef. The effectiveness of sodium ascorbate (SA) as an antioxidant was determined on both raw and cooked stored ground beef. The experimental models, i.e., standards, controls and experimental samples were similar to those described in the section on primary antioxidants. Data from the use of 150 and 200 ppm and 250 and 500 ppm are reported in Tables 7 and 8, respectively. At a concentration of 500 ppm, SA was the most effective inhibitor of MFD and lipid oxidation in both raw and cooked/stored ground beef when compared to the other three concentrations. The markers of MFD and lipid oxidation were suppressed to the greatest extent and the desirable beefy flavor markers were maintained at their highest intensities when SA was added to the beef samples. SA was effective as an inhibitor at all levels tested.

According to Sato and Herring (*39*), ascorbic acid can function as an antioxidant by interfering with the free radical mechanism due to the presence of the ene-diol portion of the molecule. However, St. Angelo et al (*5*) reported that reductone-type compounds, such as maltol, kojic acid, 3-hydroxyflavone, etc., were effective antioxidants and suggested that the alpha ketol structure may also play an important role concerning free radical mechanisms.

Rosemary: Effect on Stored Ground Beef. The effect of two different formulations of rosemary, water-base (rosemary-W) and an oil-base (rosemary-O), were tested for antioxidant activity. Preliminary studies, indicated that the oil-based preparation was more effective than the water-based preparation in retarding MFD. Therefore, less rosemary-O was used than rosemary-W (Table 9). In samples treated with 50 ppm rosemary-O, the hexanal content decreased by one-third when compared to that of the 2-day MFD sample; however, doubling the concentration to 100 ppm led to the hexanal content decreasing by over 50%. TBARS were also decreased in the sample treated with rosemary-O at a concentration of 100 ppm. These data indicated that lipid oxidation was being inhibited. The intensity value of the desirable flavor note, CBB, remained above the 2-day MFD control but slightly below the 0-day sample. The undesirable flavor notes, PTY and CBD, were high in the 50 ppm treated patties (actually less than those of the 2-day control), but the intensities decreased when patties were treated with 100 ppm rosemary-O. These data suggested that rosemary-O was able to inhibit lipid oxidation and retard MFD.

FOOD FLAVOR AND SAFETY

Table 7. Effect of Sodium Ascorbate on stored raw and cooked beef patties (N=3)

	HX	TV	TBA	CBB[a]	BM[b]	BRO[b]	BRC	SWT	PTY	CBD	SUR	BTR
Raw patties, marinated (stored) 17 hrs, 4°C.												
Standard[c]	11	164	3.31	-	3.59	2.26	2.79	1.17	0.18	0.25	0.57	0.31
(±s.e.m.)	1	50	0.44	-	0.03	0.13	0.19	0.12	0.07	0.07	0.01	0.01
Cntrl-0[d]	10	179	3.42	5.83	3.49	2.49	2.74	1.02	0.30	0.42	0.60	0.49
(±s.e.m.)	3	56	0.30		0.09	0.07	0.20	0.10	0.08	0.03	0.01	0.04
SA-150[e]	8	152	2.78	5.80	3.45	2.21	2.91	1.12	0.29	0.44	0.61	0.49
(±s.e.m.)	2	47	0.45	n.a.	0.00	0.04	0.17	0.09	0.05	0.06	0.06	0.03
SA-200	5	142	2.31	6.07	3.46	2.25	2.83	1.01	0.27	0.38	0.58	0.47
(±s.e.m.)	1	52	0.21	n.a.	0.14	0.09	0.19	0.11	0.06	0.04	0.02	0.06
Cooked patties stored 2-days, 4°C.												
Cntrl-0[d]	10	179	3.42	5.83	3.49	2.49	2.73	1.02	0.30	0.42	0.60	0.49
(±s.e.m.)	3	56	0.30	n.a.	0.09	0.07	0.20	0.10	0.08	0.03	0.01	0.04
Cntrl-2[f]	30	170	9.16	4.23	2.98	1.74	2.21	0.95	1.12	0.94	0.77	0.65
(±s.e.m.)	4	47	0.24	n.a.	0.02	0.11	0.12	0.08	0.28	0.24	0.08	0.07
SA-150	26	177	5.50	5.48	3.10	2.25	2.56	1.08	0.63	0.78	0.76	0.51
(±s.e.m.)	4	61	0.40	n.a.	0.20	0.12	0.16	0.08	0.20	0.13	0.09	0.09
SA-200	23	153	5.10	5.34	3.24	2.12	2.46	1.01	0.64	0.75	0.75	0.60
(±s.e.m.)	2	43	0.48	n.a.	0.01	0.10	0.15	0.11	0.14	0.07	0.07	0.07

[a] N=1
[b] N=2
[c] Raw, stored frozen 2-days, thawed, cooked, assayed.
[d] Control sample (0-day), stored frozen 2-days, thawed, cooked, assayed.
[e] Concentrations of additives in parts per million.
[f] Control sample (MFD), cooked, stored 2 days, rewarmed, assayed.
Abbreviations: HX, hexanal, peak area x 1000; TV, total volatiles, peak area x 1000;
TBA, 2-thiobarbituric acid reactive substances, mg MDA/kg sample; CBB, cooked beef/brothy,
all sensory values are mean intensity scores; BM, beefy/meaty; BRO, brothy; BRC,
browned/caramel; SWT, sweet; PTY, painty; CBD, cardboardy; SUR, sour; BTR, bitter.

Table 8. Effect of Sodium Ascorbate on stored raw and cooked beef patties (N=2)

	HX	TV	TBA	BM	BRO	BRC	SWT	PTY	CBD	SUR	BTR
Raw patties, marinated (stored) 17 hrs, 4°C.											
Standard[a]	17	228	5.72	3.50	2.21	2.37	1.10	0.38	0.30	0.67	0.29
(±s.e.m.)	6	64	0.12	0.24	0.23	0.27	0.04	0.05	0.04	0.02	0.03
Cntrl-0[b]	36	321	5.93	3.37	2.08	2.28	1.04	0.28	0.34	0.66	0.37
(±s.e.m.)	5	38	1.93	0.13	0.07	0.14	0.01	0.02	0.03	0.05	0.04
SA-250[c]	36	409	6.77	3.44	1.97	2.26	1.07	0.25	0.29	0.66	0.37
(±s.e.m.)	4	44	1.79	0.09	0.18	0.01	0.09	0.03	0.04	0.01	0.05
SA-500	12	267	6.13	3.56	1.91	2.28	1.14	0.29	0.38	0.58	0.44
(±s.e.m.)	3	16	0.57	0.10	0.09	0.02	0.01	0.01	0.05	0.06	0.08
Cooked patties stored 2-days, 4°C.											
Cntrl-0[b]	36	321	5.93	3.37	2.07	2.28	1.04	0.28	0.34	0.65	0.37
(±s.e.m.)	5	38	1.93	0.13	0.07	0.14	0.01	0.02	0.03	0.05	0.04
Cntrl-2[d]	62	356	9.13	3.17	1.93	2.11	1.09	0.65	0.55	0.76	0.50
(±s.e.m.)	9	12	1.05	0.11	0.20	0.11	0.02	0.11	0.06	0.01	0.02
SA-250	34	336	7.51	3.40	1.87	2.20	1.04	0.53	0.46	0.73	0.47
(±s.e.m.)	3	9	0.25	0.11	0.02	0.09	0.06	0.13	0.04	0.04	0.02
SA-500	15	277	4.05	3.39	2.01	2.15	1.15	0.40	0.44	0.64	0.46
(±s.e.m.)	1	15	0.25	0.13	0.10	0.11	0.02	0.08	0.06	0.09	0.06

[a] Raw, stored frozen 2-days, thawed, cooked, assayed.
[b] Control sample (0-day), stored frozen 2-days, thawed, cooked, assayed.
[c] Concentrations of additives in parts per million.
[d] Control sample (MFD), cooked, stored 2 days, rewarmed, assayed.
Abbreviations: HX, hexanal, peak area x 1000; TV, total volatiles, peak area x 1000;
TBA, 2-thiobarbituric acid reactive substances, mg MDA/kg sample; all sensory values are mean intensity scores; BM, beefy/meaty; BRO, brothy; BRC, browned/caramel; SWT, sweet; PTY, painty; CBD, cardboardy; SUR, sour; BTR, bitter.

Table 9. Effect of rosemary on chemical and sensory properties of cooked beef patties stored 2 days at 4° C

	Hexanal[a]	TBARS[b]	CBB[c]	PTY[c]	CBD[c]
Standard[d]	16	1.99	6.4	0.1	0.1
(±s.e.m.)	4	0.20	0	0	0
Control[e]	12	4.51	5.7	0.3	0.5
(±s.e.m.)	2	0.07	0.1	0	0
Control[f]	82	6.97	4.9	1.7	1.7
(±s.e.m.)	10	0.57	0.1	0.1	0.1
R-O[g] (50)[h]	52	7.32	5.30	1.60	1.60
R-O (100)	38	4.79	5.50	0.64	0.89
R-W[i] (125)	73	3.64	5.30	0.94	1.50
R-W (250)	38	2.35	5.15	1.20	1.25
R-W (375)	51	2.19	5.00	1.10	1.60
R-W (500)	19	3.10	3.95	1.10	1.50

[a] Area count X 1000
[b] mg MDA/kg ground meat
[c] Abbreviations: CBB, cooked beef/brothy; PTY, painty; CBD, cardboardy
[d] Fresh cooked and frozen immediately, N=16
[e] Stored overnight, cooked and assayed, 0-day sample, N=13
[f] Stored overnight, cooked, stored, assayed, 2-day MFD sample, N=13
[g] Oil-base
[h] Parts per million
[i] Water-base
   SOURCE: Adapted from ref. 14.

When rosemary-W was added to the cooked patties, as the concentration was increased from 125 to 500 ppm, both hexanal and TBARS levels decreased. The desirable flavor note, CBB, showed improvement over that of the 2-day MFD sample when the rosemary-W concentration was less than 500 ppm. However, at a concentration of 500 ppm, CBB decreased to an unacceptable level of 3.95. The undesirable flavor notes, PTY and CBD, were less in the rosemary-W group at 2-days than in the 2-day MFD samples, but were higher than the 0-day control. With the 500 ppm treatment, rosemary-W seems to effect PTY more than CBD. The data showed that rosemary-O is more effective than the water-based formulation. Consequently, smaller quantities of the oil based rosemary formulation are required for use in retarding lipid oxidation and maintaining the desirable beefy taste. A similar use of rosemary as an inhibitor of MFD in restructured beef steaks was reported recently by Stoick et al (*40*).

N-Carboxymethylchitosan. An antioxidant, obtained from a natural source, that is an inhibitor of MFD in ground beef is N-carboxymethylchitosan (NCMC). This compound is a water soluble polysaccharide derived from the chitin found in the shells of crab, shrimp, lobster and crayfish. To prepare the compound, the crustacean shells were washed with water and then treated with hydrochloric acid to make chitin. Next, hot sodium hydroxide was added to make chitosan. Glyoxylic acid was then added followed by reduction with sodium borohydride to produce the NCMC. NCMC prepared in this manner yielded a mixture of a gel form, which contained 66% free amino groups, 16% N-acetyl groups, and 18% N-carboxymethyl groups, and a soluble form, which contained 41%, 16%, and 43% respectively (*41*). When added to ground beef at 0.5%, NCMC was able to prevent MFD after storage for 2 days at 4°C. These results were determined by both chemical and sensory methods. At the 0.5% level, lipid oxidation as measured by hexanal content and TBARS numbers, was almost 100% inhibited. At dilution to the 0.06% level, inhibition was still greater than 50%. A control, chitosan, at the 1% level, was very ineffective compared to NCMC (*2*). Sensory studies showed that the desirable beefy flavor note, CBB, was very low [3.6] in patties stored for 2 days at 4°C in the control samples. The intensities of the off-flavor notes, PTY and CBD, were high, 2.1 and 1.8, respectively. However, in treated patties (0.5% NCMC), CBB remained at 5.8, whereas the off-flavor notes remained low, i. e., a PTY intensity value of 0.2 and a CBD of 0.3 (*41*). These data showed that the NCMC, i. e., the chitin derivative from a natural source, can be used to inhibit lipid oxidation and prevent MFD. The mechanism of action is thought to be through chelation of the iron, which can catalyze oxidation of the lipids. NCMC does not have any known allergic reactions or toxic side effects. A patent for the application of NCMC as a food antioxidant was issued in 1989 (*42*).

Carrageenan: A Fat Replacement in Ground Beef. Recently, Egbert et al (*43*) reported on the use of carrageenan, a hydrocolloid isolated from seaweed, as a fat replacement in ground beef. This low-fat product has been adopted by the

McDonald Corp., Oak Brook, IL and is being sold nationally as McDonald's McLean Deluxe. The authors reported on the development of a tender, juicy, low-fat (<10%), ground beef product that contained 0.5% iota carrageenan (Viscarin SD 389) and 10% water. Evaluations by a trained sensory panel showed that the lean patties had a superior beef flavor intensity value that was considered greater than the 20% or 8% fat controls. Their report included no storage studies conducted on the low-fat product. However, in a more recent report, Egbert et al. (44) showed that the low-fat patties treated with Viscarin SD 389 did undergo lipid oxidation during storage at 4°C. They attributed this increase to sodium chloride-promoted oxidation in the presence of oxygen. However, it did show that even low fat ground beef patties need antioxidants to prevent lipid oxidation, and thus MFD.

In 1988, St. Angelo et al. (2) reported on an approach to devise an acceptable and effective procedure to prevent the development of warmed-over flavor, referred to as MFD in this communication, in cooked ground beef. The method for preparing the ground beef used by St. Angelo and coworkers, had a low natural fat content ranging from 4-5% (14, 36, 38). As part of that study, several different types of compounds, i. e., chelators, free radical scavengers, Maillard reaction products and gums were examined for their antioxidant activity. Included in the gums were 7 different carrageenans. However, only one of them, identified as Carrageenan-44, actually iota carrageenan, Lactarim XP 4019, was an effective inhibitor of lipid oxidation in stored patties (2 days). Carrageenan concentrations varied from 0.25 to 1.0%. Antioxidant effectiveness was measured by TBA numbers and hexanal content. After determining (personal communication, FMC Corp., Philadelphia PA) that the Lactarim contained sodium pyrophosphate (SPP) and calcium monohydrogen phosphate (CMP), a control was run with an iota carrageenan, identified as Carrageenan-115 (Viscarin SD 389) that did not contain any SPP or CMP. Results indicated that Carrageenan-115 at the concentration tested, 0.5%, was ineffective as an antioxidant.

A second study (2) described the effect of another series of antioxidants (maltol, ascorbate-2-phosphate, Carrageenan-44 and SPP) on cooked ground beef stored at -20° and at 4°C for 3 days. These studies included sensory and chemical analyses. Results indicated that lipid oxidation was inhibited in all of the antioxidant-treated samples stored under both conditions. However, in the samples stored at 4°C, Carrageenan-44 was effective as an antioxidant but not as effective as the other antioxidants tested; it was very effective in samples stored at -20°C. In samples analyzed by sensory evaluation, those treated with Carrageenan-44 and ascorbate-2-phosphate and stored frozen had the highest intensity value for the on-flavor note, CBB [6.4], and the lowest intensity values for the off-flavor notes, PTY [0.1] and CBD [0.3]. In samples stored at 4°C, those treated with SPP and Carrageenan 44 had the highest CDD intensity [4.5 and 4.4, respectively]. Samples treated with SPP had the lowest PTY intensity value [0.8], whereas maltol had the lowest CBD intensity value [0.5]. This difference is probably due to the pyrophosphate action as a chelator and thus it would effect lipid oxidation directly. Painty has been reported to be highly correlated to hexanal (1, 5, 45). On the other hand, maltol, a reductone, is not a chelator *per se*, but presumably functions as a free radical scavenger (*2, 46*). CBD appears to evolve as a product of protein degradation via a free radical mechanism rather than from lipid oxidation (*47*). Thus, the production of the two off-flavor notes, i.e., PTY and CBD, probably evolves through different mechanisms.   Not surprising, SPP and maltol function through different mechanisms.   The intensity values for PTY and CBD in Carrageenan-44 treated samples were 2.8 and 2.5, respectively.   These values were intermediate between those of controls [4.0 and 3.5, respectively] and those from the SPP treated samples [0.8 and 1.5, respectively].

Egbert et al. (*43*) concluded by suggesting that the technology of using iota carrageenan, water, and flavor enhancers to produce a high-quality, low fat, ground-beef product should be expanded to include other fresh meat products. However, based on the results of St. Angelo et al. (*2*), the new technology should include iota carrageenan formulations that contain chelators such as pyrophosphate, or at least antioxidants should be added to the mixture when preparing low-fat food products.   By this addition, two functions can be accomplished, namely, a low-fat product is produced and its shelf-life is increased.

Conclusion

There are many compounds that are used as food grade antioxidants. Although, phenolic antioxidants are becoming suspect for health reasons, they are still are being used effectively.   However, we should not fear to explore the naturally occurring antioxidants. They too are effective and will probably be the next generation of food antioxidants.

Acknowledgements

The author thanks Dr. B. T. Vinyard for statistical evaluations, Mr. C. James, Jr. and Mrs. C. H. Vinnett for technical assistance, members of the meat sensory panel for evaluation of samples and employees of the sensory laboratory for assistance in preparing and serving the samples.

Literature Cited

1.   St. Angelo, A. J.; Vercellotti, J. R.; Legendre, M. G.; Vinnett, C. H.; Kuan, J. W.; James, Jr., C.; Dupuy, H. P. *J. Food Sci.* **1987**, *52*, 1163-1168.
2.   St. Angelo, A. J.; Vercellotti, J. R.; Dupuy, H. P.; Spanier, A. M. *Food Technol.* **1988**, *42(6)*, 133-138.
3.   Spanier, A. J.; Miller, J. A.; Bland, J. M. In *Lipid Oxidation in Food*; St. Angelo, A. J., Ed.; ACS Symposium Series No. 500; American Chemical Society: Washington, DC, 1992; pp 104-121.
4.   Dupuy, H. P.; Bailey, M. E.; St. Angelo, A. J.; Vercellotti, J. R.; Legendre, M. G. In *Warmed-Over Flavor of Meat*; St. Angelo, A. J; Bailey, M. E., Eds.; Academic Press Inc.: Orlando, FL, 1987; pp 165-191.

5.   St. Angelo, A. J.; Spanier, A. M.; Bett, K. L. In *Lipid Oxidation in Food*; St. Angelo, A. J., Ed.; ACS Symposium Series No. 500; American Chemical Society: Washington, DC, 1992; pp 140-160.

6.   Bailey, M. E.; Dupuy, H. P.; Legendre, M. G. In *The Analysis and Control of Less Desirable Flavors in Foods and Beverages*; Charalambous, G., Ed.; Academic Press, Inc.: Orlando, FL, 1980; pp 31-52.

7.   Chipault, J. R. In *Autoxidation and Antioxidants, Vol. II*; Lundberg, W. O., Ed.; Interscience Publishers, John Wiley and Sons: New York, NY, 1962; pp 477-542.

8.   Sherwin, E. R. *J. Amer. Oil Chem. Soc.*, 1972, *49*, 468-472.

9.   Hamilton R.J., In *Rancidity in Foods*, Second Edition, Allen, J. C.; Hamilton, R. J. Eds.; Elsevier Applied Science: London, 1989; pp 1-21.

10.   Nickerson, J. T. R. In *Food Processing Operations*, Vol. II, Heid, J.; Joslyn, M., Eds., Avi Publishing Company: Westport, CN, 1963; pp 218-247.

11.   Buck, D. F. *The Manufacturing Confectioner*, 1985, *65(6)*, 45-49.

12.   Dugan, L. R.; In *Principles of Food Science, Part 1. Food Chem.*; Fennema, O. R., Ed.; Marcel Dekker, Inc.: New York, NY, 1976; pp 139-203.

13.   Porter, W. L. In *Autoxidation in Foods and Biological Systems*, Simic, M. G; Karel, M., Eds.; Plenum Press: New York, NY, 1980; pp 295-365.

14.   St.Angelo, A. J.; Crippen, K. L.; Dupuy, H. P.; James, Jr., C. *J. Food Sci.* 1990, *55*, 1501-1505 & 1539.

15.   Graft, E.; Empson, K. L.; Eaton, J. W. *J. Biol. Chem.* 1987, *262*, 11647-11650.

16.   Cort, W. M. In *Ascorbic Acid: Chemistry, Metabolism, and Uses;* Seib, P. A.; Tolbert, B. M., Eds.; Advances in Chemistry Series No. 200; American Chemical Society: Washington, DC, 1982; pp 533-550.

17.   Kanner, J. In *Lipid Oxidation in Food*; St. Angelo, A. J., Ed.; ACS Symposium Series No. 500; American Chemical Society: Washington, DC, 1992; pp 55-73.

18.   Martell, A. E. In *Ascorbic Acid: Chemistry, Metabolism, and Uses,* Seib, P. A.; Tolbert, B. M., Eds.; Advances in Chemistry Series No. 200; American Chemical Society: Washington, DC, 1982; pp 153-178.

19   Gehman, H.; Osman, E. M.; *Adv. Food Res.,* 1954, *5,* 53-96.

20.   Kamiya, S. *Nippon Nogeikagaku Kaishi,* 1960, *34*, 8-12.

21.   Liao, M. L.; Seib, P. A. *J. Agric. Food Chem.* 1990, *38*, 355-366.

22.   Craig, J.; Bowers, J. A.; Seib, P. *J. Food Sci.* 1991, *56*, 1529-1531.

23.   Bauernfeind, J. C.; Pinkert, D. M.; *Adv. Food Res.,* 1970, *18,* 219-315.

24.   Seib, P. A.; Tolbert, B. M. *Ascorbic Acid: Chemistry, Metabolism, and Uses;* Advances in Chemistry Series No. 200; American Chemical Society: Washington, DC, 1982.

25.   Chipault, J. R.; Mizuno, G. R.,; Hawkins, J. M.; Lundberg, W. O. *Food Research* , 1952, *17*, 46-54.

26.   Chipault, J. R.; Mizuno, G. R.,; Lundberg, W. O. *Food Technol.*, 1956, *10*, 209-211.

27. Houlihan, C. M.; Ho, C. T. In *Flavor Chemistry of Fats and Oils,* Min, D. B.; Smouse, T. H., Eds., 1985, American Oil Chemists' Society: Champaign, IL; pp 117-143.

28. Dugan, L. R. In *Autoxidation in Foods and Biological Systems*, Simic, M. G; Karel, M., Eds.; Plenum Press: New York, NY, 1980; pp 261-282.

29. Rhee, K. S. In *Warmed-Over Flavor of Meat*; St. Angelo, A. J; Bailey, M. E., Eds.; Academic Press Inc.: Orlando, FL, 1987; pp 267-289.

30. Loliger, J. In *Rancidity in Foods*, Second Edition, Allen, J. C.; Hamilton, R. J. Eds.; Elsevier Applied Science: London, 1989; pp 105-124.

31. St. Angelo, A. J.; Legendre, M. G.; Dupuy, H. P. *Lipids,* 1980, *15*, 45-49.

32. Love, *Food Technol.* 1988, *42(6)*, 140-143.

33. Tarladgis, B. G.; Watts, B. M.;Younathan, M. T.; Dugan, Jr., L. *J. Am. Oil Chem. Soc.* 1960, *37*, 44-48.

34. Meilgaard, M.; Civille, G. V.; Carr, B. T. *Sensory Evaluation Techniques*; CRC Press, Inc.: Boca Raton, FL, 1987, Vol. 2; pp 1-23.

35. Johnsen, P. B; Civille, G. V. *J. Sensory Studies.* 1986, *1*, 99-104.

36. Spanier, A. M.; Vercellotti, J. R.; James, Jr., C. *J. Food Sci.* 1992, *57*, 10-15.

37. Koniecko, E. S. In *Handbook of Meat Analysis*, Avery Publishing Group, Inc.: Wayne, NJ; 1985, p 19.

38. Spanier, A. M.; St. Angelo, A. J.; Shaffer, G. P. *J. Agric. Food Chem.* 1992, *40*, 1656-1662.

39. Sato, K.; Herring, H. *Food Prod. Develop.* 1973, *7(9)*, 78-84.

40. Stoick, S. M.; Gray, J. I.; Booren, A. M.; Buckley, D. J.*J. Food Sci.* 1991, *56*, 597-600.

41. St. Angelo, A. J.; Vercellotti, J. R. In *Food Science and Human Nutrition*; Charalambous, G., Ed.; Elsevier Science Publishing Co. Inc.: New York, NY, 1992; pp 711-722.

42. St. Angelo, A. J.; Vercellotti, J. R. *U. S. Patent* 4,871,556, Issued October 3, 1989.

43. Egbert, W. R.; Huffman, D. L.; Chen, C.-M.; Dylewski, D. P. *Food Technol.* 1991, *45(6)*, 64-73.

44. Egbert, W. R.; Huffman, D. L.; Bradford, D. D.; Jones, W. R. *J. Food Sci.* 1992, *57*, 1033-1037.

45. Spanier, A. M.; Edwards, J. V.; Dupuy, H. P. *Food Technol.* 1988, *42(6)*, 110-118.

46. Einerson, M. A.; Reineccius, G. A. *J. Food Proc. and Pres.* 1977. *1*, 279-291.

47. Vercellotti, J. R.: Kuan, J. W.; Spanier, A. M.; St. Angelo, A. J. In *Thermal Generation of Aromas*; Parliment, T. H.; McGorrin, R. J.; Ho, C.-T.; Eds.; ACS Symposium Series No. 409; American Chemical Society: Washington, DC, 1989; pp 452-459.

RECEIVED October 28, 1992

Chapter 6

Role of Proteins and Peptides in Meat Flavor

A. M. Spanier and J. A. Miller

Southern Regional Research Center, Agricultural Research Service, U.S. Department of Agriculture, 1100 Robert E. Lee Boulevard, New Orleans, LA 70124

Human flavor perception is elicited by neurological response to the actions and interactions of food lipids, proteins, and carbohydrates. Since the main chemical component of muscle foods (meats) are proteins, this research focused on studying the impact of beef proteins and peptides on flavor. Postmortem aging, end-point cooking temperature, storage-after-cooking, factors related to meat structure and handling methods, all affect the protein and peptide composition of beef, thereby affecting meat flavor. Lipid oxidation (LO) and the free radicals generated during LO are involved both directly and indirectly in several of these changes. Experiments using a low molecular weight peptide found in beef (BMP, a beef flavor enhancer) indicate that a correlation exists between BMP loss and LO; the loss of BMP is a secondary result of LO-stimulated proteolysis. Other data presented in this manuscript led to the proposal of a taste receptor model that relates peptide structure to taste perception.

FLAVOR PERCEPTION

The perception of flavor is a fine balance between the sensory input of both desirable and undesirable flavors. It involves a complex series of biochemical and physiological reactions that occur at the cellular and subcellular level (see Chapters 1-3). Final sensory perception or response to the food is regulated by the action and interaction of flavor compounds and their products on two neural networks, the olfactory and gustatory systems or the smell and taste systems, respectively (**Figure 1**). The major food flavor components involved in the initiation and transduction of the flavor response are the food's lipids, carbohydrates, and proteins, as well as their reaction products. Since proteins and peptides of meat constitute the major chemical components of muscle foods, they will be the major focus of discussion in this chapter.

The flavor quality of food is a primary factor involved in a consumer's decision to purchase a food item. Therefore, food technologists require a thorough understanding of how flavor deteriorates if they are to prepare products that consumers will purchase repeatedly. This knowledge is particularly important in meat and meat products, since the deterioration of meat flavor is a serious and continual process (*1-4*) that involves both the loss of desirable flavor components (*4, 5*) and the formation of off-flavor compounds (*6-9*) many of which are associated with lipid oxidation (*10*).

THE EFFECT OF ISCHEMIA ON THE 3-DIMENSIONAL STRUCTURE OF MUSCLE/MEAT.

Any dialogue on meat flavor development and deterioration requires a brief discussion of muscle structure. Muscle has a highly compact and complex multicellular structural organization (**Figure 2**). Individual muscle cells contain numerous mitochondria and nuclei. They also contain contractile elements as the bulk of their structure. While the sarcoplasm of muscle (the aqueous non-organellar component) is small compared to the cytoplasm of non-muscle cells, it does have a highly evolved system of membranes called the SR/L representing an acronym for sarcoplasmic reticulum/lysosomal membrane system (*11*). The SR/L surrounds each contractile element (Fig. 9-13 in *12*; Fig. 7-10 in *13*). The close proximity of the SR/L to the contractile proteins situates the proteins in a location that is optimal for their hydrolysis by lysosomal hydrolases (*12, 13*).

Previous studies have shown that muscle lysosomal hydrolases are released early in the postmortem period due to a decrease in intracellular ATP concentrations. The decreased intracellular ATP level causes the rupture of the lysosomal membrane (*14*), releasing hydrolytic enzymes (proteases, lipases, and glycosidases) that further potentiate the weakening of membrane integrity and cellular function. Furthermore, as the acidosis increases (due to the anaerobic conditions associated with cellular death) the intramuscular pH to levels reach that which are optimal for the activity of several lysosomal thiol proteinases.

The cascade of biochemical events described above enhances the cellular necrosis and tissue breakdown, leading to what meat scientists and food technologist call meat-tenderization.   Since muscle is primarily protein in nature and since hydrolytic, and specifically proteolytic, activity increases during postmortem aging, muscle represents a remarkable pool of material for the production of flavor peptides and amino acids as well as other precursors for flavor development (*2*).

CHANGES IN BEEF SENSORY ATTRIBUTES DURING POSTMORTEM AGING:

Postmortem aging leads to an alteration in beef sensory attributes (**Figure 3**). As the meat ages, desirable flavors such as "beefy" (BEF) and "brothy" (BRO) diminish while flavors typically considered as undesirable in beef, such as "bitter"

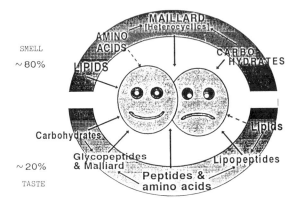

Figure 1. Representation of the 2-faces of flavor perception showing desirable and undesirable smell (odor) and taste (gustatory) input components.

Muscle Myofibril

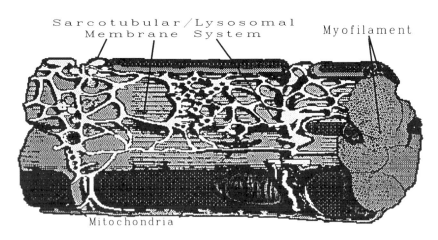

Figure 2. Representation of the three dimensional structure of skeletal muscle.

(BTR) and "sour" (SOU), increase. Most "bitter" and "sour" flavors are thought to originate from the degradation/decomposition of proteins to bitter and sour flavored peptides and amino acids.

CHANGES IN ENZYME ACTIVITY DURING POSTMORTEM AGING:
Figure 4 presents data from experiments designed to examine the effect of postmortem aging on both hydrolytic and non-hydrolytic enzyme activity in beef (2). All of the enzymes examined showed a significant rise in activity by 3.5 hours post-slaughter. Thereafter, enzyme activity decreased for the next 360 hours (15 days), but remained higher than the initial post-slaughter activity. These data indicate that during the postmortem aging process, significant enzymatic activity and particularly proteolytic activity was present for the continual production of protein-derived flavor compounds. These products of protein degradation can serve directly as flavor producing compounds and/or as potential precursors for reaction with other food components during cooking and during storage after cooking (2, 3, 5).

The intracellular pH of beef drops to 5.4 during postmortem aging process. This pH is one that is optimal for activation of several endogenous thiol proteinases such as cathepsins B and L (15). Since many products of proteolysis exhibit undesirable beefy flavors such as bitter and sour, these data (**Figure 4**) suggest that a better understanding of beef flavor development requires a good understanding of the postmortem proteolytic and hydrolytic events.

EFFECT OF POSTMORTEM AGING ON BEEF PROTEIN COMPOSITION:
The composition of extracts from beef *semimembranosus* muscle (top round cut) was examined by capillary electrophoresis (CE). The electropherogram showed several protein and peptide peaks that change during the postmortem aging process (**Figure 5**, insert). Quantitation of the area under each peak revealed that a major change in protein composition occurred between 4 and 7 days postmortem. As the peak identified as CE-3 was depleted during the postmortem aging process, another peak, CE-4, increased. Other smaller peaks such as CE-6, CE-7, and CE-11 showed an increase similar to CE-4 but to a lesser extent. CE-4 and the latter peaks are thought to arise as a result of the proteolytic degradation of a larger protein, such as CE-3. Proteolytic cleavage of any protein or peptide will result in a compositional and/or conformational change in the original protein/peptide and the target protein/peptide yielding fragments with functionalities that are most likely to be quite different from those of the parent protein/peptide.

MULTIVARIATE PRINCIPAL COMPONENTS ANALYSIS OF AGING BEEF:
Multivariate principal component or factor analysis was performed on data obtained from samples of aging beef (described above). Factor analysis was used since this method facilitates the visual examination of existing relationships (correlations) among the experimental treatments and the sensory, chemical,

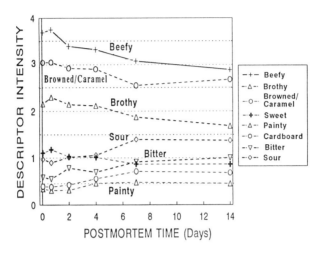

Figure 3. Descriptive sensory analysis of beef flavor attributes. Flavor attributes are examined as a function of storage time (days) postmortem.

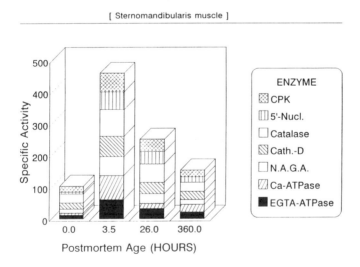

Figure 4. Specific activity of numerous enzymes during postmortem aging.

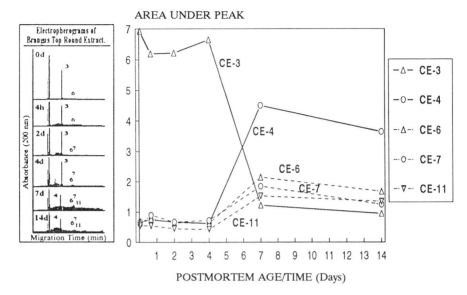

Figure 5.   Presentation of the change in levels of several beef proteins identified by capillary electrophoresis (CE). Electropherograms are shown in the insert. Numbers refer to the identification or name assigned to each peak.

and instrumental flavor descriptors (*3, 9, 16*). The bivariate factor plot (**Figure 6**) showed CE-3 to correlate well with desirable flavor descriptors such as beefy, browned/caramel, sweet and brothy (BEF, BRC, SWT, and BRO, respectively) and with beef at early postmortem ages such as 4h (hours) and 0d (days). On the other hand, examination of CE-4 and its putative fragments (CE-6, CE-7, and CE-11) showed them to cluster with the beef aged for longer periods of time such as 7d and 14d.

While lipid oxidation is reported to be the major factor responsible for the change in flavor of freshly prepared precooked beef and in the same product after storage (*1, 4-10, 16*), it is not expected to be a major factor influencing flavor during postmortem aging (*2*). This conclusion is supported by the data shown in the bivariate factor distribution seen in **Figure 6**; here the chemical and instrumental markers of lipid oxidation, such as hexanal, thiobarbituric acid reactive substances (TBARS) and total volatiles, cluster at or near the origin. This data suggests that these products of lipid oxidation have little or no influence on the final flavor perception of fresh-cooked, aging beef. The influence and importance of these lipid oxidation products is discussed below in the section on precooked and cooked/stored beef.

THERMAL INACTIVATION OF BEEF ENZYMES
Beef flavor is affected by cooking temperature and by the storage period after cooking in a manner similar to that seen during postmortem aging. Previous research (*1-3*) show that end-point cooking temperature has a significant effect on enzyme and protein composition of muscle foods (**Figure 7**). For example, while all protein levels and enzyme activity begin to diminish above 122°F

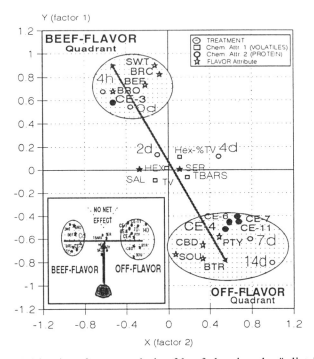

Figure 6. Multivariate factor analysis of beef showing the " distribution" of experimental treatments, volatile chemical attributes, chemical protein elements and flavor attributes in beef samples during postmortem aging. Insert represents the data redrawn with the graphical origin depicted as the fulcrum of a balance beam.

(50°C), there is a broad range of temperature-induced decreases in enzyme activity. Catalase is seen as the most thermally stable enzyme in this group, while creatine phosphokinase (or CPK) is the most thermally sensitive enzyme.

Thiol proteinases, as a group, have been reported to hydrolyze more peptide bonds than any other proteinase group including the pancreatic serine proteases of the intestine. It would be reasonable, therefore, to suggest that they can be highly effective in producing flavor proteins and peptides, particularly during postmortem aging. These enzymes deserve consideration not only because the optimum pH for their activity against most substrates is between 5.4 and 6.4, the pH range of most beef products purchased by consumers, but also because of their reaction to temperature.

Two thiol proteinases, cathepsin B and L, in particular, appear initially to be highly thermally sensitive, at least below 60°C (**Figure 7**). Above 60°C they retain more than 20% of their original activity and even display a slight increase to almost 30% of the original activity at temperatures above 70°C. This change in catheptic activity above 60°C is explained by data from earlier studies (*17*) that showed cathepsin B activity to be inhibited by myoglobin, a major soluble protein in muscle, i.e., as myoglobin levels decline (**Figure 7**), there is a rise in catheptic activity.

EFFECT OF STORAGE ON THE GENERATION AND LOSS OF FLAVOR NOTES AND PRODUCTS OF LIPID OXIDATION IN COOKED MEATS:
Precooked beef products, often referred to as "convenience" and "institutional" foods, comprise 35% of all the beef sold and consumed in America today; this represents almost $10 billion in consumer expenditures on meat. Therefore, a thorough understanding of the flavor of beef and what factors affect the flavor would be critical to continued sales in this large market.

Hornstein and Crowe (*18*) and others (*19-21*) suggested that, while the fat portion of muscle foods from different species contributes to the unique flavor that characterizes the meat from these species, the lean portion of meat contributes to the basic meaty flavor thought to be identical in beef, pork, and lamb. The major differences in flavor between pork and lamb result from differences in a number of short chain unsaturated fatty acids that are not present in beef. Even though more than 600 volatile compounds have been identified from cooked beef, not one single compound has been identified to date that can be attributed to the aroma of "cooked beef." Therefore, a thorough understanding of the effect of storage on beef flavor and on lipid volatile production would be helpful to maintain or expand that portion of the beef market.

Storage related changes in beef flavor are typically associated with lipid oxidation (*1, 4-10, 16*). Hexanal, a major lipid volatile produced from the oxidation of linoleic acid, increases during refrigerated storage (**Figure 8**). The increase in hexanal parallels the change in the aromatic sensory descriptor represented by painty (**Figure 8**). TBARS, thiobarbituric acid reactive substances, represent a chemical marker of lipid oxidation. TBARS follow a pattern of increase similar to that of hexanal and painty.

Tastes such as bitter and sour, typically considered to be undesirable in beef products, develop immediately after slaughter (**Figure 3**) and during storage (**Figure 8**). "Bitter" and "sour" flavors are present at fairly high relative-levels (31.9 and 30.5, respectively; **Figure 8**) in fresh-cooked aged meat at day 0. At the same time that these undesirable flavors increase, a decline is seen in desirable flavors such as "beefy" and "brothy" (**Figure 8**). These data demonstrate that flavor deterioration in muscle foods is a continual process that involves not only the formation of off-flavored compounds [most associated with the free radicals generated during lipid oxidation (*6*)] but also the loss of desirable flavor compounds.

EFFECT OF STORAGE ON BEEF PROTEINS:
The chemical, instrumental and sensory data presented above indicated that storage of cooked beef affects the lipid composition and concomitantly, the flavor of beef. The data also indicated that primary tastes like bitter and sour are affected by storage.

The data in the upper panel of **Figure 9** show the elution profile obtained from the separation of acid extracts of uncooked, cooked, and cooked/stored meat by size exclusion chromatography. Uncooked meat has the greatest

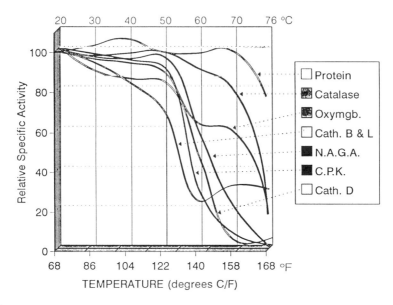

Figure 7.   Description of the enzyme activity and protein distribution determined in beef extracts cooked to various end-point cooking temperatures (adapted from 2).

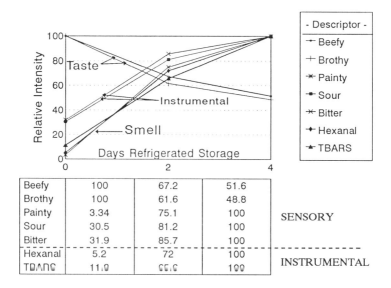

Beefy	100	67.2	51.6
Brothy	100	61.6	48.8
Painty	3.34	75.1	100
Sour	30.5	81.2	100
Bitter	31.9	85.7	100
Hexanal	5.2	72	100
TBARS	11.0	66.6	100

SENSORY

INSTRUMENTAL

Figure 8.   Change in some sensory, instrumental, and chemical descriptors in cooked ground-beef stored in a refrigerator (adapted from 6).

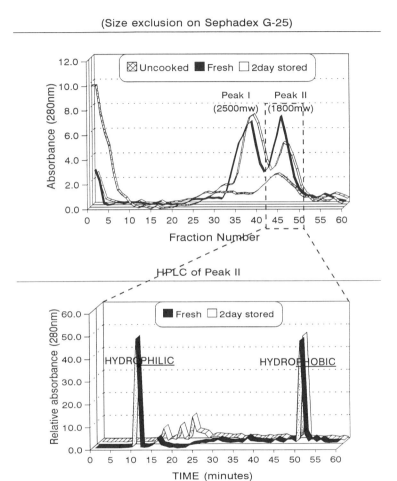

Figure 9. The effect of storage on protein and peptide composition in cooked ground beef stored in a refrigerator of 4 days (adapted from *1*). Upper graph: represents the size exclusion chromatography of acidic extracts of fresh, cooked, and cooked-stored beef. Lower graph: represents the reverse phase HPLC of peak II from the size exclusion chromatography.

amount of its proteinaceous material separating with the void volume, i.e., the 5,000 molecular weight (MW), range. On the other hand, extracts from "freshly-cooked" (60°C) meat and from 2-day "cooked-&-stored" meat have little protein in this molecular weight range (**Figure 9**, upper panel). Similar results are seen for meat cooked to an end point temperature of 77°C (not shown). Extracts from "fresh-cooked" and "cooked-&-stored" meat show 2 major peaks, one at 2,500 MW and the other at 1,800 MW. Storage of the cooked patties has no major effect on the 2,500 MW peak, while significant changes are seen in the 1,800 MW peak. The 1,800 MW peptide, therefore, appeared to be a likely candidate for further investigation (discussed below).

The lower panel of **Figure 9** shows the reverse phase-HPLC profile of the 1,800 MW peptide fractions (*1*). Both the "fresh-cooked" and "cooked-&-stored" samples resolve into separate regions, i.e., a hydrophilic region and a hydrophobic region. Hydrophilic peptides are commonly associated with flavors such as "sweet" and possibly, "meaty" and "cooked beef/brothy", whereas the hydrophobic peptides are usually associated with the more undesirable flavors like "bitter" and "sour".

The peptide fractions from "fresh-cooked" beef separate into two distinct and equal peaks. On the other hand, material from the 2-day "cooked-and-stored" sample shows a major band only in the hydrophobic region (**Figure 9**, lower panel). Based on the chromatographic elution profile, the small peaks seen in the hydrophilic region of the stored beef chromatogram are thought to be formed from the degradation of the hydrophilic peak seen in the chromatogram of the fresh meat. This change in the composition of the peptide with storage, correlates well with the observed change in flavor upon storage (*1, 22*). Whether this degradation is induced by proteolysis, by free-radical mechanisms, or both, is not known at this time. However, what is known is that, this change in peptide content is not a temperature dependent change.

EFFECT OF FREE RADICALS ON PROTEIN COMPOSITION:

The changes in the protein composition and flavor of "cooked" and "cooked-&-stored" beef seemed to be related to degradation by free radical species induced during lipid oxidation (*1-10, 16, 22, 23*). Based on the information presented in the aforementioned publication, it seemed reasonable to suggest that the appearance of "bitter" and "sour" tastes and the disappearance of "meaty" and "beefy" flavors were a result of the activity of the free radicals derived from lipid oxidation on flavor proteins (*2, 24-28*).

The formation of free radicals after lipid oxidation is known to play a key role in the deterioration of meat flavor (*8, 23*). Since proteins constitute a major portion of the muscle's composition, the relationship between chemically active radical species and decomposition of food flavor proteins and peptides needs to be studied in detail. Data has been presented showing the correlation of proteins with flavor (**Figures 5** and **6**). Data is now presented showing how soluble meat proteins change in an environment where free radicals are induced by a free-radical oxidation generating system or FROG (**Figure 10**).

The complex structure of muscle tissue as described earlier (Figure 2) make it difficult to examine directly either the targets or the mechanism(s) responsible for the changes in beef flavor. Therefore, it became necessary to develop an *in vitro* model system which permits study of this process (*6*). The model used endogenous meat lipids to prepare artificial membranes or liposomes into which a desired protein or proteins can be encapsulated or entrapped. Multilamellar liposomes (*29*) were chosen for use in this model since they come closer to imitating intact muscle than unilamellar liposomes, i.e., they provide a greater, more compact, interfacial-area for interaction between meat lipids and meat proteins. Lipid oxidation was induced by a FROG system that consisted of 3.3 mM dihydroxy-fumarate (DHF) and a mixture of 0.1 mM iron as ferric chloride with 1.0 mM adenosine diphosphate, ADP (*30*). Oxidation was initiated by the addition of the iron/ADP mixture. Reactions were stopped a mixture of 0.025% each of the antioxidant, propyl gallate (PG) and the chelator, ethylenediaminetetraacetic acid (EDTA), a mixture also added to all controls. Lipid oxidation was assessed by the measurement of thiobarbituric acid reactive substances (TBARS) and by gas chromatography (GC). Changes in the protein profile were assessed by high performance capillary electrophoresis of beef protein extracts both before and after induction of lipid oxidation using the liposome and FROG system (*6*). Electropherograms are seen in **Figure 10** (see *6* for more detail).

The upper profile depicts the liposomes containing unincubated control beef extract plus the FROG system. The lower profile depicts the same liposome + protein + FROG system after 60 minutes incubation at 37°C. Changes in the electrophoretic pattern of several proteins and protein groups are readily apparent. For example, the cluster of peaks between 2 and 3 minutes and between 3 and 4 minutes, the peaks at 4.2 and 4.4 minutes, and the peak at 5.1 minutes. The appearance of new protein and/or peptide bands which were not present prior to FROG treatment is also seen, e.g., the peaks between 8 and 9 minutes and between 12 and 13 minutes and the change in shape and migration time of the peak between 10-11 minutes. The electrophoretic peak appearing between 5.0 and 5.3 minutes is particularly intriguing since it has the same initial "shape" and "migration time" as a synthetically-prepared delicious-peptide which has been named "BMP" (see below).

EFFECT OF FREE RADICALS ON A BEEF FLAVOR PEPTIDE:
Figure 10 exhibits clearly that both the height and the shape of the peak at 5 minutes has changed significantly upon incubation-and-exposure to the FROG system (*6*). Initial semiquantitative analysis of the integrated peak areas, shows approximately a 16% reduction in peak area (lower electropherogram) suggesting that the composition of this peptide has been somehow altered. Further semi-quantitative analysis of these integrated peaks revealed that the area under the peak of the unincubated beef extract (upper electropherogram) represents 44.9% of the area of the synthetic BMP (not shown, see *6*) or

approximately 2.69 μg of presumptive octapeptide. This represents a sample containing an equivalent of 1.59 mM of octapeptide (upper electropherogram, **Figure 10**) which is greater than the 1.41 mM perception-threshold reported by Drs. Tamura and colleagues for this peptide (*31*). When these calculations are extrapolated to the sample incubated for 1 hour in the presence of the FROG system (lower electropherogram, **Figure 10**), the presumptive octapeptide has a concentration of 1.32 mM; this latter concentration is just below the 1.41 mM threshold for sensory perception of BMP.

The effect of oxidation-derived free radical activity on BMP levels has been shown to be an indirect outcome of the oxidation process. Use of the FROG-liposome system on purified BMP in the absence of other proteins (not shown) presents no change in electrophoretic behavior (*6*). On the other hand, the degradation of BMP in whole muscle extracts is believed to be due to the lipid oxidation derived free radicals affecting a protease that subsequently degrades BMP (see *6* for complete details and mechanisms). Therefore, a comprehensive understanding of factors affecting beef flavor, whether protein, lipid or carbohydrate in origin, requires a full understanding of the action and interaction of all these components within the complex matrix of muscle tissue.

BMP (BEEFY MEATY PEPTIDE):

BMP is a linear octapeptide of approximately 860 daltons molecular weight (**Figure 11**). It is a major, flavor-peptide in beef and is named BMP (beefy meaty peptide) because it is a peptide (P) which is found in beef (B) and it enhances the meaty flavor (M) already existing in the meat. BMP was originally found in papain digests of beef (*32*) and was discovered later to occur naturally in aged beef (*6, 33*). BMP production in muscle is attributed to the proteolytic digestion of a parent protein by endogenous proteinases such as cathepsin B and/or L; these proteinases share several similar characteristics with papain such as pH optimum against most substrates, proteolytic class (thiol proteinase), and reaction to inhibitors. Since papain has already been shown to produce BMP in beef (*15*) it is likely that BMP is generated naturally by cathepsin B and/or L.

To find a protein/proteins that contain BMP as part of its/their sequence, homology searches were performed using several protein databases and genetic libraries. None of the databases or libraries examined provided a single protein with a sequence homology of 100%, including the two major muscle proteins myosin and actin. Because myosin and actin together constitute approximately 70% of total muscle protein in all animals, the possible sequence homology of BMP with these two proteins was extended to include several non-beef species; these searches also yielded negative results. Because BMP is an effective flavor enhancer (1.41 mM threshold compared to monosodium glutamate, MSG, with a perception threshold of 1.56 mM; *31*), a database search was reinitiated; the new search was designed to identify proteins with less than 100% homology. This search identified only a single protein that matched BMP for four of the eight amino acids (**Table 1**). The protein, named *monellin*,

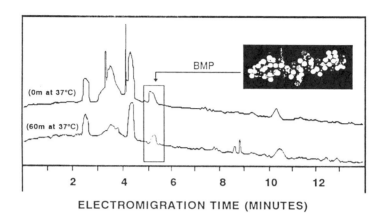

ELECTROMIGRATION TIME (MINUTES)

Figure 10. Capillary electropherogram of beef extract in the presence of liposomes and a free radical oxidation generator (FROG). Upper chromatogram is the initial unincubated protein profile while the lower profile is the protein profile after one hour incubation with the liposome + FROG system (adapted from *6*).

Lys-Gly-Asp-Glu-Glu-Ser-Leu-Ala

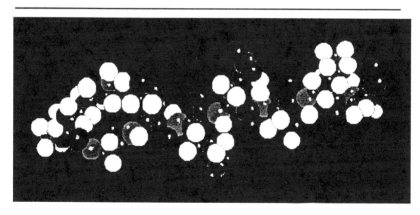

Figure 11. Computer generated molecular modeling of the structure of BMP.

Table 1. Fitting of peptides to the Tamura & Okai (Chapt. 4) model of a taste receptor[1]

| Peptide Fragment [N-terminal to C-terminal] | | | | | Taste | |
| Nakamura/Okai (Chapt. 2) equivalent | | | | Thres-hold (mM) | | |
A' (AX) Hydrophobic or Electro-positive site	B' (B) Electro-negative site	X' (X) Hydrophobic site	Ref.		1°	2°
lys-gly-	asp-glu-glu-	ser-leu-ala[2]	31/38	1.41	**savory**	sour
...-lys-gly-	tyr-glu-tyr-	gln-leu-tyr-...[3]	37	?	**sweet**	
lys-gly +	asp-glu-glu +	ser-leu-ala	31	1.41	**savory**	sour
lys-<u>glu</u> +	glu-glu +	ser-leu-ala	31	0.94	**savory**	sour
	asp-glu-glu-	ser-leu-ala[4-a]	38	?	**sour**	astrin. savory
	glu-glu-	ser-leu-ala[4-b]	38	?	**sour**	astrin. savory
	glu-	ser-leu-ala[4-c]	38	?	**sour**	astrin. savory
		ser-leu-ala	31/38	2.40	**bitter**	astrin. sour
lys-gly-	-----------------	-ser-leu-ala	38	?	**bitter**	astrin.
lys-gly-	asp		31	1.30	**sweet**	
lys-gly			31/38	1.22	salty	savory astrin.
lys-<u>glu</u> lys-asp			31		**none**	none
	asp-glu >glu-asp		31	1.69	**savory**	salty
	glu-glu		31	3.41	**savory**	salty
asp-	glu-glu		31	n.d.	sweet	
monosodium glutamate (MSG)			31	1.56	**savory**	

[1] Adapted from references 31, 37, and 38. [2] BMP amino acids 1-8 from 32. [3] Monellin's amino acids 4-11 from 37. [4] sour: a>b>c; savory: a=b>c; astringent: a<b<<c

is an intensely sweet tasting protein present in the sap of "Serendipity berries," the fruit of the West African plant, *Dioscoreophyllum comminsii*. Monellin has a molecular weight of 10,700 and is completely free of carbohydrate (*35-37*). Amino acids number 1, 2, 4, and 7 of BMP are identical to amino acids number 4, 5, 7 and 10 of *monellin* (**Table 1**). Based on these similarities and on the information compiled in Table 1, it is hypothesized that BMP and *monellin* may act at the same receptor site. The suggestion that BMP and *monellin* share and/or compete for the same receptor site requires that one or more residues are similar and that these residues bind to the same receptor. The first two sets of sequence data in **Table 1**, if fitted to the mechanism proposed by Drs. Nakamura and Okai (Chapter 2), yield strong support for this hypothesis.

Expansion or enhancement of the proposed mechanism of Nakamura and Okai is shown in **Figure 12**. Their model is based on the demonstration that several synthetically prepared di- and tri- peptide fragments composed of basic or acidic amino acids, produced individual tastes such as salty (*lys-gly*), sweet (*lys-gly-asp*), sour (*asp-glu-glu*) and bitter (*ser-leu-ala*; *31*). The expanded mechanism we propose is shown in Figure 12 and is based on the data tabulated in Table 1 (*31, 38*).

The expanded theory, indicates that "savory" (*umami*) shares a separate niche within the same receptor as "sweet" and "bitter" (**Figure 12**). "Sour" is also postulated to share the same receptor, but the data on the latter are insufficient at this time. The model proposes that a "savory/*umami*" flavor is imparted by the amino acids reacting at site "B'" or with "B'-A'", the electronegative and the electronegative - hydrophobic/electropositive site, respectively (**Figure 12**). This suggestion is based on the data indicating that most peptides containing "*glu*" yielding a "savory" flavor similar to monosodium glutamate (MSG). Certain "glu-containing" peptides such as "*lys-glu*," on the other hand, exhibit no flavor at all (**Table 1**). A "*glu*-containing" peptides with an N-terminal "*asp*" ("*asp-glu-glu*") gives a "sweet" taste even though the model says that this peptide should have give a "savory" flavor. Therefore, these data suggest that there may be a stronger connection between "savory/*umami*" and "sweet" than realized. These data also explain some of the similarities between BMP and *monellin*.

The intensely sweet taste of *monellin* can be explained in a second way. The intense sweetness of *monellin* can be envisioned as being dependent upon the molecular size of *monellin*. The size of *monellin* is thereby large enough to span two or more individual receptors. In this way, amino acid residues #4-11 would enhance the sweet taste produced by another residue(s) further along the *monellin* molecule. These other residues that might impart a sweet taste could include amino acid residue #16-19 (*asp-lys-leu-phe*) or #33-37 (*lys-leu-leu-arg-phe*) on subunit 1 of *monellin*. Experiments examining the taste of *monellin* both with and without residue #4-11 should confirm or deny the above hypothesis.

Additional support for the single receptor model (**Figure 12**) comes from experiments that utilized various peptide mixtures. For example, when the N-terminal di- tri- or tetra- peptide are removed from BMP the remaining peptide

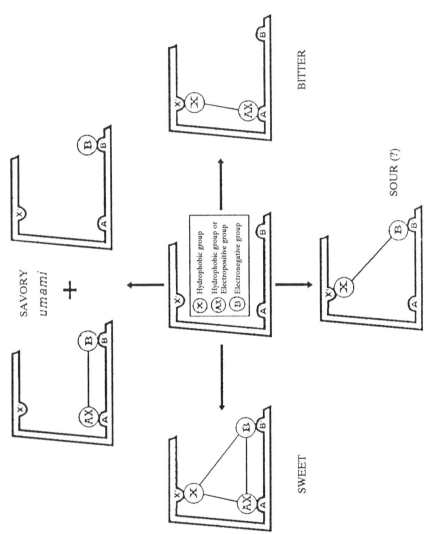

Figure 12. Proposed model of a taste receptor (adapted from Chapter 2).

fragments yield a "sour" taste (**Table 1**). The remaining C-terminal hexa-, penta- and tetra- peptide fragments of BMP give a mixed flavor impression that is predominantly "sour" (60%, 55%, and 44%, respectively) yet has a mildly "savory / *umami*" (10%, 10%, and 5%, respectively) taste and a moderate "astringency" (15%, 18%, and 32%, respectively; *38*). When the N-terminal dipeptide of BMP, i.e., *lys-gly* [by itself primarily "salty" with mild "savory" and "astringent" (5%)], is added back to the remaining hexapeptide fragment [by itself "sour"] the final perception is one of "savory/*umami*;" the perception threshold is similar to the original intact octapeptide (**Table 1**; additional information regarding flavor peptides may be found in chapter 10; *34*).

SUMMARY

The long-term goal of food technologists and biotechnologists is to design foods (specifically proteins) with a desired function or quality. Unlike the alchemist of yesteryear who tried to change lead into gold, the goal of the modern protein alchemist requires a detailed understanding of the relationship between peptides and the activation of receptors. Part of this understanding comes from knowledge of several phenomenon related to protein functionality such as the rules governing structure and function, the basis for a given reaction mechanism, and the important attributes of the protein/peptide. The data presented in this communication show that the production and deterioration of flavor proteins and peptides is a natural consequence of slaughter, cooking and storage involving scission by free radicals, thermal denaturation, and proteolysis.

While we are a long way from unraveling the relationships between protein structure and taste perception, some of the scientific and technical achievements over the last three decades allow study to be approached at the molecular level. The continued success of technological and biotechnological innovation as related to protein/flavor peptide engineering is rapidly permitting overproduction of flavor protein(s) as well as their redesign for improved resistance to proteolytic denaturation/degradation, free radical oxidation, and enzyme activation. As our understanding of taste perception and the functionality of enzymes and proteins continues, *de novo* synthesis of flavor peptides should become feasible. All these new innovations will lead to the production of better tasting, high quality muscle food products with improved stability and shelf-life at different cooking and cooking/storage situations.

LITERATURE CITED

1   Spanier, A.M.;Edwards, J.V.;Dupuy, H.P. *Food Tech.* **1988** *42(6)*:110-118.
2   Spanier, A.M.;McMillin, K.W.;Miller, J.A. *J. Food Sci.* **1990** *55(2)*:318-326.
3   Spanier, A.M.;Vercellotti, J.R.;James, C., Jr. *J. Food Sci.* **1992** *57(1)*:10-15.
4   St. Angelo, A.J.;Verecllotti, J.R.;Dupuy, H.P.;Spanier, A.M. *Food Tech.* **1988** *42(6)*:133-138.

5   Drumm, T.D.;Spanier, A.M. *J. Agric. Food Chem.* **1991** *39(2)*:336-343.
6   Spanier, A.M.;Miller, J.A.;Bland, J.M. In, *Lipid Oxidation in Food.* St.
       Angelo, A.J., Ed.;ACS Symposium Series No. 500;American Chemical
       Society:Washington, **D.C.**; **1992** Chapt. 7, pp **104-119.**
7   Timms, M.J.;Watts, B.M. *Food Technol.* **1958**, *12*:240-243.
8   Bailey, M.E. *Food Technol.* **1988**, *42(6)*:123-126.
9   St. Angelo, A.J.;Spanier, A.M.;Bett, K.L. In, *Lipid Oxidation in Food.* St.
       Angelo, A.J., Ed.;ACS Symposium Series No. 500;American Chemical
       Society:Washington, **D.C.;1992** Chapt. 9, pp **140-160.**
10   *Lipid Oxidation in Food.* St. Angelo, A.J., Ed.;ACS Symposium Series No.
       500;American Chemical Society:Washington, **D.C.:1992.**
11   Bird, J.W.C.;In, *Lysosomes in Biology and Pathology. Vol 4.* Dingle,
       J.T.,Dean, R.T., Ed.;American Elsevier Publishing Co.:New York, **NY**;
       1975 pp **75-109.**
12   Bird, J.W.C.;Schwartz, W.N.;Spanier, A.M.;*Acta Biol. Med. Germ.* **1977**
       36:1587-1604.
13   Bird, J.W.C.;Spanier, A.M.;Schwartz, W.M.;In, *Protein Turnover and
       Lysosomal Function.* Segal, H., Doyle, D., Eds.;Academic Press:New
       York, **NY**; **1978** pp **589-604.**
14   Spanier, A.M.;Dickens, B.F.;Weglicki, W.B.;*Am. J. Physiol.* **1985** *249
       (Heart Circ. Physiol. 18)*:H20-H28.
15   *Proteinases in Mammalian Cells and Tissues.*Barrett, A.J., Ed., Research
       monographs in Cell and Tissue Physiology. Volume 2;North-Holland
       Publishing Co.:Amsterdam, **The Netherlands**; **1977.**
16   Spanier, A.M.;St. Angelo, A.J.;Shaffer, G.P. *J. Agric. Food Chem.* **1992**
       40(9):1656-1662.
17   Spanier, A.M.;Bird, J.W.C. *Muscle & Nerve.* **1982** *5*:313-320.
18   Hornstein, I.;Crowe, P.F. *J. Agric. Food Chem.* **1960** 8:494-498.
19   Sink, J.D. *J. Food Sci.* **1979** *44*:1-5.
20   Wasserman, A.E. *J. Food Sci.* **1979** *44*:6-11.
21   Wasserman, A.E.;Talley, F. *J. Food Sci.* **1968** *33*:219-223.
22   Spanier, A.M. In, *Developments in Food Science. 29. Food Science and
       Human Nutrition.* Charalambous, G., Ed.;Elsevier Science Publishers
       B.V.:Amsterdam; **1992** pp **695-709.**
23   Asghar, A.;Gray, J.I.;Buckley, D.J.;Pearson, A.M.;Booren, A.M. *Food
       Tech.* **1988** *42(6)*:102-109.
24   Davies, K.J.A. *J. Biol. Chem.* **1987** *262*:9895-9901.
25   Davies, K.J.A.;Delsignore, M.E. *J. Biol. Chem.* **1987** *262*:9908-9914.
26   Davies, K.J.A.;Delsignore, M.E.;Lin S.W. *J. Biol. Chem.* **1987b** *262*:9902-
       9907.
27   Davies, K.J.A.;Goldberg, A.L. *J. Biol. Chem.* **1987a** *262*:8220-8226.
28   Davies, K.J.A.;Goldberg, A.L. *J. Biol. Chem.* **1987b** *262*:8227-8234.
29   Cohen, C.M.;Weissman, G.;Hottstein, S.;Awasthi, Y.C.;Srivastava, S.K.
       Biochem. **1976**, *15(2)*:452-460.

30   Mak, I.T.;Misra, H.P.;Weglicki, W.B. *J. Biol. Chem.* **1983**, *258*:13733-13737.

31   Tamura, M.;Nakatsuka, T.;Tada, M.;Kawasaki, Y.;Kikuchi, E.;Okai, H. *Agric. Biol. Chem.* **1989** *53(2)*:319-325.

32   Yamasaki, Y.;Maekawa, K. *Agric. Biol. Chem.* **1978**, *42*:1761-1765.

33   Spanier, A.M.;Miller, J.A. *ACS International Congress* **1989**, *AGRD 382* (abstract).

34   Kuramitsu, R.;Tamura, M.;Nakatani, M.;Okai, H. In, *Molecular Approaches to the Study of Flavor Quality*. Spanier, A.M., Okai, H., Tamura, M., Eds.;ACS Symposium Series No. 52_;American Chemical Society:Washington, **D.C.**; **1993**, Chapt. 7, pp **xy-zz**.

35   Bohak, Z.;Shoei-lung, L. *Biochimica et Biophysica Acta.* **1976**, *427*:153-170.

36   Morris, J.A.;Martenson, R.;Deibler, G.;Cagan, R.H. *J. Biol. Chem.* 1973 *248*:534-536.

37   Hudson, G.;Biemann, K. *Biochem. Biophys. Res. Commun.* **1976**, *71(1)*:212-220.

38   Yamasaki, Y.;Maekawa, K. *Agric. Biol. Chem.* **1980**, *44(1)*:93-97.

RECEIVED February 8, 1993

QUALITY ANALYSIS AND RESEARCH APPLICATIONS TOWARD PRODUCTION OF QUALITY FOODS

Chapter 7

Modern Statistics and Quantitative Structure–Activity Relationships in Flavor

John R. Piggott and S. J. Withers

Department of Bioscience and Biotechnology, University of Strathclyde, 131 Albion Street, Glasgow G1 1SD, Scotland

Partial least squares regression analysis (PLS) has been used to predict intensity of 'sweet' odour in volatile phenols. This is a relatively new multivariate technique, which has been of particular use in the study of quantitative structure-activity relationships. In recent pharmacological and toxicological studies, PLS has been used to predict activity of molecular structures from a set of physico-chemical molecular descriptors. These techniques will aid understanding of natural flavours and the development of synthetic ones.

Structure-activity relationships evolved from the assumption that the structure of a molecule is related in some way to its biological activity. Such relationships have been the backbone of numerous pharmacological and toxicological studies for many years. It is not surprising that the field of pharmaceutical science has lead to the identification and implementation of new techniques, to aid in defining structure-activity relationships.

Quantitative Structure-Activity Relationships

Statistical and computational methods have been used to quantify structure-activity relationships leading to quantitative structure-activity relationships (QSAR). The concept of QSAR can be dated back to the work of Crum, Brown and Fraser from 1868 to 1869, and Richardson, also in 1869. Many notable papers were published in the period leading up to the twentieth century by men such as Berthelot and Jungfleisch in 1872, Nernst in 1891, Overton in 1897 and Meyer in 1899 (1). Professor Corwin Hansch is now regarded by many as the father of QSAR, because of his work in the development of new and innovative techniques for QSAR. He and his co-workers produced a paper that was to be known as the birth of QSAR, and was entitled: "Correlation of biological activity of phenoxyacetic acids with Hammett substituent constants and partition coefficients" (2).

This important piece of work had four main facets: first, it showed that structure could be quantitatively related to biological activity; secondly, it introduced the concept of activity being described by more than one parameter; thirdly, the two-stage variation of activity with the water-octanol partition coefficient, denoted by $\log P$, was quantified using a quadratic equation; and fourthly, it introduced the concept of the

0097–6156/93/0528–0100$06.00/0

hydrophobic substituent constant, denoted by π. The logarithm of the partition coefficient (logP) of low molecular weight organic compounds is a physicochemical parameter used extensively in structure-biological activity studies to model interactions of the compounds with non-polar phases *in vitro* and *in vivo*. The hydrophobic substituent constant expresses the relative free-energy change on moving a derivative from one phase to another.

The paper by Hansch and co-workers, and many other publications of that school, generated great interest in QSAR research world-wide. A major development occurred when discrepancies were observed between measured logP values and those calculated using hydrophobic substituent constants (3). However, a "reductionist" fragmental constant approach for the calculation of logP did not show these discrepancies (4), and there appeared to be a fundamental weakness in the Hansch (π) approach. In response, the Hansch group developed their own fragmental constant approach for the calculation of logP, called the "constructionist" technique (5). The "reductionist" technique uses a statistical approach in order to analyse a large number of measured logP values, which give the best values for particular molecular fragments. So, the logP values are calculated by summation of these fragment values, with the addition of a few correction factors. The "constructionist" technique relies on the use of a small number of fundamental fragments which are derived from very precise logP measurements, and uses a correspondingly larger set of correction factors. This technique is now available as part of Pomona College's MEDCHEM software package. This computer package not only allows the calculation of logP, but also of other properties.

QSAR in Olfaction

QSAR is not only applicable to pharmacological and toxicological studies, but also may be applied to determine the odour quality of, for example, an artificial flavouring. It has been argued that QSAR could be applied more successfully to odour-active compounds, since the general sites of action can be identified (6). Also metabolic activity need not be considered since odour-active compounds are usually direct acting, and the biological end-point is well-defined. The recognition of an odour by the human olfactory system was considered to be a bimolecular process (7), involving the interaction of an airborne molecule with a receptor system, which takes place at the interfaces of peripheral nerve cells, located within the mucous layer of the olfactory epithelium. This interaction generates a response characterised by its quality, that is, its informational structure, and its intensity.

The relationship between odour quality and chemical structure is of considerable practical and theoretical interest. A number of methods have been used to determine quantitatively the relationships between the structure of a molecule and its odour quality (7). Though quantitative results were not obtained, a number of interesting theories were presented in that the intermolecular interaction in olfaction involved electrostatic attraction, hydrophobic bonding, van der Waals forces, hydrogen bonding, and dipole-dipole interactions. Hydrophobic interactions also appeared to be a major force for substrate binding in olfaction. It had previously been shown that lipophilicity and water solubility were factors that significantly influenced the odour thresholds of the pyrazines (8).

A method based on gas-liquid chromatography for establishing solubility factors of solutes has been applied to QSAR in olfaction (9). The hypothesis, concerning the recognition of odorous substances by receptor cells, was that the intermolecular forces involved in this phenomenon were similar to those involved in solutions. In other words, these forces are only of van der Waals and hydrogen-bonding types, and not the highly specific "lock and key" type as generally encountered in pharmacology. However, subsequent work has addressed the problem of olfactory perception at a molecular level (10). The experiment involved the cloning and characterisation of eighteen different members of a very large multigene family. This family encoded seven transmembrane domain proteins, whose expression is restricted to the olfactory epithelium. It was surmised that the members of the novel gene family were likely to encode the individual odorant receptors. Observations made suggested a model whereby receptors that bind distinct structural classes of odorant are encoded by specific genes. Therefore, the activation of receptors, distinct from each other, with similar structures of odour-active molecules, could elicit different odours, since perceived odour will depend upon higher order processing of primary sensory information. This model, unlike previous stereochemical models, does not necessarily predict that similar molecular structures will have similar odours. In essence, a "lock and key" mechanism is being suggested between the odorous molecule and its receptor.

Gas-chromatography on two stationary phases, Carbowax-20M and OV-101, was used to determine the retention indices of sixty disubstituted pyrazines (11). Sensory evaluation was used to determine the odour characteristics and odour thresholds of these compounds. It was observed that on a polar Carbowax-20M column, pyrazines with a non-polar group had a smaller $\Delta \Delta I$ value than the pyrazines with at least one polar group. From the sensory evaluation, it was determined that odour threshold was more significantly influenced by the kind of substituent, rather than the position of the ring substituent, and furthermore, the odour thresholds appeared to decrease with increasing lipophilicity. A two-term regression equation was used to model the logarithm of the odour threshold, where one term $\Sigma \delta I_R$ contained the empirical parameters for each of the substituents on the pyrazine ring. The other term $\Delta \Delta I$ contained a descriptor (Kovats retention index). The structure-activity relationship appeared to be satisfactory since calculated and observed values appeared to be in fairly good agreement.

However, although this equation was effective in modelling the odour thresholds of the disubstituted pyrazines, two main weaknesses have been identified (12): the first was that it was difficult to draw physical meaning from the descriptor $\Delta \Delta I$, since it was not clear which aspects of the molecular structure determined the odour threshold. The second weakness was discovered when pyrazine itself and thirteen mono-substituted pyrazines were added to the original set. The calculated and observed odour threshold values were no longer in agreement. This result indicated that the model was insufficient for more heterogeneous data sets.

Odour threshold techniques have been combined with chromatographic techniques to probe the structural characteristics of odour-active molecules, using an alcohol data set and a pyrazine data set (12). Two stages were involved: first of all, regression equations were developed, for the two sets of compounds involved, to model the logarithm of odour thresholds of these compounds. The descriptors were grouped into several general types, which are as follows: topological descriptors, which include molecular connectivity indices, atom, bond and path counts; geometrical descriptors such as shape, surface area, length and breadth; electronic descriptors such as partial charges, energies and dipole moments; combination descriptors such as charged partial surface area descriptors (CPSA); and physical property descriptors such as octanol-water partition coefficient (logP) and Kovats retention indices. Each quantitative structure-retention relationship and quantitative structure-property relationship was determined using a software system called ADAPT, Automated Data Analysis and Pattern Recognition Toolkit (13, 14). The process was executed in four steps: the first step was entry and storage of the molecular structure and the associated Kovats retention index, and odour threshold values. Secondly, three-dimensional molecular models were developed using molecular mechanics equations. Thirdly, molecular structure descriptors were calculated; and finally the model was generated using multiple linear regression techniques. Kovats retention indices or the log of the odour threshold were the independent variables.

Multiple linear regression (MLR) is the method most commonly used in QSAR of odours (*15*). It is used for modelling quantitative relationships between a y-variable, and a block of x-variables. This popular and simple statistical tool imparts advantages as well as disadvantages to data analysis. Multiple linear regression is advantageous in that regression analysis software packages are readily available with many helpful features. It is also a technique that is familiar to most scientists, particularly in the form of simple linear regression. The resulting regression models have good predictive ability and so quantitative estimates of biological activity may be made, and individual contributions of particular terms identified. However, MLR also has disadvantages which are derived partly from the statistical assumptions, and partly from the fitting of a function. It is assumed, before carrying out multiple linear regression, that the descriptors are uncorrelated, and that they model homogeneous sets of compounds. For example, for a homologous group, logP, a reflection of hydrophobicity, was strongly correlated with odour intensity (*16, 17*). However, when an attempt was made to model larger and more heterogeneous data sets, using multiple linear regression, the results were not as satisfactory.

Principles and Applications of Partial Least Squares (PLS) Regression Analysis

In any experimental design, two facts must be considered: the first is that enough data are needed to cover the complexity of the system, and the second is that the appropriate analytical tools should be available to deal effectively with the data, so that relevant information is not overfitted, over-simplified or ignored.

Multiple linear regression (MLR), although a popular technique, does not meet the requirements of the experimental design described above. MLR can only deal with one dependent variable at a time and assumes that all variables are orthogonal (uncorrelated), and that they are all completely relevant to the experiment. When dealing with an experimental system for the first time, it is not always possible to predict which variables will be relevant to the experiment, and which will not. So a technique is needed that can reconcile such uncertainties.

The method which satisfies these conditions is partial least squares (PLS) regression analysis, a relatively recent statistical technique (*18, 19*). The basis of the PLS method is that given k objects, characterised by i descriptor variables, which form the X-matrix, and j response variables which form the Y-matrix, it is possible to relate the two blocks (or data matrices) by means of the respective latent variables u and t in such a way that the two data sets are linearly dependent:

$$u(a) = d(a)t(a,k) + h(a,k)$$

where u and t are linear combinations of the descriptors and response variables respectively, d is the coefficient of the inner relation between u and t, h are the residuals that represent the non-systematic part of the relationship, and a is the dimensionality of the model. Ideally, the maximum relation between the latent variables is realised when each object becomes associated to pairs of u and t values that form a linear plot in the u/t space.

PLS falls in the category of multivariate data analysis whereby the X-matrix containing the independent variables is related to the Y-matrix, containing the dependent variables, through a process where the variance in the Y-matrix influences the calculation of the components (latent variables) of the X-block and vice versa. It is important that the number of latent variables is correct so that overfitting of the model is avoided; this can be achieved by cross-validation. The relevance of each variable in the PLS-method is judged by the modelling power, which indicates how much the variable participates in the model. A value close to zero indicates an irrelevant variable which may be deleted.

Partial least squares (PLS) regression analysis can be described as a predictive method, which can handle more than one dependent variable, and is not critically influenced by correlations between the independent variables. This technique can exist in two forms: PLS1 and PLS2. PLS1 regression predicts a single y-variable from a block of x-variables, and so, resembles multiple linear regression. PLS2 regression predicts a whole block of y-variables from a block of x-variables. PLS2 is most easily implemented as an extension of the orthogonalized PLS1 algorithm.

PLS is related to principal components analysis (PCA) (20). This is a method used to project the matrix of the X-block, with the aim of obtaining a general survey of the distribution of the objects in the molecular space. PCA is recommended as an initial step to other multivariate analyses techniques, to help identify outliers and delineate classes. The data are randomly divided into a training set and a test set. Once the principal components model has been calculated on the training set, the test set may be applied to check the validity of the model. PCA differs most obviously from PLS in that it is optimized with respect to the variance of the descriptors.

Other statistical techniques which are important in describing data are canonical correlation analysis (CCA) (21), and redundancy analysis (22). Canonical correlation analysis is analogous to PLS2, in that it constructs linear combinations of the variables from the X-block that correlate maximally with the Y-block, and vice versa. It is required that the number of objects is higher than the number of variables in X and Y. So, CCA and PLS2 estimate a small number of factors or dimensions in order to express the systematic variations common to the two data matrices, the X-block and the Y-block. The major difference between the two techniques is that CCA is a purely correlative technique, while PLS2 gives a predictive direction from X to Y. Maximum redundancy analysis calculates components in the X-block which predict the variables in the Y-block in an optimal way, optimal here being defined as explaining the maximum proportion of variance in the Y-block.

PLS has been successfully applied to quantitative structure-activity relationships in pharmacology and toxicology. PLS has been used for QSAR between polycyclic aromatic hydrocarbons and a receptor in rat liver, by defining the linear relation between the X-block containing the descriptor variables, and the Y-block containing the dependent variable, that is the logarithm of the receptor binding affinity of the compounds (23). The analysis showed which of the descriptors played an important role in binding affinities. PLS was useful in QSAR analysis using molecular orbital (MO) indices as descriptors (24). There are many problems associated with molecular orbital indices, due to the fact that the number derived from each molecule is high and they are usually strongly correlated with each other. Also, their relevance to the measured activity is not known prior to the experiment. The PLS method was not deterred by such co-linearities and so the model satisfied the experimental system from both a predictive and an interpretive point of view.

PLS was advantageous when studying the relationship of the toxicity of thirty triazines on *Daphnia magna* (25), and in a comparison between Hansch analysis and PLS analysis, using the same data set, it was shown that the multivariate approach of PLS provided more useful models than the Hansch type approach (26).

A relatively recent development in QSAR research is molecular reference (MOLREF). This molecular modelling technique is a method that compares the structures of any number of test molecules with a reference molecule, in a quantitative structure-activity relationship study (27). Partial least squares regression analysis was used in molecular reference to analyse the relation between X- and Y-matrices. In this paper, forty-two disubstituted benzene compounds were tested for toxicity to *Daphnia magna*. The results from the PLS analysis showed that MOLREF was able to quantify the relationship between the molecular structures and the resulting toxicity. MOLREF is carried out by first choosing a molecule, from a calibration set, which spans the structure space. Next, an appropriate number of atomic variables are included such as x,y,z-co-ordinates, van der Waals radius and atomic charge. The distances between the atoms in the reference molecule and the nearest neighbour atoms in the test molecule are calculated, and the co-ordinates of these neighbours are stored. Using the PLS model it is possible to generate the structure of molecules with higher or lower Y-values than the most active molecules found in the calibration set.

PLS does not appear to have been applied to QSAR of flavours, and although much progress has been made in the field of flavour chemistry, a greater insight into odour quality could be derived by the concept of applying many physico-chemical descriptors to the appropriate molecules.

QSAR of Odour Quality of Phenols

To test the potential of PLS to predict odour quality, it was used in a QSAR study of volatile phenols. A group of trained sensory panelists used descriptive analysis (*28*) to provide odour profiles for 17 phenols. The vocabulary consisted of 44 descriptive terms, and a scale from 0 (absent) to 5 (very strong) was used. The panel average sensory scores for the term 'sweet' were extracted and used as the Y-block of data, to be predicted from physico-chemical data.

A set of physico-chemical descriptors was gathered by direct measurement or calculation. The capacity factor, log k, was determined by reversed phase high performance liquid chromatography (*29*), and Kovats indices on OV101 and Carbowax-20M by gas chromatography. The molecular weight, dipole moment, ionization potential, electron energy and heat of formation were calculated using the MOPAC program in the Interchem molecular modelling package (*30*). Zero-order and first-order connectivities, and first-order connectivity/n, were finally calculated (*31*).

With the molecular descriptors as the X-block, and the sensory scores for 'sweet' as the Y-block, PLS was used to calculate a predictive model using the Unscrambler program version 3.1 (CAMO A/S, Jarleveien 4, N-7041 Trondheim, Norway). When the full set of 17 phenols was used, optimal prediction of 'sweet' odour was shown with 1 factor. Loadings of variables and scores of compounds on the first two factors are shown in Figures 1 and 2 respectively. Figure 3 shows predicted 'sweet' odour score plotted against that provided by the sensory panel. Vanillin, with a sensory score of 3.3, was an obvious outlier in this set, and so the model was recalculated without it. Again 1 factor was required for optimal prediction, shown in Figure 4.

While far from perfect, there is clearly a good correlation between predicted and measured scores, demonstrating the potential of PLS in this type of study. In terms of the model proposed by Buck and Axel (*10*), this study is attempting to predict the strength of interaction between the test molecules and the receptors giving rise to the perception of a 'sweet' odour. To improve the model, and particularly to accommodate vanillin, further descriptors would appear to be necessary. The greatest value of such a model would obviously lie in its ability to predict sensory scores from calculated descriptors only, eliminating the need to obtain or synthesize authentic samples of compounds. This would allow the model to suggest which molecules might be important in a natural flavour, and thus which compounds should be sought by flavour researchers.

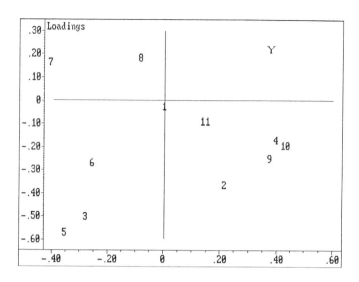

Figure 1. Loadings of molecular descriptors and sensory 'sweet' score on two PLS factors. 1 = log k, 2 = Kovats index on OV101 and (3) Carbowax-20M, 4 = molecular weight, 5 = dipole moment, 6 = ionization potential, 7 = electron energy, 8 = heat of formation, 9 = zero-order connectivity, 10 = first-order connectivity, 11 = first-order connectivity/n; Y = sensory 'sweet' score.

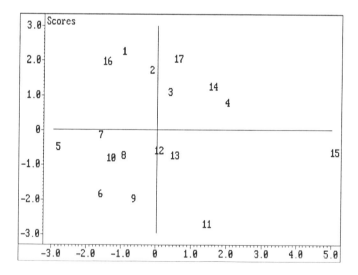

Figure 2. Scores of 17 volatile phenols on two PLS factors. 1 = eugenol, 2 = iso-eugenol, 3 = 4-methyl-guiacol, 4 = 4-ethyl-guiacol, 5 = 4-hydroxy-benzaldehyde, 6 = 2,4-dihydroxy-benzaldehyde, 7 = 2,3-dihydroxy-benzaldehyde, 8 = 2,5-dihydroxy-benzaldehyde, 9 = 2-hydroxy-5-methoxy-benzaldehyde, 10 = 4-hydroxyacetophenone, 11 = ethyl-4-hydroxy-benzoate, 12 = 4-hydroxy-phenethyl alcohol, 13 = ortho-vanillin, 14 = vanillin, 15 = syringaldehyde, 16 = 3,4-dimethyl-phenol, 17 − 2,4,6-trimethyl-phenol.

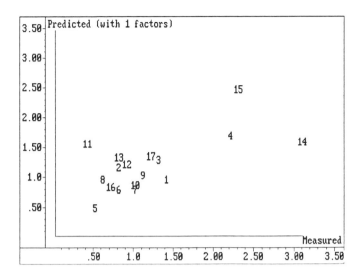

Figure 3. 'Sweet' score predicted by PLS model plotted against measured sensory score for 17 volatile phenols.

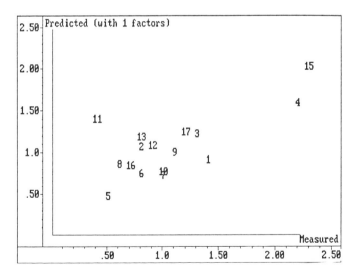

Figure 4. 'Sweet' score predicted by PLS model plotted against measured sensory score for volatile phenols, omitting vanillin.

Acknowledgments

The authors wish to thank Chris Gotts for collecting the sensory data; Peter Bladon for assistance in calculating molecular descriptors.

Literature Cited

1 Dearden, J.C. In *QSAR: Quantitative Structure-Activity Relationships in Drug Design*; Fauchere, J.L., Ed.; Alan R. Liss: New York, 1989; pp 7-14.
2 Hansch, C.; Maloney, P.P.; Fujita, T.; Muir, R.M. *Nature* **1962**, *194*, 323-351.
3 Hansch, C.; Anderson, S.M. *J. Org. Chem.* **1967**, *32*, 2583-2586.
4 Nys, G.G.; Rekker, R.F. *Chimica Therapeutica* **1973**, *8*, 521-528.
5 Leo, A.; Jow, P.Y.C.; Silipo, C.; Hansch, C. *J. Med. Chem.* **1975**, *18*, 865-868.
6 Hopfinger, A.J.; Mazur, R.H.; Jabloner, H. In *Analysis of Foods and Beverages*; Charalambous, G., Ed.; Academic Press: New York, 1984; pp 324-351.
7 Ohloff, G. *Experientia* **1986**, *42*, 271-279.
8 Fors, S.M.; Olofsson, B.K. *Chem. Senses* **1985**, *10*, 287-296.
9 Laffort, P.; Patte, F. *J. Chromatog.* **1987**, *406*, 51-74.
10 Buck, L.; Axel, R. *Cell* **1991**, *65*, 175-187.
11 Mihara, S.; Masuda, H. *J. Agric. Food Chem.* **1988**, *36*, 1242-1247.
12 Edwards, P.A.; Anker, L.S.; Jurs, P.C. *Chem. Senses* **1991**, *16*, 447-464.
13 Jurs, P.C.; Chou, J.T.; Yuan, M. In *Computer-Assisted Drug Design*; Olson, E.C., Ed.; ACS Symposium Series 112; American Chemical Society: Washington, D.C., 1979; pp 103-129.
14 Stuper, A.J.; Brugger, W.E.; Jurs, P.C. *Computer Assisted studies of Chemical Structure and Biological Function*; Wiley: New York, 1979.
15 Pike, D.J. In *Statistical Procedures in Food Research*; Piggott, J.R., Ed.; Elsevier Applied Science: London, 1986; pp 75-100.
16 Greenberg, M.J. *J. Agric. Food Chem.* **1979**, *27*, 347-352.
17 Greenberg, M.J. In *Odor Quality and Chemical Structure*; Moskowitz, H.R.; Warren, C.B., Eds.; ACS Symposium Series 148; American Chemical Society: Washington, D.C., 1981; pp 177-194.
18 Gerlach, R.W.; Kowalski, B.R.; Wold, H.O.A. *Anal. Chim. Acta* **1979**, *112*, 417-421.
19 Martens, M.; Martens, H. In *Statistical Procedures in Food Research*; Piggott, J.R., Ed.; Elsevier Applied Science: London, 1986; pp 293-359.
20 Piggott, J.R.; Sharman, K. In *Statistical Procedures in Food Research*; Piggott, J.R., Ed.; Elsevier Applied Science: London, 1986; pp 185-216.
21 Schiffmann, S.; Beeker, T. In *Statistical Procedures in Food Research*; Piggott, J.R., Ed.; Elsevier Applied Science: London, 1986; pp 255-291.
22 Israels, A.Z. *Applied Stochastic Models and Data Analysis* **1986**, *2*, 121-130.
23 Johnels, P.; Gilner, M.; Norden, B.; Toftgard, R.; Gustafsson, *Quantitative Structure-Activity Relationships* **1989**, *8*, 83-89.
24 Cocchi, M.; Menziani, M.C.; Rastelli, G.; De Benedetti, P.G. *Quantitative Structure-Activity Relationships* **1990**, *9*, 340-345.
25 Tosato, M.L.; Cesareo, D.; Marchini, S.; Passerini, L.; Pino, A. In *QSAR: Quantitative Structure-Activity Relationships in Drug Design*; Fauchere, J.L., Ed.; Alan R. Liss: New York, 1989; pp 417-420.
26 Galassi, S.; Mingazzini, M.; Vigano, L.; Cesareo, D.; Tosato, M.L. *Ecotoxicol. Environ. Saf.* **1988**, *16*, 158-169.
27 Alsberg, B. *Chemometrics and Intelligent Laboratory Systems* **1990**, *8*, 173-181.
28 Powers, J.J. In *Sensory Analysis of Foods*; Piggott, J.R., Ed.; Elsevier Applied Science: London, 1988; pp 187-266.
29 Kaliszan, R.; Blain, R.W.; Hartwick, R.A. *Chromatographia* **1988**, *25*, 5-7.
30 Breckenridge, R.; Bladon, P. INTERCHEM. Department of Chemistry, University of Strathclyde, Glasgow, Scotland, 1991.
31 Kier, L.B.; Hall, L.H. *Molecular Connectivity in Chemistry and Drug Research*; Academic Press: New York, 1976.

RECEIVED October 28, 1992

Chapter 8

Approaches to Mapping Loci That Influence Flavor Quality in Raspberries

Alistair Paterson, John R. Piggott, and Jianping Jiang

Food Science Laboratories, Department of Bioscience and Biotechnology, University of Strathclyde, 131 Albion Street, Glasgow G1 1SD, Scotland

Flavour quality is important in the raspberry and related soft fruits. Using multivariate statistics, it has been possible to resolve varietal and seasonal differences and processing effects in raspberry flavour into variations in fruit composition. Such approaches transform varietal flavour variation into quantitive traits that can be mapped genetically to chromosomal loci (*qtls*). Physical mapping of plant genomes has been facilitated by new polymerase chain reaction technologies that use oligonucleotide primers of arbitrary sequence. With such techniques, the presence of physical genetic markers in parental and progeny strains can be screened using rapid automated protocols. Such approaches can readily be used in diploid raspberries and related berry crops with simple genomes. In varieties with more complicated cytogenetics, genetic mapping of both phenotypic and physical markers is more complex.

Molecular biology is a reductionist science that has dominated modern genetics. In the past decade, molecular genetics has reached maturity and strategies developed to explore how molecules interact can now be employed to elucidate the genetic basis of variation in plants. Quantitive variations in certain compositional characters in fruit are perceived by the human observer in an integrative manner as taste and aroma. The genetic coding of such variation can be localised on plant chromosomes to *quantitive trait loci* (*qtl*) that have major influences upon fruit character, notably flavour and composition (*1*). These loci must also, in principle, be present in animal genomes, although, in domesticated animals, choice of feed, growth conditions, and slaughter conditions are likely to have the dominant influence.

A prerequisite for using such strategies to analyze the regulation of flavour in fruits and vegetables, is an understanding of which variations in a complex mixture of organic compounds are important to the human observer. Use of multivariate statistics allows the analyst to identify single or multiple compounds with a strong influence on varietal character, or producing flavour changes following processing, against a background of others that contribute to overall flavour, but are not central to varietal character (*2*).

Mapping complex genomes has been invigorated by the development of rapid molecular genetic techniques. The presence of individual plant genes can now be

0097–6156/93/0528–0109$06.00/0

determined, and subsequent inheritance followed, using a physical "tag" in the form of a restriction fragment length polymorphisms (3). Developments in polymerase chain reaction (PCR) technology have also facilitated such approaches: using short DNA primers, of arbitrary sequence, populations of DNA fragments similar to RFLPs can be generated in rapid and reproducible protocols (4, 5).

The Raspberry and Related Hybrid Fruits

Raspberry species were widely distributed through Northern Europe and the Americas, Asia, and East and Southern Africa, and belong to the subgenus *Idaeobatus* of the genus *Rubus*. They are closely related to, but differentiated from, the blackberries (*Eubatus*), cloudberries (*Chamaemorus*), and various arctic berries (*Cylactis*) (6). Each produces distinctive fruits, appreciated by consumers, and a tremendous range of hybrid cultivars has been generated for evaluation and growth by horticulturalists.

Rubus genomes are relatively simple and contain seven chromosomes present in ploidies (n) ranging from two to twelve (6). Crossing between members of the genus has proven relatively straightforward and a number of progeny have achieved success, notably the loganberry. This was thought to arise from an accidental crossing between the European raspberry "Red Antwerp" (*R. idaeus*) and the Western American blackberry "Auginbaugh" (*R. ursinus*), discovered in the garden of a Judge Logan, at Santa Cruz, California in 1881. Crane and Thomas (7, 8) showed that this fruit was a hexaploid (6n), with 42 chromosomes, subsequently shown to be an allopolyploid with two different pairs of genomes (4n) from *R. ursinus* and a pair of identical genomes (2n) from *R. idaeus*.

Other important North American crosses include the Young- and Boysenberries (7n) (9) and in Scotland crossings between the Oregon blackberry "Aurora" (8n) and tetraploid raspberries (4n) have generated the important Tay- and Tummelberries (hexaploid; 6n) (6). In Finland, the arctic raspberry (*R. arcticus*) has been crossed into *R. idaeus* and the diploid "Heija" has enabled commercial exploitation of the special aroma character of this fruit (10).

Analyses of Raspberry Flavour

To achieve widespread success, the raspberry and related fruits have required major breeding programmes, in Scotland carried out at the Scottish Crops Research Institute at Invergowrie. Major research targets in recent years have included developing berries suitable for machine-harvesting and freezing. Flavour has, to an extent, been a secondary selection character in breeding being largely related to acceptability with expert assessments of quality based upon rejection of defects.

The "impact compound" that provides the primary stimulus for fruit character in the raspberry is the ketone, 1-(p-hydroxphenyl)-3-butanone (11). Other important flavour contributors are *cis*-3-hexen-1-ol, α - and β - ionones, and α - irone (12, 13). In *R. arcticus* the characteristic aroma is considered to be from mesifurane (10). It has, however, been reported that steam distillates of raspberries can be assessed for aroma content using a colorimetric procedure and 80% of aroma is accounted for by geraniol, nerol, linalool, α - terpineol and the ionones (13).

Despite this detailed knowledge of raspberry aroma compounds, major flavour problems still emerge in commercial production. For many years the Scottish crop was dominated by Glen Clova, released in the early 1970s, which produced a good yield, and had good disease resistance, but gave fruit subject to deleterious flavour and colour changes associated with chilled storage and thermal processing. Such flavour faults can prove a major handicap in the current retail market. Moreover, the aroma of modern raspberries is notably sensitive to heat, and freezing and thawing, yet to satisfy consumer demand for premium fruit, and successfully compete against reduced labour costs in Eastern European fruit, character must be retained.

Matching of Compositional and Sensory Data on Flavour

Determining compounds responsible for varietal character and flavour changes is thus of more than passing interest. Use of multivariate statistics has made it possible to discern relationships between instrumental quantifications of individual fruit flavour compounds and sensory data, obtained from panels of experienced observers. A decade ago such strategies were the realm of the specialist, but with the appearance of suitable statistical packages for the microcomputer, such techniques are more approachable. Principal component analysis (PCA) (*14*), partial least square regression (PLS) (*15*) and redundancy analysis (RA) (*16*) have proven of particular value in allowing the experimenter to determine which of many compounds, have a major influence on an individual varietal character or flavour changes from processing.

Principal component analysis calculates linear combinations of variables that explain the maximum possible proportion of variance in data and generates orthogonal latent variables describing relationships among variables and between objects. This yields relationships among groups of variables in data sets and shows relationships between samples (*14*). Partial least square regression, commercially available in an accessible form, searches for a small number of orthogonal latent factors that describe the interrelationships between two blocks of variables. Factors that are extracted from the regressor matrix (*X*) are relevant to prediction of the regressand (*Y*) and factors in one matrix not relevant to the other are eliminated. Redundancy analysis, not previously extensively used in food science, projects *Y* (sensory) variables on to *X* (compositional) variables, and performs principal component analyses on the totality of *Y*-variables.

Elucidation of Important Features in Raspberry Taste

Flavour is of increasing importance when food is sufficiently abundant for consumers to exert choice. Sensory analysis, using trained laboratory panels, has been developed to profile fruit flavours, and describe relationships between products with a marked degree of confidence but is time-consuming, requiring dedicated observers who appreciate the nuances of individual character. Many, if not most, consumers, however, do not discriminate between fruit flavours. In dried orange juices, sweetness has been shown to be the major factor determining preference; in canned juices, sourness; and in frozen juices the interaction between sweetness and sourness is the significant factor (*17*).

Three major sugars (% w/v) contribute to sweet taste in the raspberry: fructose (17 - 40), glucose (14.5 - 33) and sucrose (0 - 10.8) giving total contents of 31.5 - 73.0 g l^{-1}. Raspberry juices are, unfortunately, remarkably sour in taste, through the presence of 10.5 - 26.7 g l^{-1} citric and malic acids. In a recent study of varietal and seasonal influences on raspberry taste, in this laboratory (*18*), it was concluded that 26 growth trial cultivars fell into three classes with respect to acidity: low (10.5 - 13.5); intermediate (15 - 30) and high (20.1 - 26.7 g l^{-1}). Principal component analysis of 14 compositional variables and 6 sensory attributes gave indicated the relationships between compositional and sensory variables (Figure 1a) and cultivars studied (Figure 1b) which allowed conclusions to be drawn with respect to flavour characters, their compositional origin, and influence on fruits.

Although the three statistical techniques PCA, PLS, and RA gave different sets of information, it was clear that sweetness and sourness and related compositional factors were of central importance in flavour. Contents of total sugars, sweet taste, sugar/acid ratio gave positive loadings and high negative loadings were obtained for sour taste, citric and total acid content, indicating malic acid had little influence on flavour as concluded previously (*19*). Utilizing PLS and RA, 55% and 77%, respectively, of total variance of sensory variables was explained by compositional variables. The correlation between such results and those of earlier studies give confidence in this methodology.

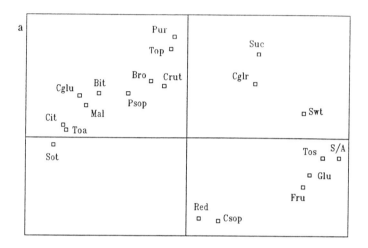

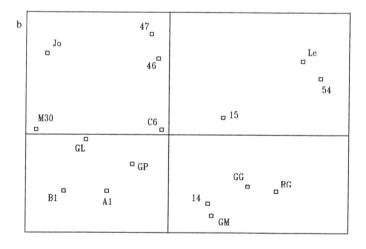

Figure 1. Analysis of the relationships between (a) compositional and sensory variables and (b) cultivars in terms of these variables. (a) Loadings of the 6 sensory and 14 compositional variables (Table I) on first (horizontal) and second (vertical) principal components. (b) Sample scores of the first (horizontal) and second (vertical) principal components of raspberry cultivars (Table II).

Table I. Compositional and sensory variables in principal component analysis of flavour in raspberry cultivars

Composition	Symbol	Composition	Symbol
Fructose	Fru	Sugar/acid ratio	S/A
Glucose	Glu		
Sucrose	Suc	Cyanidin-3-sophoroside	Csop
Total sugars	Tos	Cyanidin-3-glucosylrutinoside	Cglr
		Cyanidin-3-glucoside	Cglu
Malic acid	Mal	Cyanidin-3-rutinoside	Crut
Citric acid	Cit	Pelargonidin-3-sophoroside	Psop
Total acids	Toa	Total pigments	Top

Sensory attributes

Red	Red
Brown	Bro
Purple	Pur
Sweet taste	Swt
Sour taste	Sot
Bitter taste	Bit

Table II. Raspberry cultivars in study

Named varieties	Symbol	Cultivars	Symbol
Comox	Co	M30	M30
Glen Clova	GC	14/106	14
Glen Garry	GG	16A10	A1
Glen Lyon	GL	16C6	C6
Glen Moy	GM	19B11	B1
Glen Prosen	GP	31/7	31
Malling Joy	Jo	32RG3	RG
Malling Jewel	MJ	54/39	34
Malling Landmark	La	55/15	15
Malling Leo	Le	55/46	46
		55/47	47
		55/48	48
		789C1	C1

Important Features in Raspberry Flavour

Compositional analyses of flavour compounds in fresh and processed fruits are often of limited value because it is clear that compounds with very low aroma thresholds can have dominant effects upon fruit flavour. Moreover after processing, the compounds with the primary influence on flavour may change. Quantification of individual aroma compounds is also problematic requiring high resolution gas chromatographs linked to ion-trap or related detectors (HRGC/MS) although flame-ionisation detection is often more convenient.

Nearly 200 aroma compounds have been identified in fresh raspberries, many in trace quantities (20, 21), although Jiang (18) quantified only 62 compounds. Principal component analysis of volatile compound data showed that the first component was dominated by nine compounds, and the second by eight. Incorporating sensory data showed that *grassy*, correlated with *cis*-3-hexen-1-ol, had a high negative loading and *fruity (citric)* was associated with three compounds. In contrast, Williams (22) reported that apple flavour notes could be related to many compounds.

Redundancy analysis was able to explain 47% of total variance in flavour in relation to the 62 aroma compounds in the first two components; PLS explained only 36%. Neither method, however, correlated either raspberry ketone or linalool with important aroma notes, suggesting concentrations of these impact compounds are not important in determining varietal character.

Definition of taste and aroma character in sensory terms, and assigning this to precise variation in fruit composition, allows experiments with sensory panels to be limited to defining compositional components correlated with character changes perceived important by observers. Subsequent experimentation can then be effected by automated chromatography.

Quantitive Trait Loci

Studies of raspberry flavour have centered on understanding how quantitive variation in individual volatile and non-volatile compounds influence perceived quality. Such variation can be seasonal, environmental, or of genetic origin; both seasonal and varietal influences are important. Interaction between sources of variation is also observed. Seasonal effects in raspberry aroma volatiles appears to be primarily associated with variation in *cis*-3-hexenol content (*grassy* note), whereas genetic variation is stable. In strawberries, genotype determines ratios of individual sugars and organic acids in fruits (23).

A number of aspects of genetic variation in tomato flavour have been shown by Tanksley and his colleagues (1) to originate from regulation of levels of phenotype expression by *quantitive trait loci*. These can be mapped in the diploid genome by physical methods such as restriction fragment length polymorphism (RFLP) analysis. This has been carried out using random genomic DNA clones, isolated from the organism being studied, in hybridizations utilizing radiolabelled DNA (24). While inexpensive for a small number of samples, such strategies are tedious when monitoring the progeny of crosses is required. Thus attention has turned to use of new approaches, made feasible by development of polymerase chain reaction (PCR) technology, which does not require radioisotopes, and can be routinely performed by affordable automated equipment to yield data that can be digitized.

Polymerase chain reactions are initiated by binding oligonucleotide primers to opposing strands of the DNA and synthesizing the intervening sequence using a thermostable DNA polymerase. This effects an amplification of a predefined sequence, generating significant quantities of duplex DNA in experiments lasting a few hours. Random primers (typically decamers), not corresponding to any known sequence, have also been utilized to generate physical genetic maps, and alternate mixtures of primers generates complementary data. Moreover as organisms diverge, the similarities between PCR product patterns is reduced, indicating changes in genomic sequence. Analyses using Random Amplified Polymorphic DNA (RAPD markers) (4) have been of value in genetic fingerprinting in cultivated sub-populations of cocoa (*Theobroma*), generating certain fragments that are shared between and others that differentiate varieties (25).

Physical Mapping of Quantitive Trait Loci in Raspberries

Mapping *qtls* in raspberries is an exciting concept. It is clear that compositional factors determining fruit character can be identified, utilizing sensory analysis and multivariate statistics. It is also clear that the cytogenetics of the *Rubus* genus may be varied and complex, as in many fruit. The strategy of Tanksley and his coworkers (*1*) in mapping *qtls* in the tomato was dependent upon backcrossing progeny obtained from a cross between a modern cultivar and a wild-type tomato, and is dependent upon progeny being diploid. In the raspberry, crosses between diploid cultivars are possible and a wide range of genes have already been identified (6). This fruit may thus prove a valuable model system in developing research in an area that has proven recalcitrant.

Acknowledgments

The authors are very grateful to the Director, and staff of the Scottish Crops Research Institute for provision of fruit and helpful discussions, and the University of Strathclyde Research and Development Fund, Agricultural and Food Research Council and Chivas Bros (Keith) Ltd for support.

Literature Cited

1 Paterson, A.H.; Lander, E.S.; Hewitt, J.D.; Peterson, S.; Lincoln, S.E.; Tanksley, S.D. *Nature* **1988**, *335*, 721-726.
2 Piggott, J.R. *J. Sensory Stud.* **1990**, *4*, 261-272.
3 Prince, J.D.; Tanksley, S.D. *Proc. Roy. Soc. Edin.* **1992**, *99B*, 23-29.
4 Williams, J.G.K.; Kubelik, A.R.; Livak, K.L.; Rafalski, J.A.; Tingey, S.V. *Nucleic Acids Research* **1991**, *18*, 6531-6535.
5 Welsh, J.; McClelland, M. *Nucleic Acids Research* **1991**, *18*, 7213-7218.
6 Jennings, D.L. *Raspberries and Blackberries: Their Breeding Diseases and Growth*; Academic Press: New York, 1988.
7 Crane, M.B.J. *Genet.* **1940**, *40*, 109-118.
8 Thomas, P.T.J. *Genet.* **1940**, *40*, 119-128.
9 Thompson, M.M. *Amer. J. Botany* **1961**, *48*, 667-673.
10 Hiirsalmi, H.; Sako, J. *Acta Horticult.* **1976**, *60*, 151-157.
11 Nursten, H.E. In *The Biochemistry of Fruits and their Products*; Hulme, A.C., Ed.; Academic Press: London, 1970; p 253.
12 Kallio, H.J. *Food. Sci.* **1976**, *41*, 563-566.
13 Latrasse, A.; Lantin, B.; Mussillon, P.; Sanis, J.L. *Lebensmittel Wissensch. Technol.* **1982**, *15*, 19-21.
14 Piggott, J.R.; Sharman, K. In *Statistical Procedures in Food Research*; Piggott, J.R., Ed.; Elsevier Applied Science: London, 1986; pp 181-232.
15 Martens, M.; van der Berg, E. In *Progress in Flavour Research 1984*; Adda, J., Ed.; Elsevier Science Publishers BV: Amsterdam, 1985; pp 131-148.
16 Liu, Y. *Multivariate Methods for Relating Different Sets of Variables with Applications in Food Research*. PhD Thesis, University of Bristol, UK, 1990.
17 Ennis, D.M.; Keeping, L.; Chin-Ting, J.; Ross, N.J. *Food Sci.* **1979**, *44*, 1011-1016.
18 Jiang, J. *Variation in Raspberry Composition and Sensory Qualities as Influenced by Variety, Season and Processing*. PhD Thesis, University of Strathclyde, UK, 1991.

19 Green, A. In *The Biochemistry of Fruits and their Products*; Hulme, A.C., Ed.; Academic Press: London, 1971.
20 Nursten, H.E.; Williams, A.A. *Chem. Indust.* **1967**, *12*, 486-497.
21 de Vincenzi, M.; Badellino, E.; Di Folco, S.; Dracos, A.; Magliola, M.; Stacchini, A.; Stachinin, P.; Silano, V. *Food Additives and Contaminants* **1989**, *6*, 235-267.
22 Williams, A.A. In *Geruch- und Geshmackstoffe*; Drawert, F., Ed.; Verlag Hans Carl: Nurnberg, 1975; pp 141-151.
23 Shaw, D.V.J. *Amer. Soc. Hort. Sci.* **1988**, *113*, 770-774.
24 Bonierbale, M.W.; Plaisted, R.L.; Tanksley S.D. *Genet.* **1988**, *120*, 1095-1103.
25 Wilde, J.; Waugh, R.; Powell, W. *Theor. Appl. Genet.* **1992**, *83*, 871-877.

RECEIVED December 30, 1992

Chapter 9

Supercritical Fluid Extraction of Muscle Food Lipids for Improved Quality

Milton E. Bailey[1], Roy R. Chao[1], Andrew D. Clarke[1], Ki-Won Um[1], and Klaus O. Gerhardt[2]

[1]Department of Food Science and Human Nutrition, 21 Agriculture Building and [2]Department of Biochemistry, 4 Agriculture Building, University of Missouri, Columbia, MO 65211

Supercritical CO_2 (SC-CO_2) was used to reduce the lipid of meat and the cholesterol of meat and beef tallow. Lipids can be removed quantitatively from dried muscle foods by SC-CO_2, but relatively high temperatures are needed. The use of SC-CO_2 in conjunction with ethanol, adsorbents and multi-separators also reduced the cholesterol of beef tallow. SC-CO_2 was also used to concentrate volatile flavor compounds from beef and pork fat. The volatile components in various extraction fractions were identified and quantitated.

Meat products provide approximately 36% of the energy and many of the required nutrients in the diet. They also contribute more than 50% of the total fat, 75% of the saturated fatty acids and essentially all of the cholesterol. Consumers have been advised to reduce their dietary fat and cholesterol levels for health reasons, and the utilization of red meats has suffered greatly in the past few years. Consumption of red meat products world-wide has been reduced considerably compared to a high consumption of 3.1 million tons per year in 1986-1988 (1).

Along with the reduced consumption of meat products, the production of rendered animal fats has also declined. Beef tallow consumption has increased greatly since 1950 due to its distinctive flavor and stability as a frying medium. In 1985, 460,000 metric tons of beef tallow were used in the U.S. for baking and frying, but this figure was reduced to 289,000 metric tons in 1990 and is probably less this year. The decreased production of beef tallow has made it unprofitable for renderers to continue the processing procedures required to produce edible tallow (1).

Processed meat products with a minimum of fat and cholesterol appear to be in high demand. As for edible beef tallow, reduced usage has increased

0097–6156/93/0528–0117$06.25/0

research efforts to preserve the flavor characteristics while reducing the cholesterol and saturated fatty acids. These quality improvements will require specific processing procedures and supercritical fluid extraction may be a useful technic.

Warmed-Over Flavor

Another processing procedure that could involve supercritical fluid extraction with CO_2 is the preparation of flavor concentrates from meat lipids for use in mixtures of other natural precursors for the preparation of synthetic meat flavor additives that serve both as antioxidants that prevent warmed-over flavor (WOF) in cooked meat during storage and enhance the flavor of the natural products.

WOF is a problem associated with the use of precooked meat products such as roasts and steaks. The term WOF was first used by Tims and Watts (2) to describe the rapid development of oxidized flavors in refrigerated cooked meats. Published evidence indicates that the predominant oxidation catalyst is iron from myoglobin and hemoglobin, which becomes available following heat denaturation of the protein moiety of these complexes. The oxidation of the lipids results in the formation of low molecular weight components such as aldehydes, acids, ketones and hydrocarbons which may contribute to undesirable flavor.

Harris and Lindsay (3) reported that panelists were able to detect off-flavors in reheated fried chicken after 2 hr of refrigeration. The high concentration of polyunsaturated fatty acids in muscle phospholipids supplies readily oxidizable lipids, and these components play an important role in the development of WOF (4, 5). The rising demand for precooked convenience foods has increased the importance of controlling WOF.

The most practical method for preventing WOF in meat products is to add antioxidants prepared from natural precursors such as sugars and amino acids by heating them to produce constituents that not only act as antioxidants but serve to enhance meaty flavor as well. The resulting Maillard products have been known to have antioxidant activity in lipid systems (6-8). It is assumed that the antioxidative property of the Maillard reaction is associated with the formation of low molecular weight reductones and high molecular weight melanoidins (6, 7, 9-13).

While the above reactions are responsible for meaty flavors, animal fat plays an important role in the formation of the characteristic species flavor of cooked meat (14). More than 100 compounds have been identified in heated beef fat, including aldehydes, n-alkanes, n-alkenes, fatty acids, ketones, lactones and heterocyclics (15-18). Of these compounds, the lactones, methyl ketones and free fatty acids appear to be the most important for desirable meat flavor.

Supercritical Fluid Extraction

Supercritical fluid extraction (SFE) is a fluid extraction process that has been applied on both a processing scale and for analytical chemical applications.

The process involves the use of supercritical fluids rather than liquids as solvents. A fluid is in the supercritical state when its pressure and temperature exceed the physical properties which defines its critical point. Carbon dioxide is by far the most widely used supercritical solvent. Many other selected fluids have potential use for SFE technologies.

The unique feature of supercritical fluids as solvents is that their solvating strength is directly related to their densities, which can be easily varied as a function of pressure and temperature. Above the critical point, the densities of supercritical fluids increase with increased pressure and decrease with increasing temperatures. Their properties are similar to those of both liquids and gases. The densities and solvating power can approach that of a liquid, whereas the viscosity is intermediate and diffusivity is much closer to properties of gases (*19*).

Liquid carbon dioxide is a non-polar solvent much like hexane, but in its supercritical state the polarity increases with increase in pressure. By controlling pressure (or by the addition of a polar co-solvent) the selectivity of SC-CO$_2$ can be adjusted for preferential extraction of compounds based on their polarity (*19*).

Many authors of technical articles in trade journals have described the advantages of SFE in a broad range of applications such as coffee, decaffeination, spice extraction and lipid purification; and the processing principles have been well established and practiced in a number of industries as indicated by McHugh and Krukonis (*20*) in a recent article. The general advantages of using SC-CO$_2$ include: low critical pressure and temperature (73.8 bar/31.06°C), low viscosity, high diffusivity, good solvent for many organics, high relative volatility, non-toxic, non-flammable, and ready availability. There are likewise many disadvantages for using SFE including high cost non-specific extracting qualities and low capacity for removing low-boiling compounds.

Four aspects of research involving the use of SFE for the improvement of quality of muscle food products are briefly discussed. These include supercritical CO$_2$ extraction of lipids from fresh ground beef and from dried muscle foods; the extraction and separation of lipid and cholesterol from beef tallow; supercritical CO$_2$ extraction of flavor volatiles from beef and pork lipids for use as additives in synthetic meat flavors; and identification and quantitation of flavor volatiles extracted with SC-CO$_2$.

Experimental

SFE Equipment. The extraction apparatus (Figure 1) used was manufactured by the Superpressure Division of Newport Scientific, Inc. (Jessup, MD) and modified by adding two more separators. Detailed extraction conditions were described by Chao et al. (*21*). For each test run, approximately 200 g of frozen and thawed ground beef sample containing about 18% fat or 100 g of animal fat was loaded into the extraction vessel and extracted with supercritical CO$_2$.

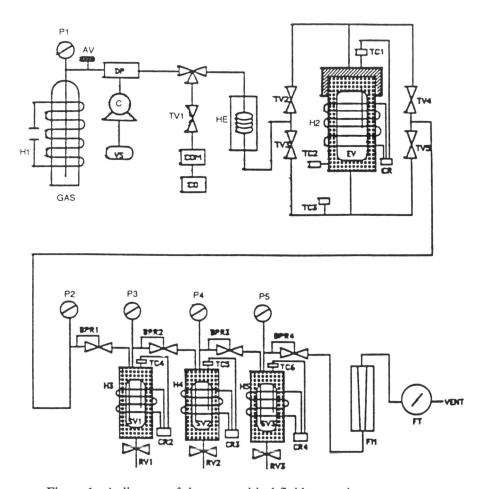

Figure 1. A diagram of the supercritical fluid extraction system.
P1 = cylinder pressure gauge; C+DP = diaphragm compressor;
EV = extraction vessel; HE = water bath; TV = two-way through
valve; TC = thermocouple; BPR = back pressure regulator;
SV = separation vessel; RV = pressure relief valve; AV = adjusting
valve; C+VS = variable speed compressor; FT = gas flow totalizer.

The extraction pressures and temperatures were varied, ranging from 100 to 310 bar and 30 to 50°C. The soluble components were generally collected in a separator at 34.5 bar/40°C. Edible beef tallow of commercial grade was obtained from Anderson Clayton/Humko Products, Inc. (Memphis, TN) and melted at 50°C, mixed, subdivided into smaller plastic containers with 200 g each in weight, and stored at -27°C until used. The procedures mentioned above were followed for the extraction of 100 g beef tallow.

Determination of Cholesterol. For meat extraction, the procedures for determining the cholesterol of extracted lipid samples were described by Chao et al. (*21*). For edible beef tallow extraction, the preparation of samples for cholesterol content was based on the AOAC (*22*) method Section 28.110. The prepared sample was then injected into a Supelco SPB-1 fused silica capillary column of 30 meters x 0.32 mm i.d. in a Varian Model 3700 gas chromatograph equipped with dual flame ionization detectors. The initial holdup time was 4 min at 270°C and then programmed to a temperature of 300°C at a ramp rate of 10°C/min. Helium flow rate and split ratio were 1.5 ml and 50:1, respectively, while the injector/detector temperature was 310°C.

GLC/MS Analysis of Volatile Compounds. The volatile compounds of heated pork fat were determined by the direct sampling capillary chromatogrpahic method of Suzuki and Bailey (*23*)

Fat Analysis. Fat was determined by AOAC (*22*) methods.

Results

Supercritical CO_2 Extraction of Lipids from Ground Beef and from Dried Muscle Foods. Clarke (*24*) found that cumulative yield curves of total lipids from fresh ground beef containing 18% fat extracted at 103, 172 and 310 bar/50°C revealed that the yields increased with increased pressure, which is the dominant factor for the extraction process. This result was confirmed by extraction studies at several pressures and temperatures. He found that extraction was very erratic, which was related to the presence of high moisture content of the samples which co-extracts with the lipid and somewhat limits the use of this methodology for processing high moisture foods such as fresh meat (*24*) or fish (*25*).

Most studies of SC-CO_2 quantitative extraction of lipids from muscle foods have been limited to dried materials with moisture levels below 10%. King et al. (*26*), Froning (*27*) and Hardardottir and Kinsella (*28*) were able to quantitatively remove lipid from muscle foods and data for these studies are shown in Table I. This table also includes data from study of dried pork. Maximum removal of lipid from dried beef and luncheon meat was obtained at 300-345 bar/45-80°C (*26, 27*) and at 275 bar/80°C for trout (*28*).

Table I. Supercritical Carbon Dioxide Efficiency
for Removing Fat from Dried Muscle Foods

Food	Pressure, Bar	Temperature, °C	Fat Extracted, %	Moisture, %
pork[a]	310	50	71.9	3.9
beef powder[b]	300	45	66.9	2.5
beef chunks[b]	300	45	97.9	1.7
luncheon meat[c]	345	80	98.9	1.8
imported ham[c]	345	80	97.3	2.3
trout[d]	275	80	97.3	<2.0
sardine meat powder[e]	255	40	57.4	33.0

[a]Clarke (24).
[b]Froning (27).
[c]King et al. (26).
[d]Hardardottir & Kinsella (28).
[e]Fujimoto et al. (25).

Data cited by Clarke (24) indicate that fatty acid content may be a good indication of lipid fractionation by SFE. He found that saturated fatty acids were reduced by 72% and unsaturated fatty acids reduced by 63% compared to the unextracted control. Cholesterol was reduced by 76% to an average of 41.6 mg/100 g in the extracted meat residue (24).

The cholesterol was reduced another 30% to 27.9 mg/100 g tissue by using 3% ethyl alcohol as an entrainer or co-solvent with the SC-CO_2, but higher temperatures may be required to completely remove cholesterol from dried meat. The use of entraining solvents, however, is not an advantage in the processing of meats for human consumption and defeats the purpose of using SC-CO_2 and higher temperatures should be avoided to maintain the functional properties of the muscle proteins (24).

Extraction and Separation of Lipid and Cholesterol from Beef Tallow with SC-CO_2. The cumulative yields of the fractions for various operating conditions used to extract 100 g of edible beef tallow are shown in Figure 2. The SC-CO_2 used to extract all lipids charged in the extractor at 345 bar and 241 bar were 10 and 22 kg, respectively. At 138 bar, 22% of the total beef tallow was extracted after 20 kg CO_2 was passed. These results indicate the high dependency of triglyceride solubility on the applied pressure and temperature and confirm the greater extraction efficiencies at the higher pressures.

A variety of experiments were carried out to determine the influence of chemical and physical parameters on the extractability of cholesterol. Results of a study to determine the importance of pressure on the extraction of cholesterol are given in Figure 3.

The maximum concentration of cholesterol extracted occurred after the passage of 2.5 kg of CO_2 at 138 bar/50°C. This was 2.5 times the amount extracted at 345 bar/50°C and approximately 2.3 times the amount extracted at 241 bar/150°C. Extraction at lower pressures increased the solubility of cholesterol but decreased the weight of total lipid extracted per unit weight of CO_2 used. Most of the cholesterol was removed after extracting with 20 kg CO_2 at 138 bar/40°C where only 22% of the total lipid had been removed.

Multiple Separators for More Efficient Separation of Cholesterol from Beef Tallow. Since the solvent power of CO_2 depends upon its density, a step-wise reduction of separation pressure will alter the CO_2 density so that the soluble components of beef tallow can be separated and collected in different fractions.

In one study, the three separators (Figure 1) were adjusted at 170 bar/40°C (S1); 102 bar/40°C (S2); and 34.5 bar/40°C (S3) during extraction of beef tallow at 345 bar/40°C. Table II is a tabulation of the yields of lipid and cholesterol and the cholesterol concentration from the fractions collected from the three separators during continuous extraction. Although the weight of the lipid extracted in each separator with progressive fractions (6, 12 and 18 kg) of CO_2 remains about the same from each separation, the concentration of cholesterol in these fractions decreased with increased CO_2 weight.

Table II. Weights of Lipid and Cholesterol of Beef Tallow Fractions Extracted at 345 Bar/40°C and Fractionated by Separation at Different Pressures[a]/40°C

CO_2 kg	Separator	Wt. Lipid, g	Cholesterol Conc., mg/100 g	Wt. Cholesterol, mg
6	S1	10.2	128	13.0
	S2	4.1	171	6.9
	S3	7.1	433	30.7
12	S1	11.2	92	10.3
	S2	5.1	126	6.4
	S3	10.3	376	38.8
18	S1	11.9	49	5.8
	S2	4.4	75	3.3
	S3	10.7	272	29.2

[a]S1 = 172 bar; S2 = 103 bar; S3 = 34.5 bar.

The lipid collected in separator (S3) at 34.5 bar/40°C during the run was 37% of the total lipid extracted, whereas the amount of cholesterol separated at 34.5 bar/40°C was 70% of the cholesterol extracted. This amounts to two-fold concentration of the cholesterol at the lower pressure. Use of multiple

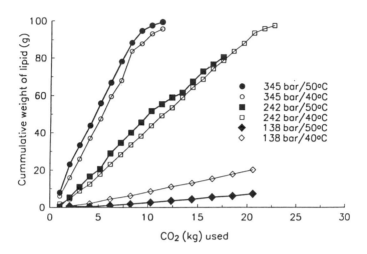

Figure 2. Cumulative yields of beef tallow lipids extracted with SC-CO_2 at different pressures and temperatures.

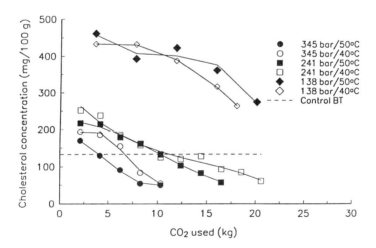

Figure 3. Yields of cholesterol extracted at different pressures and temperatures plotted against increment weight of CO_2 used.

separators allowed one to produce relatively small quantities of lipid containing high amounts of cholesterol. It would be possible to reduce the amount of cholesterol in beef tallow and other muscle lipids by this method by about 65% by collecting fractions from S1 and S2 which contain 60% of the lipids.

Ethanol as a Co-Solvent with SC-CO$_2$ for Extracting Cholesterol from Beef Tallow. Figure 4 contains data for comparing the extraction of lipid and cholesterol with 5% ethanol as a co-solvent (entrainer) compared to extracting with SC-CO$_2$ alone.

Extraction with SC-CO$_2$-ethanol (5%) at 138 bar/40°C progressively removed more lipid and cholesterol than SC-CO$_2$ alone. Twelve kilograms of SC-CO$_2$-ethanol (5%) extracted 100 mg of cholesterol and 39 g of lipid. The remaining 61 g of lipid contained 30 mg of cholesterol, thus reducing the cholesterol concentration from 130 mg/100 g lipid to 49 mg/100 g lipid, or by 63%. The same amount of SC-CO$_2$ without ethanol extracted 10 g of lipid and 30 mg of cholesterol, for a reduction in cholesterol concentration of 23%.

Cholesterol Extraction from Beef Tallow by SC-CO$_2$ in the Presence of Adsorbents. The utilization of SC-CO$_2$ extraction in the presence of adsorbers enhances the fractionation of solutes dissolved in the supercritical fluid. This procedure has been reviewed by King (*29*).

Figure 5 has data on the extraction of cholesterol in the presence of a 1:1 mixture of adsorbents and beef tallow with a SC-CO$_2$ extraction pressure of 345 bar/40°C. The adsorbents were activated alumina (AA), silica gel (SG) and molecular sieve (MS), a synthetic silico-aluminate (1:1) zeolite. The data reveal a continuous decrease in cholesterol concentration during progressive extraction with SC-CO$_2$.

Extraction at 345 bar/40°C of a sample mixed with AA (1:1) removed lipid with the least amount of cholesterol. The average concentration of cholesterol extracted in the presence of AA at 345 bar/40°C (separated at 34.5 bar/40°C) was 60 mg/100 g of lipid extracted. At fractions beyond 6 kg of CO$_2$ used where one-half of the lipid remained, the average concentration of cholesterol in the lipid was less than 40 mg/100 g of lipid extracted. This was a 70% reduction in the concentration of cholesterol in the extracted lipid.

Concentration and Identification of Flavor Volatiles in Heated Beef Fat by SC-CO$_2$ Extraction. Um et al. (*30*) studied the flavor intensities of lipids separated in different fractions of SC-CO$_2$ extracts at two pressures from heated beef tallow. The tallow was heated at 100°C for 2 hr and extracted at 207 bar/50°C and 345 bar/50°C. Six 1 kg fractions of CO$_2$ were used to extract 100 g of tallow at each pressure and separated at 34.5 bar/40°C.

The individual fractions (F1 through F6) were evaluated for beefy odor intensity by 10 trained panelists who used a nine-point structured scale with descriptive anchors at 1 (no odor) and 9 (very strong beefy odor). Results of these analyses are shown in Figure 6 for beef tallow extracted with SC-CO$_2$ at 207 bar/50°C. Lipids extracted in fraction F1 had very strong beefy odor which decreased in subsequent fractions.

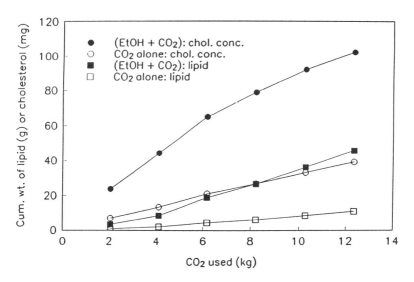

Figure 4. Cumulative yields of cholesterol and lipid extracted from beef tallow with SC-CO_2 at 138 bar/40°C with and without 5% ethyl alcohol as solvent modifier.

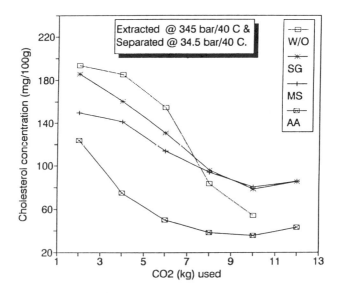

Figure 5. Extraction of cholesterol from beef tallow with SC-CO_2 at 345 bar/40°C in presence of activated alumina (AA), molecular sieve (MS), silica gel (SG) and without adsorbent (W/O).

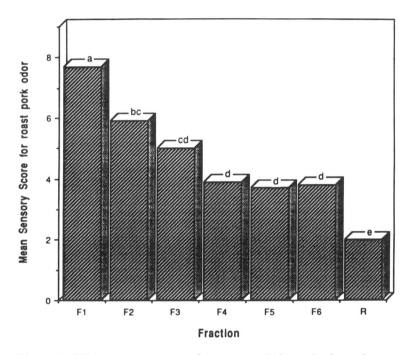

Figure 6. Mean sensory scores for roast pork intensity in various fractions of heated pork fat extracted with SC-CO_2 at 207 bar/50°C. Means headed by different letters indicate significantly (P < 0.05) different values.

Volatile Compounds from Heated Beef Fat. Um et al. (30) studied the volatile lipids in fraction F1 from extractions at 207 bar/50°C and 345 bar/50°C using the qualitative GLC/MS method of Suzuki and Bailey (23). They identified 71 compounds in fraction F1 of the extracts and quantified 66 compounds including 17 hydrocarbons, 4 terpenoids, 15 aldehydes, 3 ketones, 4 phenols, 10 carboxylic acids, 6 esters and 7 lactones.

Many of the hydrocarbons were similar to those identified in butter fat by Urbach and Stark (31), and some of these were concentrated up to 6-fold in fraction F1 compared to the unextracted control.

Among the terpenoids, phyte-1-ene, neophytadiene, phyte-2-ene and farnesol were all highly concentrated in fraction F1 compared to the non-extracted control. These compounds were previously associated with beef flavor by Larick et al. (32) and by Peterson and Chang (33). They were also identified in butter fat by Urbach and Stark (31) and in lamb by Suzuki and Bailey (23). Concentration of some of these compounds correlate highly with grassy flavor of beef.

2-Nonenal, 2-decenal and 2-undecenal were concentrated appreciably in fraction F1 of the 207 bar/50°C extract, whereas aldehydes associated with lipid oxidation such as pentanal, hexanal and octanal were not concentrated by the extraction method used, probably because these constituents continue to accumulate following extraction or because low volatiles are not extracted well by the SC-CO$_2$ method used.

Of the fatty acids, nonanoic and decanoic acid were the most abundant in the concentrate of fraction F1. Ha and Lindsay (34) postulated that many of the volatile acids present in beef tallow contributed to the desirable flavor of deep-fried potatoes.

Lactones are key components in beef aroma (35). All seven lactones were concentrated in fraction F1 compared to the non-extracted control. δ-Tetra-decalactone and α-hexadecalactone were present in the greatest concentration, and were present in the highest concentration of any other constituent in fraction F1.

Volatile Compounds from Heated Pork Fat. Pork fat was heated at 160°C for 1 hr, fractionated at 207 bar/50°C and at 345 bar/50°C, and analyzed as described by Um et al. (30). The fractions (Figure 6) were analyzed by sensory analysis and the results were similar to those described for beef tallow. Fraction F1 was observed to have the strongest "porky" odor, and the residue had the least.

Different fractions from the extracts of pork fat were analyzed for volatile compounds by the GLC/MS method of Suzuki and Bailey (23). Fifty-six compounds consisting of 13 hydrocarbons, 15 aldehydes, 4 ketones, 3 alcohols, 6 phenols, 9 carboxylic acids, 1 ester and 5 heterocyclics were identified and quantified in extracts from pork fat heated at 160°C for 1 hr.

Analytical data of volatile compounds extracted from heated pork fat with SC-CO$_2$ at 207 bar/50°C are given in Table III and similar data on volatiles extracted at 345 bar/50°C are presented in Table IV. Most volatiles in

Table III. Concentrations of Volatile Compounds in Heated Pork Fat
Extracted with SC-CO_2 at 207 bar/50°C

Peak	Compound[a]	Concentration, ppm		
		Control	F1[b]	Residue
Hydrocarbon: n-alkanes				
3[c]	heptane	2.03 (0.01)[d]	5.29 (0.76)[e]	1.89 (0.57)[d]
6	octane	0.35 (0.02)[d]	2.32 (1.37)[d]	0.32 (0.38)[d]
9	nonane	0.28 (0.01)[de]	1.65 (0.82)[df]	<0.05[e]
18	decane	0.22 (0.01)[d]	2.94 (1.41)[e]	<0.05[d]
26	undecane	0.47 (0.03)[de]	1.78 (0.83)[ef]	<0.05[d]
33	dodecane	0.35 (0.15)[d]	3.46 (1.30)[e]	0.16 (0.15)[d]
40	tridecane	<0.05[d]	4.12 (0.92)[e]	<0.05[d]
48	tetradecane	0.47 (0.07)[d]	2.80 (0.90)[e]	0.15 (0.14)[d]
49	pentadecane	0.25 (0.02)[d]	34.15 (7.22)[e]	0.43 (0.07)[d]
53	hexadecane	0.27 (0.05)[d]	7.88 (1.17)[e]	<0.05[d]
Hydrocarbon: branched, cyclic, alkenes				
5	1,2,4-trimethyl (1A,2B,4A)-cyclopentane	0.40 (0.07)[d]	4.88 (2.89)[d]	<0.05[d]
20	2-ethyl-1,4-hexadiene	0.57 (0.09)[d]	5.37 (1.90)[e]	<0.05[d]
45	3,7-dimethyl-undecane	0.27 (0.02)[d]	6.76 (0.82)[e]	<0.05[d]
Aldehydes				
3[c]	pentanal	2.03 (0.01)[d]	5.29 (0.76)[e]	1.89 (0.57)[d]
7	hexanal	2.53 (0.06)[d]	11.01 (6.74)[d]	1.35 (1.28)[d]
10	heptanal	0.46 (0.01)[d]	3.29 (1.08)[e]	0.21 (0.22)[d]
12	2-heptenal (Z)	0.80 (0.30)[d]	8.63 (1.10)[e]	<0.05[d]
13	benzaldehyde	0.19 (0.01)[d]	3.22 (0.18)[e]	<0.05[d]
19	octanal	0.64 (0.04)[d]	10.84 (0.61)[e]	0.21 (0.22)[d]
21	2-octenal	0.61 (0.02)[d]	7.68 (1.97)[e]	<0.05[d]
27	nonanal	1.62 (0.20)[d]	27.96 (0.88)[e]	2.58 (1.09)[d]
30	2-nonenal	<0.05[d]	2.41 (0.64)[e]	<0.05[d]
34	decanal	0.46 (0.28)[d]	12.78 (3.23)[e]	0.18 (0.19)[d]
35	2,4-nonadienal	0.21 (0.06)[d]	9.66 (1.06)[e]	<0.05[d]
38	2-decenal (Z)	3.63 (0.16)[d]	51.70 (3.07)[e]	<0.05[d]
41	2,4-decadienal (E,Z)	1.30 (0.07)[d]	22.67 (3.11)[e]	0.39 (0.49)[d]
42	2,4-decadienal (E,E)	3.34 (0.01)[d]	112.93 (28.56)[e]	0.31 (0.36)[d]
44	2-undecenal	1.78 (0.13)[d]	53.42 (12.42)[e]	0.28 (0.01)[d]

Continued on next page

Table III. Continued

Peak	Compound[a]	Concentration, ppm		
		Control	F1[b]	Residue

Ketones

8	2-heptanone	0.63 (0.05)[d]	2.74 (0.81)[e]	<0.05[d]
24	2-nonanone	<0.05[d]	2.56 (1.77)[d]	<0.05[d]
54	2-pentadecanone	8.29 (0.18)[d]	536.81 (60.35)[e]	0.56 (0.37)[d]
55	2-heptadecanone	0.53 (0.09)[d]	54.89 (0.30)[e]	0.42 (0.52)[d]

Alcohols

2	1-butanol	<0.05[d]	3.85 (5.44)[d]	<0.05[d]
17	heptanol	0.68 (0.06)[d]	10.69 (2.58)[e]	0.26 (0.29)[d]
22	1-octanol	<0.05[d]	2.82 (0.32)[e]	<0.05[d]

Phenols

15	phenol	1.44 (0.40)[d]	4.19 (0.71)[e]	<0.05[d]
25	3-methyl-phenol	0.21 (0.01)[d]	2.57 (1.93)[d]	<0.05[d]
29	2,5-dimethyl-phenol	0.32 (0.06)[d]	7.26 (2.81)[e]	<0.05[d]
32	2,3-dimethyl-phenol	2.03 (1.22)[d]	7.02 (0.31)[e]	0.48 (0.61)[d]
36	3-(1-methylethyl)-phenol	0.33 (0.08)[d]	3.34 (1.29)[e]	0.58 (0.15)[d]
52	2,6-bis-(1,1-dimethylethyl)-4-ethyl-phenol	1.06 (0.15)[de]	6.33 (3.17)[ef]	<0.05[d]

Acids

1	acetic acid	6.58 (0.64)[d]	19.58 (7.41)[d]	1.16 (1.58)[d]
4	2-methyl-propionic acid	<0.05[d]	18.36 (12.30)[e]	<0.05[d]
14	hexanoic acid	0.58 (0.00)[d]	9.36 (3.85)[e]	0.53 (0.18)[d]
23	heptanoic acid	1.40 (0.08)[d]	12.72 (2.64)[e]	<0.05[d]
28	2-ethyl hexanoic acid	1.06 (0.31)[d]	4.59 (0.01)[e]	<0.05[f]
31	octanoic acid	0.23 (0.05)[d]	1.57 (0.55)[e]	0.27 (0.06)[d]
39	nonanoic acid	5.38 (2.27)[d]	18.75 (20.48)[d]	<0.05[d]
43	decanoic acid	<0.05[d]	10.22 (2.82)[e]	<0.05[d]
51	dodecanoic acid	<0.05[d]	4.43 (2.99)[d]	<0.05[d]

Esters

47	2-methyl-propionic acid, methyl-ester	0.24 (0.02)[d]	4.31 (0.53)[e]	0.21 (0.22)[d]

Table III. Continued

Peak	Compound[a]	Concentration, ppm		
		Control	F1[b]	Residue
Furans, furanones and lactones				
11	dihydro-2(3H)-furanone	0.40 (0.09)[d]	3.46 (2.21)[d]	<0.05[d]
16	2-pentyl furan	<0.05[d]	5.52 (7.80)[d]	<0.05[d]
46	dihydro-5-pentyl-2(3H)-furanone (γ-nonalactone)	0.51 (0.09)[d]	2.55 (0.50)[e]	0.15 (0.14)[d]
50	δ-decalactone	0.62 (0.12)[d]	11.84 (1.79)[e]	0.21 (0.21)[d]
Others				
37	benzothiazole	<0.05[d]	2.08 (1.20)[d]	<0.05[d]

[a]Identified by mass spectrometry.
[b]Fraction F1: extraction pressure = 207 bar/50°C; separation pressure = 34.5 bar/40°C.
[c]Peak contains more than one volatile.
[def]Means in same row bearing unlike superscripts are significantly different ($P < 0.05$).

Table IV. Concentrations of Volatile Compounds in Heated Pork Fat
Extracted with SC-CO$_2$ at 345 bar/50°C

Peak	Compound[a]	Concentration, ppm		
		Control	F1[b]	Residue
Hydrocarbon: n-alkanes				
3[c]	heptane	2.04 (0.01)[d]	2.68 (0.07)[e]	1.02 (0.22)[f]
6	octane	0.35 (0.02)[d]	0.79 (0.11)[e]	0.43 (0.13)[d]
9	nonane	0.28 (0.01)[d]	0.33 (0.01)[d]	0.29 (0.09)[d]
18	decane	0.22 (0.01)[d]	1.04 (0.09)[e]	0.24 (0.04)[d]
26	undecane	0.47 (0.03)[d]	0.63 (0.01)[e]	0.60 (0.01)[de]
33	dodecane	0.35 (0.15)[d]	0.67 (0.02)[d]	0.32 (0.10)[d]
40	tridecane	<0.05[d]	0.57 (0.06)[e]	<0.05[d]
48	tetradecane	0.47 (0.07)[d]	0.46 (0.02)[d]	0.46 (0.17)[d]
49	pentadecane	0.25 (0.02)[d]	4.10 (0.25)[e]	0.62 (0.29)[d]
53	hexadecane	0.28 (0.05)[d]	1.07 (0.08)[e]	0.26 (0.03)[d]
Hydrocarbon: branched, cyclic, alkenes				
5	1,2,4-trimethyl (1A,2B,4A)-cyclopentane	0.39 (0.07)[d]	1.06 (1.00)[d]	<0.05[d]
20	2-ethyl-1,4-hexadiene	0.67 (0.09)[d]	1.23 (0.04)[e]	0.61 (0.14)[d]
45	3,7-dimethyl-undecane	0.27 (0.02)[d]	0.88 (0.04)[e]	0.13 (0.10)[d]
Aldehydes				
3[c]	pentanal	2.04 (0.01)[d]	2.68 (0.07)[e]	1.02 (0.22)[f]
7	hexanal	2.51 (0.06)[d]	4.57 (0.41)[e]	2.23 (0.30)[d]
10	heptanal	0.46 (0.01)[d]	1.18 (0.08)[e]	0.57 (0.18)[d]
12	2-heptenal (Z)	0.80 (0.30)[d]	2.89 (0.01)[e]	0.80 (0.27)[d]
13	benzaldehyde	0.19 (0.01)[d]	0.59 (0.03)[d]	<0.05[d]
19	octanal	0.64 (0.04)[d]	1.88 (0.02)[e]	0.93 (0.18)[d]
21	2-octenal	0.60 (0.02)[d]	2.11 (0.05)[e]	0.56 (0.11)[d]
27	nonanal	1.62 (0.20)[d]	4.43 (0.53)[e]	2.15 (1.17)[de]
30	2-nonenal	<0.05[d]	0.45 (0.15)[e]	<0.05[d]
34	decanal	0.46 (0.28)[d]	1.97 (0.15)[e]	0.51 (0.01)[d]
35	2,4-nonadienal	0.21 (0.06)[d]	1.69 (0.14)[e]	0.65 (0.10)[f]
38	2-decenal (Z)	3.63 (0.16)[d]	8.19 (0.21)[e]	2.09 (1.50)[d]
41	2,4-decadienal (E,Z)	1.31 (0.07)[d]	3.34 (0.15)[e]	0.83 (0.40)[d]
42	2,4-decadienal (E,E)	3.34 (0.01)[d]	14.18 (1.34)[e]	2.06 (1.80)[d]
44	2-undecenal	1.78 (0.13)[d]	10.03 (6.24)[d]	0.90 (0.61)[d]

Table IV. Continued

Peak	Compound[a]	Concentration, ppm		
		Control	F1[b]	Residue
Ketones				
8	2-heptanone	0.63 (0.05)[d]	1.01 (0.47)[d]	0.77 (0.15)[d]
24	2-nonanone	<0.05[d]	<0.05[d]	<0.05[d]
54	2-pentadecanone	8.29 (0.18)[d]	65.17 (7.69)[e]	1.34 (0.44)[d]
55	2-heptadecanone	0.53 (0.09)[d]	5.11 (1.25)[e]	0.23 (0.18)[d]
Alcohols				
2	1-butanol	<0.05[d]	0.61 (0.79)[d]	<0.05[d]
17	heptanol	0.68 (0.05)[d]	1.66 (0.10)[e]	0.58 (0.10)[d]
22	1-octanol	<0.05[d]	<0.05[d]	<0.05[d]
Phenols				
15	phenol	1.44 (0.40)[d]	3.85 (0.17)[e]	<0.05[f]
25	3-methyl-phenol	0.21 (0.01)[d]	0.47 (0.05)[e]	0.15 (0.08)[d]
29	2,5-dimethyl-phenol	0.32 (0.06)[d]	1.01 (0.15)[e]	<0.05[d]
32	2,3-dimethyl-phenol	2.03 (1.22)[d]	1.44 (0.06)[d]	0.67 (0.49)[d]
36	3-(1-methylethyl)-phenol	0.33 (0.08)[d]	0.46 (0.03)[d]	0.46 (0.10)[d]
52	2,6-bis-(1,1-dimethylethyl)-4-ethyl-phenol	1.06 (0.15)[d]	2.83 (0.19)[e]	0.06 (0.14)[d]
Acids				
1	acetic acid	6.58 (0.64)[d]	8.51 (0.82)[d]	3.37 (1.31)[e]
4	2-methyl-propionic acid	<0.05[d]	5.22 (0.93)[e]	<0.05[d]
14	hexanoic acid	0.58 (0.00)[d]	2.04 (0.07)[e]	0.46 (0.20)[d]
23	heptanoic acid	1.40 (0.08)[d]	4.06 (0.38)[e]	0.75 (0.68)[d]
28	2-ethyl hexanoic acid	1.06 (0.31)[d]	1.68 (0.02)[d]	1.09 (0.15)[d]
31	octanoic acid	0.23 (0.05)[d]	2.18 (0.50)[e]	0.25 (0.06)[d]
39	nonanoic acid	5.38 (2.27)[d]	13.35 (2.67)[e]	<0.05[d]
43	decanoic acid	<0.05[d]	1.60 (0.27)[e]	<0.05[d]
51	dodecanoic acid	<0.05[d]	1.14 (0.04)[e]	<0.05[d]
Esters				
47	2-methyl-propionic acid, methyl-ester	0.24 (0.02)[d]	0.61 (0.05)[e]	0.14 (0.12)[d]

Continued on next page

Table IV. Continued

Peak	Compound[a]	Concentration, ppm		
		Control	F1[b]	Residue
Furans, furanones and lactones				
11	dihydro-2(3H)-furanone	0.40 (0.09)[d]	2.09 (0.05)[d]	<0.05[d]
16	2-pentyl furan	<0.05[d]	<0.05[d]	<0.05[d]
46	dihydro-5-pentyl-2(3H)-furanone (γ-nonalactone)	0.51 (0.09)[d]	0.65 (0.09)[e]	<0.05[d]
50	δ-decalactone	0.62 (0.12)[d]	1.84 (0.19)[e]	<0.05[f]
Others				
37	benzothiazole	<0.05[d]	0.73 (0.24)[e]	<0.05[de]

[a]Identified by mass spectrometry.
[b]Fraction F1: extraction pressure = 345 bar/50°C; separation pressure = 34.5 bar/40°C.
[c]Peak contains more than one volatile.
[def]Means in same row bearing unlike superscripts are significantly different (P < 0.05).

fraction F1 from these extracts were significantly ($P < 0.05$) concentrated compared to the control or the residue. Among hydrocarbons, pentadecane was 100-fold more concentrated in fraction F1 in the extract at 207 bar/50°C compared to the control. n-Alkanes and n-alkenes are readily formed from lipid hydroperoxides by β-scission of alkoxy radicals and have been implicated in pork flavor.

The concentrations of aldehydes extracted in fraction F1 at 207 bar/150°C were 3 to 8 times higher than those extracted at 345 bar/50°C. The concentrations of 2,4-dienals which are responsible for the deep-fat fried odor of many fats and oils (*14*) were very high compared to similar compounds found in beef fat, and they were concentrated up to 34-fold in fraction F1 compared to non-extracted control. 2,4-Decadienal (E,E) is an apparent precursor of 2-pentylpyridine, and it has also been found that the 2,4-dienals contribute to fatty, oily and tallowy odor of fats (*36*).

Ketones were also more readily extracted at 207 bar/50°C than at 345 bar/150°C. 2-Pentadecanone was concentrated 65-fold in fraction F1 in the lower pressure extract.

Hexanoic, heptanoic, nonanoic and decanoic acids were the most predominant fatty acids in heated pork fat. Nonanoic acid was the most abundant. Fraction F1 for the 207 bar/50°C extract was highly concentrated compared to other fractions.

Lactones are minor constituents in pork fat compared to beef fat, but δ-decalactone was highly concentrated in the F1 fraction of the 207 bar/50°C extract. The furans, furanones and thiazoles were undoubtedly formed from Maillard reaction precursors and this type of volatile would be more prevalent in the lipid fraction of cooked pork compared to pork fat alone.

Conclusions

High extraction pressure (>300 bar) is the predominant factor for increasing the extractability of lipid from muscle foods, although greater quantities of lipid are extracted at higher temperatures (50-80°C) compared to lower temperatures such as 35 or 40°C.

Extraction of lipids from muscle foods is more predictable if the food has less than 10% moisture. High recoveries of glycerides can be obtained from dry meat products by extracting with SC-CO_2 above 300 bar, but cholesterol is extracted less effectively. Cholesterol is extracted from beef tallow more efficiently at lower pressures than at higher pressures (<300 bar).

Use of ethanol as a co-solvent with SC-CO_2 enhances the extraction of cholesterol from dried muscle foods. High extraction pressure coupled with multiple separation vessels for gradual reduction of SC-CO_2 density reduces the separation time necessary for separating cholesterol from beef tallow using less CO_2 than extractions at lower pressure. Ethanol is a good co-solvent used with SC-CO_2 for separating cholesterol from other lipids in beef tallow.

Cholesterol can be extracted more effectively from beef tallow with the addition of adsorbents such as activated alumina, silica gel or molecular sieve to the extraction chamber, but flavor constituents are removed as well.

Flavor volatiles from beef and pork fat can be concentrated up to 30-fold by extracting with small quantities of SC-CO_2 at low pressure.   SC-CO_2 concentrated volatile fractions from heated beef tallow have greater numbers of terpenoids, more high molecular weight ketones, more lactones, more esters, more phenols and more branched cyclic and unsaturated aldehydes than similar extracts from heated pork fat, but the latter has more 2,4-dienals and higher concentrations of aldehydes.

Acknowledgments

Some of the research data cited or used in this manuscript was supported by grants from the Missouri Beef Council and the National Live Stock and Meat Board.

The contribution to the preparation of this manuscript by Louise Noland is truly appreciated.

Literature Cited

1. Dotson, K. *Inform.* **1992,** *3,* 152.
2. Tims, M.; Watts, B.M. *Food Technol.* **1958,** *12,* 240-243.
3. Harris, N.D.; Lindsay, R.C. *J. Food Sci.* **1972,** *37,* 19-22.
4. Bailey, M.E.; Dupuy, H.P.; Legendre, M.G. In *The Analysis and Control of Less Desirable Flavors in Foods and Beverages;* Charalambous, G., Ed.; Academic Press: NY, 1980; pp 31-52.
5. Pearson, A.M.; Gray, J.I. In *The Maillard Reaction in Food and Nutrition;* Walter, G.R.; Feather, M.S., Eds.; ACS Symposium Series 215; ACS: Washington, DC, 1983; pp. 287-300.
6. Hodge, J.E.; Rist, C.E. *J. Am. Oil Chem. Soc.* **1953,** *75,* 316-322.
7. Evans, C.O.; Mosa, H.A.; Cooney, P.M.; Hodge, J.E. *J. Am. Oil Chem. Soc.* **1958,** *35,* 84-88.
8. Yamaguchi, N.; Nakoo, Y.; Koyama, Y. *Nippon Shokuhin Kogyo Gakkaishi* **1964,** *11,* 184-189.
9. Yamaguchi, N. *Nippon Shokuhin Kogyo Gakkaishi* **1969,** *16,* 140-144.
10. Yamaguchi, N.; Fujimaki, M. *Nippon Shokuhin Kogyo Gakkaishi* **1974,** *21,* 280-284.
11. Mills, F.D.; Hodge, J.E.; Parks, L.W. *J. Org. Chem.* **1981,** *46,* 3597-3693.
12. Yamaguchi, N.; Oda, T. *Nippon Shokuhin Kogyo Gakkaishi* **1984,** *31,* 577-580.
13. Bailey, M.E. In *Meat Flavor;* Shahidi, F., Ed.; Elsevier: Amsterdam, 1992; in press.
14. Hornstein, I.; Crowe, P.F. *J. Agric. Food Chem.* **1960,** *8,* 494-498.
15. Watanabe, K.; Sato, Y. *Agric. Biol. Chem.* **1968,** *32,* 191-196.
16. Ohnishi, S.; Shibamato, T. *J. Agric. Food Chem.* **1984,** *32,* 987-992.
17. Umano, K.; Shibamato, T. *J. Agric. Food Chem.* **1987,** *35,* 14-18.
18. Ha, J.K.; Lindsay, R.C. *J. Food Sci.* **1991,** *56,* 1197-1202.

19. Stahl, E.; Quirin, K.W.; Gerard, D. *Dense Gases in Extraction and Refining;* Springer-Verlag: Berlin, 1988; pp 1-29.
20. McHugh, N.A.; Krukonis, V.J. *Supercritical Fluid Extraction Principles and Practice;* Butterworths: Boston, MA, 1986.
21. Chao, R.R.; Mulvaney, S.J.; Bailey, M.E.; Fernando, L.N. *J. Food Sci.* **1991,** *56,* 183-187.
22. AOAC. *Official Methods of Analysis, 14th Edition;* Association of Official Analytical Chemists: Washington, DC, 1984.
23. Suzuki, J.; Bailey, M.E. *J. Agric. Food Chem.* **1985,** *33,* 343-347.
24. Clarke, A.D. In *Proceedings of the Reciprocal Meats Conference;* 1991, Vol. 44; pp 101-107.
25. Fujimoto, K.; Endo, Y.; Cho, S.Y.; Watabe, R.; Suzuki, Y.; Konno, M.; Shoji, K.; Ara, K.; Saito, S. *J. Food Sci.* **1989,** *54,* 265-268.
26. King, J.W.; Johnson, J.H.; Friedrich, J.P. *J. Agric. Food Chem.* **1989,** *37,* 951-954.
27. Froning, G.W., University of Nebraska, personal communication, 1991.
28. Hardardottir, I.; Kinsella, J.E. *J. Food Sci.* **1988,** *53,* 1656-1658.
29. King, J.W.; Eissler, R.L.; Friedrich, J.P. In *Supercritical Fluid Extraction and Chromatography, Techniques and Applications;* Charpentier, B.A.; Sevenants, M.R., Eds.; ACS Symposium Series 366; ACS: Washington, DC, 1987; pp. 63-88.
30. Um, K-W.; Bailey, M.E.; Clarke, A.D.; Chao, R.R. *J. Agric. Food Chem.* **1992,** accepted for publication, September issue.
31. Urbach, G.; Stark, W. *J. Agric. Food Chem.* **1975,** *23,* 20-24.
32. Larick, D.K.; Hedrick, H.B.; Bailey, M.E.; Williams, J.E.; Hancock, D.L.; Garner, G.B.; Morrow, R.E. *J. Food Sci.* **1987,** *52,* 245-252.
33. Peterson, R.J.; Chang, S.S. *J. Food Sci.* **1982,** *47,* 1444-1453.
34. Ha, J.; Lindsay, R.C. *J. Am. Oil Chem. Soc.* **1991,** *68,* 294-298.
35. Yamato, T.; Kurata, T.; Kato, H.; Fujimaki, M. *Agr. Biol. Chem.* **1970,** *34,* 88-94.
36. MacLeod, G.; Ames, J.M. *Flavor and Fragrance J.* **1986,** *1,* 91-104.

RECEIVED October 28, 1992

Chapter 10

New Usage of Aspartic Acid and Glutamic Acid as Food Materials

Rie Kuramitsu[1], Masahiro Tamura[2], Masaru Nakatani[2], and Hideo Okai[2]

[1]Department of Chemistry, Faculty of General Education, Akashi College of Technology, Uozumi, Akashi, Hyogo 674, Japan
[2]Department of Fermentation Technology, Faculty of Engineering, Hiroshima University, Higashihiroshima, Hiroshima 724, Japan

The role of aspartic acid and glutamic acid was investigated in BMP (Beefy Meaty peptide, Lys-Gly-Asp-Glu-Glu-Ser-Leu-Ala) isolated from enzymatic digests of beef soup. The taste of BMP was affected by the sequence of acidic fragment. Sodium ion uptake of acidic dipeptides and their taste, when mixed with sodium ion, were dependent on the component and/or sequence of dipeptides containing acidic amino acids.

From the era of Aristotle, it is traditional recognition that tastes consist of four factors, that is, sweet, bitter, sour and salty. In 1924, Dutch scientist, Henning, proposed that these four tastes were considered to form a "taste tetrahedron" and all tastes might be expressed as a mixture of these four tastes. However, the typical compounds he used as four fundamental tastes had no common chemical feature. His "principle of taste", therefore, could not reveal the mystery of taste. Sucrose was selected as sweet taste, strychnine as bitter taste and NaCl as salty taste. It is impossible to find some relationship among such saccharide, alkaloid and inorganic salts. Not such ideal approach, for last ten years, some groups have tried to reveal the construction of tastes systematically using chemicals in the same category such as peptides or saccharide derivatives.

Ionic Taste

Nakamura *et al.* (*1*) reported properties of sweetness and bitterness, and their correlation in this volume. Their proposal is based on the stoichiometrical combine at molecular level between sweet or bitter compounds and taste receptors (*2*). We would like to discuss other tastes, including sourness and saltiness, which may be categorized as "ionic taste". We classify these tastes into three groups; (1) "acidic taste" is produced by electronegative groups, (2) "alkaline taste" is produced by electropositive groups, (3) "Neutral salt taste" is produced by both electronegative and electropositive groups. There are two essential reasons for our above proposal. The first is the remarkable difference in exhibition of tastes between ionic taste and sweetness or bitterness. A mixture of sweet and bitter compounds provides only a simple taste as a mixture of the both tastes without any novel taste. On the other hand, a mixture of

0097–6156/93/0528–0138$06.00/0

neutral salts usually produces a taste which is different from those of original ones. The second is that the threshold values of ionic compounds or their mixture are always around 1 mM, regardless their chemical structure. In the case of bitterness or sweetness, we know that many compounds have threshold values less than 1 mM. Some aspartame analogs produced a sweetness 1,200 times stronger than that of sucrose. Tamura *et al.* (4) prepared a compound producing a bitterness more than 300 times stronger than that of caffeine (threshold value of caffeine is 1 mM).

Acidic Taste. It has been believed that acidic taste depends on proton concentration. If this theory is true, acidic taste has to become more potent as pH goes down. As listed in Table I, each acid requires various concentration to get the same pH. Sensory analysis of these acids, which were then carried out to measure their sour potency is listed in Table II. Sour potency of acetic acid was used as standard. Score 1 means that sour potency is at threshold value (0.63 mM) while scores 2, 3, 4 and 5 correspond to sour potency of 1.25 , 2.5 , 5.0 and 10 mM acetic acid. Sour intensities varied even though all acid showed the same pH. The result in Table II was arranged based on the threshold value (Table III). The threshold value of each acid is almost identical at about 1 mM when standardizing by molarity. In other words, the strength of acidity is not in direct proportion to the proton concentration, but to the concentration of electronegative groups. In order to confirm sure of these thoughts, strength of each acid was compared at the same concentration. Expected results were obtained as shown in Table IV.

Table I. Concentration and pH of Acids

Acid	Concentrations (mM) at each pH				
	3.0	3.5	4.0	4.5	5.0
HCl	4.6 mM	1.2 mM	0.032 mM	0.013 mM	0.006 mM
H_2SO_4	1.3 mM	0.052 mM	0.007 mM	0.003 mM	0.001 mM
HCOOH	16 mM	1.3 mM	0.032 mM	0.015 mM	0.010 mM
CH_3COOH	35 mM	12 mM	1.5 mM	0.082 mM	0.035 mM
Citric Acid	2.5 mM	0.071 mM	0.12 mM	0.056 mM	0.028 mM
Aspartic Acid	17 mM	0.93 mM	0.040 mM	0.020 mM	0.010 mM
Glutamic Acid	38 mM	6.3 mM	0.08 mM	0.043 mM	0.015 mM

Table II. Taste of Acids

Acids	Scores at each pH				
	3.0	3.5	4.0	4.5	5.0
HCl (Score)	1	1	0	0	0
H_2SO_4 (Score)	1	0	0	0	0
HCOOH (Score)	3	1	0	0	0
CH_3COOH (Score)	5	3	1	0	0
Citric Acid (Score)	2	1	0	0	0
Aspartic Acid (Score)	4	1	0	0	0
Glutamic Acid (Score)	4	2	1	0	0

Table III. Acids and Hydrogen Ion Concentration
Near the Threshold Value

Acid	Concentration (mM)	pH	$[H^+]$ (mM)
HCl	1.2	3.5	1.0
H_2SO_4	1.3	3.0	0.32
HCOOH	1.3	3.5	0.32
CH_3COOH	1.5	4.0	0.1
Citric Acid	0.71	3.5	0.32
Aspartic Acid	0.93	3.5	0.1
Glutamic Acid	0.80	4.0	0.1

Table IV. Taste of Acids

Acids	Scores at each Concentration (mM)					
	0.63 mM	1.25 mM	2.5 mM	5.0 mM	10 mM	20 mM
HCl	1	2	3	4	5	6
H_2SO_4	2	3	4	5	6	7
HCOOH	1	3	4	5	6	7
CH_3COOH	1	2	3	4	5	6
Citric Acid	1	3	4	5	6	7
Aspartic Acid	1	2	3	4	5	6
Glutamic Acid	1	1	2	3	4	5

Sodium Chloride

Recently, diet low in NaCl has been indispensable for health conscious people in modern civilized countries. At present, an average Japanese takes about 12 g of sodium chloride a day. Patients who need a low sodium chloride diet are suggested to reduce intake to 3 g a day. It is not difficult to reduce sodium chloride intake to 10 g a day under attentive health care. However, it is hard to reduce sodium chloride intake to less than 8 g a day as long as our daily life depends on natural foods.

Some attempts to reduce sodium chloride intake have been carried out. The first is to use some sodium chloride substitutes. Pottasium chloride is widely used for this purpose. However, pottasium chloride is not thought as perfect sodium chloride substitute because it contains bitter taste. Okai and his associates have synthesized several salty peptides (3). These peptides are expected to be good for hypertension, gestosis, diabetes mellitus and other deseases because they contain no sodium ions. These peptides, however, are not expected to be used as sodium chloride substitutes immediately because of the difficulty of in their synthesis and their cost. Although they are struggling to establish a new synthetic method of peptides in a mass production system with reasonable costs and to improve the salty potency of peptides, they have not dissolved this problem. Since the threshold value of ionic taste is around 1 mM regardless their kinds, it seems to be very difficult to prepare an artificial sodium chloride substitute produces a saltiness more than hundreds times of that of NaCl. The second is to obtain a compound which has an enhancing effect of the saltiness of sodium chloride. Tamura *et al.* reported that glycine ethyl ester hydrochloride has an enhancing effect of the saltiness of NaCl (*4*).

Sodium Ion. The excessive intake of sodium ion coming from other than NaCl should be noticed, though reduced intake of NaCl is now a matter of great concern. Monosodium glutamate (MSG), for instance, is a subject of discussion. Since MSG effectively provides *umami* taste, it has been very popular as a Japanese seasoning. In the United States, MSG has currently been marked as a cause of "Chinese restaurant syndrome". In addition, beef, liver, blood and their processed foods contains a large amount of sodium ion. Sine sodium ion combines with aspartic acid and glutamic acid residues in protein, study of affinity of acidic amino acids to sodium ion has to be set out first.

Acidic Peptides. We studied four sodium salts of acidic dipeptides composed of aspartic acid and/or glutamic acid. As shown in Table V, the pH's before adding NaOH were varied although peptides (Asp-Glu and Glu-Asp) contains the same amino acids. The difference in pH value should be caused by the difference of their amino acid sequence. The pH's of the sample solutions were then brought up to 6 with sodium hydroxide and the solutions were tasted. All four samples produced both the salty and the *umami* tastes. The threshold values, however, were different from sample to sample. The amounts of sodium ion, needed to bring the sample solution to 6, were also all different. It would be easily understandable that Asp-Asp having the lowest pH demands the highest amount of sodium ion in this group. Other three peptides required almost the same amount of sodium ion no matter peptides showed different pH's before NaOH was added.

Table V. Taste of Acidic Dipeptides at pH 6

Compound	pH before adjusting to 6	Na^+/Peptide (mM/mM)	Taste	Threshold Value (mM)
Asp-Asp	3.10	2.00	Salty>*Umami*	4.79
Asp-Glu	3.19	1.69	Salty>*Umami*	1.25
Glu-Asp	3.40	1.65	Salty>*Umami*	3.14
Glu-Glu	3.41	1.74	Salty>*Umami*	2.73

These peptides produced the same taste. The interesting point was that Asp-Asp, took the largest amount of sodium ion, produced the weakest taste. The taste intensity of Asp-Asp was about 1/4 of that of Asp-Glu which produced the strongest taste. These results showed that sodium ion intake capacity and taste intensity depend on amino acid sequence of acidic peptides. We have to carefully think about these facts when we construct the taste of foods and carry out sodium ion diet.

Sensory analysis of mixed solutions of MSG and each of four dipeptides was carried out (Table VI). The *umami* taste of MSG was not changed when it was mixed with peptides containing aspartic acid. On the other hand, Glu-Glu, produced the second strongest taste intensity at pH 6, clearly decreased the *umami* intensity of MSG. Taste intensities of acidic peptides are thus influenced by the condition in which they are placed. This result also shows that we must be very careful when we study sodium ion diet.

Table VI. Taste Score of Mixed Solutions of MSG and Acidic Dipeptides

Peptides	MSG and Acidic Dipeptides at each Concentration (mM)[a]				
	0.32 mM	1.25 mM	2.5 mM	5 mM	10 mM
Asp-Asp	0	0	1	3	5
Asp-Glu	0	0	1	2	4
Glu-Asp	0	0	1	3	5
Glu-Glu	0	0	0	1	3
MSG[b]	0	0	1	3	5

[a]Molar ratio of MSG and acidic dipeptide were the same (1:1 M/M). For example, 10 mM MSG and dipeptide were contained in the 10 mM sample.
[b]Taste without acidic dipeptides.

Beefy Meaty Peptide (BMP)

The function of acidic peptides was examined by using the peptide which was isolated from beef soup and sequenced as Lys-Gly-Asp-Glu-Glu-Ser-Leu-Ala by Yamasaki and Maekawa (5) and Tamura *et al.* (6). According them, the peptide solution tasted just like beef soup. Spanier (7) named this peptide as Beefy Meaty Peptide (BMP) based on its taste.

Synthesis of BMP. Figure 1 shows diagramatical explanation of the synthetic method of BMP. On the upside, eight amino acids are positioned. Amino groups are on the left side of the vertical lines while carboxyl groups on the right side and side chain groups on the oblique lines. The horizontal lines between vertical lines shows synthesized peptide. Boc-Ser-OH and H-Leu-OMe were condensed and obtained Boc-Ser-Leu-OMe (1). Compound 1 was converted to the hydrazide (2). Compound 2 was condensed with H-Ala-OBzl by azide method to give the tripeptide (3). Boc-Ser-Leu-Ala-OMe was treated with HCl solution in dioxane to remove amino protecting group and tripeptide methyl ester hydrochloride (4) was obtained. Compound 4 and Boc-Glu(OBzl)-OH were condensed by the mixed anhydride (MA) method to give tetrapeptide derivative (5). The peptide was elongated in the same manner to give protected hexapeptide (13). All protecting groups were removed by hydrogenation. The purity of the octapeptide (14) was confirmed by amino acid analysis and HPLC analysis.

In the synthesis of acidic peptide such as BMP, aspartic acid and glutamic acid residues were stepwisely elongated. In case of the practical synthesis of peptide containing a many acidic amino acid residues, multi reaction procedure for long duration may be required.

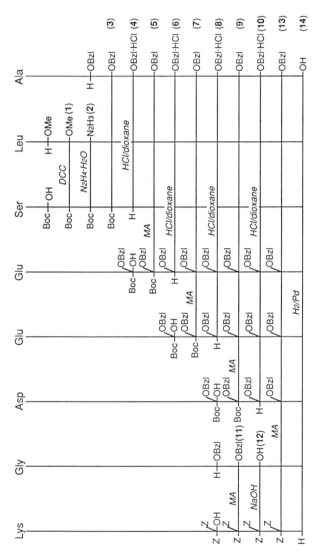

Fig. 1. Synthesis of BMP (Reproduced with permission from ref. 6. Copyright 1989 Japan Society for Bioscience, Biotechnology, and Agrochemistry.)

Table VII. Esterification of Boc-Glu-OH with HOBMCl

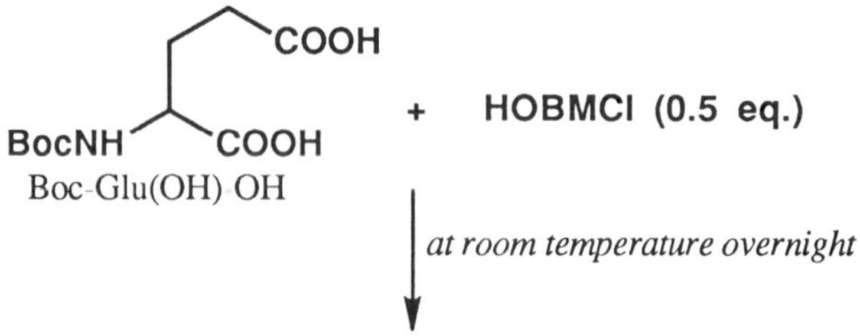
COOH
BocNH
COOH
Boc-Glu(OH)-OH
+ HOBMCl (0.5 eq.)
at room temperature overnight

	Ratio of (α:γ) in the presence of basic catalyst		
Solvent	K_2CO_3	DCHA[a]	CHA[b]
H_2O	0:100	-	-
DMF[c]	67:33	-	-
MeOH	63:38	-	-
CH_2Cl_2	57:43	20:80	67:33

[a]Dicyclohexylamine.
[b]Cyclohexylamine.
[c]*N,N*-Dimethylformamide.

Table VIII. Esterification of Boc-Phe-Glu-OH with HOBMCl

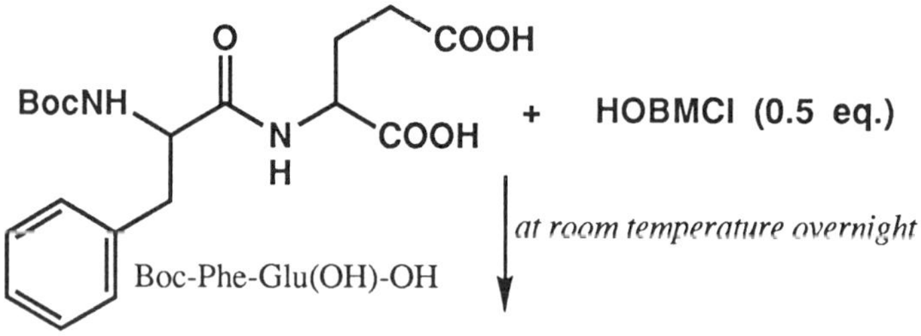
O
COOH
BocNH
N
H
COOH
Boc-Phe-Glu(OH)-OH
+ HOBMCl (0.5 eq.)
at room temperature overnight

	Ratio of (α:γ) in the presence of basic catalyst		
Solvent	K_2CO_3	DCHA	CHA
H_2O	50:50	-	-
DMF	33:67	-	-
MeOH	50:50	-	-
CH_2Cl_2	50:50	25:75	50:50

The innovative esterification reagent, HOBMCl was developed in our laboratory in a few years ago (*8-11*). As shown in Table VII, HOBMCl reacts under milder condition than that for the conventional reagents. Boc-Glu(OH)-OH was converted to benzyl ester by adding a half equivalent of HOBMCl. "α-Ester" rich product was catalyzed by K_2CO_3 and "γ-ester" was selectively produced in aqueous solution, though its yield was somewhat poor.   On the other hand, when using dicyclohexylamine or cyclohexylamine as a basic catalyst, the yield ratio of each isomer was inversely changed.   Boc-Phe-Glu(OH)-OH was converted to benzyl ester in the same way. In this reaction, the yield ratios were nearly same regardless of the kinds of basic catalyst and of medium used (Table VIII). In case of Boc-Phe-Asp(OH)-OH, the yield ratios of benzyl ester isomers were varied depending on the kind of basic catalyst used (Table IX). Furthermore, the ratio of α-ester and ω-ester from peptide containing Asp was totally different from the case of the peptide containing Glu. The reaction of the peptide which includes Asp or Glu is varied not only by the amino acids contents, but by their sequence. Synthesis of peptide would have many difficulties. This means at the same time that acidic peptides have amply varied properties by their chemical structures, offering active interest in the mechanism of taste.

Table IX.   Esterification of Boc-Phe-Asp-OH with HOBMCl

Boc-Phe-Asp(OH)-OH   + HOBMCl (0.5 eq.)

at room temperature overnight

Solvent	Ratio of (α:γ) in the presence of basic catalyst		
	K_2CO_3	DCHA	CHA
H_2O	60:40	-	-
DMF	50:50	-	-
MeOH	50:50	-	-
CH_2Cl_2	29:71	83:17	17:83

Taste of BMP.   In order to investigate the mechanism of the taste of BMP, tastes of peptide fragments of BMP were then investigated.   As shown in Table X, Lys-Gly·HCl, N-terminal fragment, produced the salty taste including *umami* taste. Lys-Gly-Asp, N-terminal tripeptide, produced sweetness. The middle part of the peptide, Asp-Glu-Glu produced a sourness, while the C-terminal fragment, Ser-Leu-Ala produced a weak bitterness. Through these analytical pursuit, the taste of the delicious peptide should be produced as the integrated taste of these peptide fragments. Further investigation of the taste of each fragment was then carried out.

Table X.   Tastes of BMP and its Fragments

Compound	Taste	Threshold Value (mM)
Lys-Gly-Asp-Glu-Glu-Ser-Leu-Ala	*Umami*>Sour	1.41
Lys-Gly·HCl	Salty>*Umami*	1.22
Lys-Gly-Asp	Sweet	1.30
Asp-Glu-Glu	Sour	------
Ser-Leu-Ala	Bitter	2.40

SOURCE: Reproduced with permission from ref. 6. Copyright 1989 Japan Society
for Bioscience, Biotechnology, and Agrochemistry.

Taste of Dipeptides Containing Lys and/or Gly.   Since Lys-Gly·HCl
produces the saltiness, we prepared some dipeptides composed of Lys and/or Gly.
The results are listed in Table XI. Gly-Lys, of which the amino acid sequence is
opposite to the salty peptide Lys-Gly·HCl, produced a weakly sweet taste instead of
the salty taste. Dipeptide composed of only Lys or Gly did not any taste.

Table XI.   Tastes of Lys-Gly and Analogs

Compound	Taste	Threshold Value (mM)
Lys-Gly·HCl	Salty>*Umami*	1.22
Lys-Lys·HCl	Tasteless	-----
Gly-Lys·HCl	Sour>Sweet	5.48
Gly-Gly	Tasteless	-----

SOURCE: Reproduced with permission from ref. 6. Copyright 1989 Japan Society
for Bioscience, Biotechnology, and Agrochemistry.

Taste of Peptides Containing Basic and Acidic Amino Acids.   We
obtained some interesting results from the sensory test of peptides composed of basic
and acidic amino acids. As shown in Table XII, peptides, in which the basic amino
acid at the N-terminal and the acidic amino acid at the C-terminal, produced no taste.
On the contrary, peptide in which Glu is at N-terminal produced both sour and weak
umami taste while peptides in which Asp is at N-terminal produced only sour taste.
However, we could not observe any strong saltiness or *umami* taste. No investigators
have pointed out that tastes of peptides containing acidic amino acids vary as locations
of acidic amino acids changes. We should carry out further study and establish the
most effective location of acidic amino acid in peptides in order to design peptides
which produce good tastes.

Taste of Orn-β-Ala Derivatives.   We reported in the previous paper (3) that
ornithyl-β-alanine monohydrochloride produced the salty taste as ornithyltaurine. We
prepared two tetrapeptides composed of ornithine and β-alanine. Both tetrapeptides,
however, produced mainly an astringent taste instead of the salty or the *umami* taste
(Table XIII).

Table XII. Tastes of Lys-Gly-Asp and Analogs

Compund	Taste	Threshold Value (mM)	Quality Salty	Quality Umami
Lys-Gly-Asp	Sweet	1.30	0	0
Lys-Asp	Tasteless	----	0	0
Lys-Glu	Tasteless	----	0	0
Orn-Asp	Tasteless	----	0	0
Orn-Glu	Tasteless	----	0	0
Asp-Lys	Sour	1.56	0	0
Glu-Lys	*Umami*>Sour	3.12	0	2
Asp-Orn	Sour	3.12	0	0
Glu-Orn	*Umami*>Sour	3.12	0	2
NaCl	Salty	3.12	5	0
MSG	*Umami*	1.56	0	5

SOURCE: Reproduced with permission from ref. 6. Copyright 1989 Japan Society for Bioscience, Biotechnology, and Agrochemistry.

Table XIII. Tastes of Orn-Orn-β-Ala-β-Ala and Orn-β-Ala-Orn-β-Ala

Compound	Taste	Threshold Value (mM)	Quality Salty	Quality Umami
MSG	*Umami*	1.56	0	5
NaCl	Salty	3.20	5	0
Orn-β-Ala·HCl	Salty>*Umami*	1.25	4	1
Orn-Orn·2HCl	*Umami*	1.50	0	1
β-Ala-β-Ala	Tasteless	-----	0	0
Orn-Orn-β-Ala-β-Ala	Astringency>Sweet	3.09	0	0
Orn-β-Ala-Orn-β-Ala	Astringency>Sweet	2.25	0	0

SOURCE: Reproduced with permission from ref. 6. Copyright 1989 Japan Society for Bioscience, Biotechnology, and Agrochemistry.

Re-formation of the Taste of BMP. According to the above results, the taste of BMP might produced by the combination of the basic amino acid (Lys) at N-terminal and acidic amino acids at the middle part. To confirm this idea, we prepared a mixed solution of the N-terminal dipeptide (Lys-Gly), the acidic tripeptide at the middle part (Asp-Glu-Glu) and C-terminal tripeptide (Ser-Leu-Ala) and examined the taste. We also studied tastes of a mixture in which a basic dipeptide fragment was replaced by Orn-β-Ala, a salty dipeptide, and a mixture in which Glu-Glu replaced an acidic tripeptide fragment. The result are shown in Table XIV. All of the combinations produced the same character of the taste as BMP. It means that the taste of BMP is mainly produced by the combined effect of the N-terminus basic dipeptide and the acidic tripeptide of the middle part. However, taste strength of the mixture became more potent when Glu-Glu replaced acidic tripeptide of BMP (Asp-Glu-Glu).   The taste of BMP is clearly influenced by acidic peptide fragment.

Table XIV. Sensory Test on the Mixture of N-Terminal, Middle Part and C-Terminal Fragments of BMP

Combination	Taste	Threshold Value (mM)
Lys-Gly-Asp-Glu-Glu-Ser-Leu-Ala	*Umami*/Sour	1.41
Lys-Gly + Asp-Glu-Glu- + Ser-Leu-Ala	*Umami*/Sour	1.41
Orn-b-Ala + Asp-Glu-Glu + Ser-Leu-Ala	*Umami*/Sour	1.41
Lys-Gly + Glu-Glu + Ser-Leu-Ala	*Umami*/Sour	0.94

SOURCE: Reproduced with permission from ref. 6. Copyright 1989 Japan Society for Bioscience, Biotechnology, and Agrochemistry.

Conclusions

Change of taste behavior of acidic peptides may cause various changes of the taste of mixture in which acidic peptides are contained. When we attempt to utilize the ionic taste, we must carefully think about the behavior of acidic peptides. These studies of taste of acidic peptide should give us useful information about the role of aspartic and glutamic in flavor enhancing and in designing taste of foods or sodium ion diet.

Literature Cited

(1)   Nakamura, K.; Okai, H. in *Molecular Approaches to the Study of Food Quality*, Spanier, A. M; Okai, H; Tamura, M., Eds; ACS Symposium Series; American Chemical Society: Washington, D.C., in press.
(2)   Shinoda, I.; Okai. H. *J. Agric. Food Chem.*, **1985**, *33*, 792.
(3)   Tada, M.; Shinoda, I.; Okai, H. *J. Agric. Food Chem.*, **1984**, *32*, 992.
(4)   Tamura, M.; Seki, T.; Kawasaki, Y.; Tada, M.; Kikuchi, E.; Okai, H. *Agric. Biol. Chem.*, **1989**, *53*, 1625.
(5)   Yamasaki, Y.; Maekawa, K. *Agric. Biol. Chem.*, **1978**, *53*, 1623.
(6)   Tamura, M.; Nakatsuka, T.; Tada, M.; Kawasaki, Y.; Kikuchi, E.; Okai, H. *Agric. Biol. Chem.*, **1989**, *53*, 319.
(7)   Spanier, A. M. in *Molecular Approaches to the Study of Food Quality*, Spanier, A. M; Okai, H; Tamura, M., Eds; ACS Sympodium Series; American Chemical Society: Washington, D.C, in press.
(8)   Kouge, K.; Koizumi, T.; Okai, H.; Kato, T. *Bull. Chem. Soc. Jpn.*, **1987**, *60*, 2409.
(9)   Kouge, K.; Soma, H.; Katakai, Y.; Okai, H.; Takemoto, M.; Muneoka, Y. *Bull. Chem. Soc. Jpn.*, **1987**, *60*, 4343.
(10) Kouge, K.; Katakai, Y.; Soma, H.; Kondo, M.; Okai, H.; Takemoto, M.; Muneoka, Y. *Bull. Chem. Soc. Jpn.*, **1987**, *60*, 4351.
(11) Azue, I.; Okai, H.; Kouge, K.; Yamamoto, M.; Koizumi, T. *Chem. Express*, **1988**, *3*, 21.

RECEIVED January 29, 1993

Chapter 11

Convenient Synthesis of Flavor Peptides

Masaru Nakatani and Hideo Okai

Department of Fermentation Technology, Faculty of Engineering, Hiroshima University, Higashihiroshima, Hiroshima 724, Japan

For the purpose of synthesizing flavor peptides or proteins in large scale, we developed "protein recombination method" and "enzymatic synthesis using chemically modified enzyme". "Protein recombination method" was applied to the synthesis of C-terminal portion of β-casein and its analog. Chymotrypsin was chemically modified by Z-DSP in aqueous solution. It was stable for organic solvents. Using this modified enzyme, we succeeded in the synthesis of Inverted-Aspartame-Type Sweetener "Ac-Phe-Lys-OH" in one step.

Today, it is well-known that peptides or proteins exhibit various kinds of taste. Our group has been researching on the relationship between taste and structure of peptides, BPIa (Bitter peptide Ia, Arg-Gly-Pro-Pro-Phe-Ile-Val) *(1)* as a bitter peptide, Orn-β-Ala·HCl (OBA), Orn-Tau·HCl as salty peptides*(2)*, and "Inverted-Aspartame-Type Sweetener" (Ac-Phe-Lys-OH) as a sweet peptide*(3)*. The relationship between taste and chemical structure was partly made clear. Since commercial demand for these flavor peptides is increasing, we need to develop new synthetic methods which can prepare these peptides in large scale. We developed the following two methods: (1) protein recombination method as a chemical method, (2) enzymatic synthesis using chemically modified enzyme as a biochemical method.

Protein Recombination Method

For protein recombination method, starting materials are free peptides or proteins which are existing inexpensive, in large quantities in nature. We can also obtain peptides by solid-phase peptide synthesis or gene manipulation. By this method, we need to leave the functional group on the side chain unprotected as long as they don't make any problems. Another word, we only have to protect the amino group and the thiol group on the side chain. This protected peptide is coupled with another protected peptide, and product is next deblocked (Figure 1). This is an ideal route. But for realization of this method, there are following three problems:
 1. selective blocking of side chain of basic amino acid in amine component

2. selective blocking or activation of acidic amino acids in acid component
3. selection of coupling method without racemization.

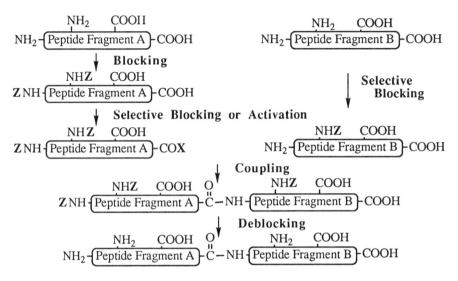

Fig. 1.   Protein Recombination

Problem of Amine Component. If basic amino acids exist in amine component, selective blocking of amino group on the side chain is necessary. But it was impossible by conventional methods. We attempted to develop a new method.

Selective Blocking of Amino Group on Side Chain of Basic Amino Acids. In 1987, we developed a new water-soluble active ester reagent, *p*-hydroxy-phenyldimethylsulfonium methyl sulfate (HODSP) *(4-6)*, and acylating reagents, Z-DSP {[*p*-(benzyloxycarbonyloxy)phenyl]dimethylsulfonium methyl sulfate}, Boc-DSP{[*p*-(t-butoxycarbonyloxy)phenyl]dimethylsulfonium methyl sulfate}, Fmoc-DSP{*p*-(9-fluorenylmethyloxycarbonyloxy)phenyl]dimethylsulfonium methyl sulfate} and so on *(7)*. These reagents can be used in aqueous solution because of high solubility due to the presence of a counter anion in the molecules. Making the best use of this character, Z-DSP was allowed to react with lysine by controlling the pH of the reaction mixture in aqueous media. The amino group on the side chain was selectively benzyloxycarbonylated in good yield when pH of the reaction mixture was 11.3.

Application to Selective Blocking of Amino group on Side Chain of Peptides. We investigated to selectively protect the amino group on the side chain in peptides using Z-DSP. Phenylalanyllysine, ornithylphenylalanine and others were selected for this model. Z-DSP was allowed to react with these peptides in water in the same manners as described before. As shown in Table I, the result definitely shows that we succeeded in selective blocking of the amino group on the side chain of these peptides.

Selection of Coupling Method. For protein recombination, coupling method must not cause racemization. Most suitable one seems to be azide method. If azide method is employed, the problem is esterification of peptide fragment. If esterification of peptide fragment is possible and it was applied to azide method, we will be get the object. A new reagent for the preparation of benzyl ester, p-hydroxyphenylbenzylmethyl-sulfonium chloride (HOBMCl), was introduced by our group *(9)*. This reagent converts N-acyl amino acids into N-acyl amino acid benzyl esters under mild conditions. We thought that this reagent has a possibility of esterification of peptide fragment. Actually as shown in Table II, many kinds of N-protected peptides were converted into benzyl esters in satisfactory yields. And we examined whether this esterification cause racemization. As a model peptide, Gly-Ala-Leu was synthesized by azide method *via* benzyl ester prepared by this reagent . We confirmed that no racemization occurred in this reaction *(10)*.

Application to the Synthesis of Flavor Peptide. As shown in figure 2, we applied this method to the synthesis of C-terminal portion of β-casein (Arg^{202}-Gly-Pro-Phe-Pro-Ile-Ile-Val^{209}) and its analog (Arg-Gly-Pro-D-Phe-Pro-Ile-Ile-Val) *(10)*. This synthetic peptide by this method possessed a bitter taste 250 times that of caffeine, and it was the same result as which Kanehisa et al. had reported 8 years ago*(11)*. And it is interesting that this analog produced the bitterness only about one twentieth as strong as that of original octapeptide. These results showed that protein recombination method is an effective strategy in synthesis of flavor peptides.

Enzymatic Synthesis Using Chemically Modified Enzyme

It is advantageous if enzymatic synthesis could be carried out in organic solvents. In order to avoid denaturation, we thought to prepare more hydrophobic enzyme than native one. We made good use of water-soluble acylating reagents (Z-DSP, Boc-DSP) described above, and tried to modify the amino groups of enzyme.

Preparation of Chemically Modified Chymotrypsin. Chemically modified chymotrypsin *(12)* was prepared by following methods: Z-DSP (2.5-20 equivalent to the amino groups of chymotrypsin) was added to the buffer solution (Na$_2$B$_4$O$_7$-HCl, pH 8) of chymotrypsin (14mg). The solution was allowed to stand for 16 hours in refrigerator. And the solution was centrifuged (10.000rpm, 10min). The solution was dialyzed, and obtained chemically modified chymotrypsin by lyophilization.

Characterization of Modified Chymotrypsin. Measurement of the degree of modification of prepared chymotrypsin was determined by measuring the unreacted amino groups of the modified products with trinitrobenzenesulfonic acid *(13)*. It was found that the degrees of benzyloxycarbonylation increased as the amount of the reagent increased (Table III). Activities of the modified and unmodified chymotrypsin were determined by measuring the hydrolysis rates of Ac-Tyr-OEt, and methyl 2,3-di-O-(L-phenylalanyl)-α-D-glucopyranoside, which was synthesized in our laboratory and found to be a suitable substrate for chymotrypsin. When the degrees of acylation were 35%, the modified chymotrypsin retained its activity of hydrolysis about 80% compared with native one. The best degree of modification was 35%. Less modified or more modified ones were inferior. We measured pH Dependence (Figure 3) and temperature dependence (Figure 4) of 35% modified chymotrypsin, and confirmed that those of modified chymotrypsin were not changed compared with those of native one.

Table I. Selective Protection of Peptides Containing
            Lysine or Ornithine in Water

$$\text{Phe-Lys} \quad \xrightarrow[\text{NaOH, in water}]{\text{Z-DSP, pH controled}} \quad \text{Phe-Lys(Z)}$$

Reagents	Peptides	pH	Yield(%)
Z-DSP(1.6eq)	Phe-Lys	11.3	75
	Phe-Orn	11.3	89
	Lys-Phe	11.3	90
	Orn-Phe	11.3	84

Table II. Preparation of N-Protected Peptide
             Benzyl Esters Using HOBMCl

N-protected peptides	solvent	yield (%)
Z-Phe-Gly-OH	DCM	83
Z-Val-Tyr-OH	DCM-DMF	46
Z-Pro-Leu-OH	DCM	85
Z-Gly-Ala-OH	DCM	57
Z-Lys(Z)-Pro-OH	DCM	91
Boc-Ala-Glu-OH	DCM-DMF	66 *)
Boc-Gly-Gly-OH	DCM	89
Boc-Ile-Val-OH	DCM	64
Boc-Pro-Phe-OH	DCM	79
Boc-Thr-Pro-OH	DCM	53
Boc-Gln-Pro-OH	DCM-DMF	92
Boc-Ser-Ile-OH	DCM	74
Boc-Lys(Boc)-Val-OH	DCM-DMF	24
Boc-Met-Pro-OH	DCM	71
Boc-Phe-Asp-OH	DCM-DMF	57 *)
Boc-Pro-Pro-Pro-OH	DCM	80
Boc-Val-Val-Val-Pro-Pro-OH	DCM-DMF	57

*) Dibenzyl ester

DCM - dichloromethane

DMF - N,N- dimethylformamide

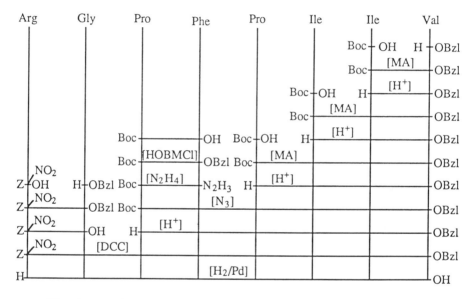

Fig. 2. The Synthetic Route of C-Terminal Portion of β-Casein
(Arg-Gly-Pro-Phe-Pro-Ile-Ile-Val) and Its Analog

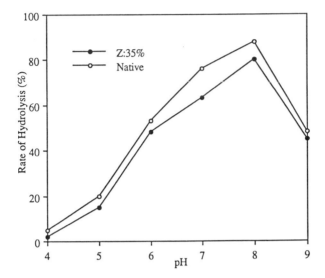

Fig. 3. pH Dependence of Modified (Z:35%) and Native
Chymotrypsin (Reproduced from ref. 8 with permission.
Copyright 1992 Japan Society for Bioscience,
Biotechnology, and Agrochemistry.)

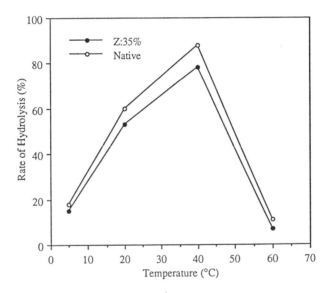

Fig. 4. Temperature Dependence of Modified (Z:35%) and
Native Chymotrypsin (Reproduced from ref. 8 with permission.
Copyright 1992 Japan Society for Bioscience, Biotechnology,
and Agrochemistry.)

Table III. Determination of Modified Amino
Groups of Chymotrypsin

Z-DSP (eq.)	number of free amino groups	degree of modification (%)
0	17	0
42.5	15.4	10
85	11.1	35
175	9.6	44
340	6	65

SOURCE: Adopted from ref. 8.

Synthesis of Z-Tyr-Gly-NH$_2$ in Aqueous-DMF Solvent Media. Using this modified enzyme, we carried out the synthesis of "Z-Tyr-Gly-NH$_2$", which has never been formed in 100% aqueous system *(14)*, and compared with native chymotrypsin on the effect of organic solvent. To a solution of Z-Tyr-OH (315mg) in Tris buffer (pH 6.7, 0.5ml), which contained N,N-dimethylformamide (DMF) (0-100%), was added a solution of H-Gly-NH$_2$·HCl (11mg) and native or modified chymotrypsin (2mg) in the same Tris buffer. The mixture was incubated at 20°C for 24 hours and heated at 100°C for 15 minutes. The products were isolated by HPLC (ODS column, 278nm, 50% acetonitrile). Native chymotrypsin inactivated when concentration of DMF was 50%, while chemically modified chymotrypsin kept its activity even up to 80% (Table IV).

Table IV. Effect of DMF Concentration on Synthesis of
Z-Tyr-Gly-NH$_2$ by Native and Modified Chymotrypsin

DMF (%)	synthetic rate of Z-Tyr-Gly-NH$_2$ (%)	
	native chymotrypsin	modified chymotrypsin (Z:35%)
0	0	0
10	0	14
20	13	24
30	16	19
40	25	43
50	5	33
60	0	27
70	0	24
80	0	25
100	0	0

SOURCE: Adopted from ref. 8.

Table V. Synthetic Reactions of Ac-Phe-Lys-OH in
Various Water-Organic Solvents

Solvent / Conc. (%)	DMF	n-BuOH	AcOEt	CH$_2$Cl$_2$	Hexane	Acetone
10	-	-	-	+	-	+
30	-	-	-	+	-	+(1%)
50	-	-	-	-	-	-
70	-	-	.	-	+	-
90	-	-	+(10%)	-	+(5%)	-

SOURCE: Adopted from ref. 8.
-: Ac-Phe-Lys-OH was not detected.
+: Ac-Phe-Lys-OH was detected by thin layer chromatography
(): Yields of Ac-Phe-Lys-OH were measured by high performa.nce
liquid chromatography .

One Step Synthesis of Ac-Phe-Lys-OH. Aso reported that Inverted-Aspartame-Type Sweetener "Ac-Phe-Lys-OH", chemically synthesized by our group (3), was prepared using native chymotrypsin (14). This method demands five steps, and has no merit compared with chemical synthesis that we had reported. Therefore we investigated more effective synthesis of Ac-Phe-Lys-OH using chemically modified chymotrypsin. It is obvious that amino acids, especially lysine, are not suitable for nuceophiles because the amino groups of amino acids are ionized in aqueous solution. If in organic solvent, this ionization decreases extremely. So we attempted this one step synthesis of Ac-Phe-Lys-OH using this enzyme in the organic-aqueous solvent system. To a solution of Ac-Phe-OH (2.07mg), and Lys·HCl (5-100 equivalent to Ac-Phe-OH) in phosphate buffer (pH=6.7), organic solvents (DMF, n-butanol, benzene, acetone, dichloromethane, hexane and ethyl acetate) was added. And a solution of modified chymotrypsin (2mg) in 0.1mM HCl was added to the solution. The mixture was incubated at 37°C for 24 hours and the enzyme was filtered off by MOLCUT L (MILLIPORE). Synthesis of Ac-Phe-Lys-OH was followed by thin layer chromatography. In the case of DMF or n-butanol, Ac-Phe-Lys-OH was not detected, contrary to our expectation. But in the case of benzene, acetone, dichloromethane, hexane and ethyl acetate, synthesis of the objects was detected by thin layer chromatography in some cases (Table V). And quantitative analysis was carried out by ion exchange high performance liquid chromatography. Through a series of experiment, maximum yield was 10% in the case of 90% ethyl acetate-10% water solvent system.

Conclusions

For the purpose of synthesis of flavor peptides or proteins in large scale, we developed "protein recombination" as a chemical method. C-terminal portion of β-casein and analog were synthesized by using this method effectively. We also developed "enzymatic synthesis using chemically modified enzyme" as a biochemical method.

Inverted-Aspartame-Type Sweetener, Ac-Phe-Lys-OH, was synthesized in one step in the organic-aqueous solvent media. Maximum yield was 10%. But considering collection of enzyme and raw materials, we think that this synthesis is able to be industrialized. We also think that this modified technique can be applied to the preparation of another modified enzyme. There are a few small problems in these two methods. But we hope that these two methods will be made full use as synthetic strategies of flavor peptides or the other functional peptides.

Literature Cited

1   Fukui, H.; Kanehisa, H; Ishibashi, N.; Miyake, I.; Okai, H. *Bull. Chem. Soc. Jpn.*, **1983**, *56*, 766-769.
2   Tada, M.; Shinoda, I.; Okai, H. *J. Agric. Food Chem.*, **1984**, *32*, 992-996.
3   Nosho, Y.; Seki, T.; Kondo, M.; Ohfuji, T.; Tamura, M.; Okai, H. *J. Agric.Food Chem.*, **1990**, *38*, 1368-1373.
4   Kouge, K.; Koizumi, T.; Okai, H.; Kato, T. *Bull. Chem. Soc. Jpn.*, **1987**, *60*, 2409-2418.
5   Kouge, K.; Soma, H.; Katakai, Y.; Okai, H.; Takemoto, M.; Muneoka, Y. *Bull. Chem. Soc. Jpn.*, **1987**, *60*, 4343-4349.
6   Kouge, K.; Katakai, Y.; Soma, H.; Kondo, M.; Okai, H.; Takemoto, M.; Muneoka, Y. *Bull. Chem. Soc. Jpn.*, **1987**, *60*, 4351-4356.
7   Azuse, I.; Tamura, M.; Kinomura, K.; Okai, H.; Kouge, K.; Hamatsu, F.; Koizumi, T. *Bull. Chem. Soc. Jpn.*, **1989**, *62*, 3103-3108.
8   Kawasaki, Y; Murakami, M; Dosako, S; Azuse, I; Nakamura, T; Okai, H. *Biosci. Biotech. Biochem.*, **1992**, *56*, 441.
9   Mukaiyama, N.; Kouge, K.; Nakatani, M.; Okai, H. *Chem. Express*, **1991**, *6*, 985-988.
10 Nakatani, M. Hiroshima University, unpublished data.
11 Kanehisa, H.; Miyake, I.; Okai, H.; Aoyagi, H.; Izumiya, N. *Bull. Chem. Soc. Jpn.*, **1984**, *57*, 819-822.
12 Azuse, I.; Tamura, M.; Kinomura, K.; Kouge, K.; Kawasaki, Y.; Okai, H. *Chem. Express*, **1989**, *4*, 539-542.
13 Habeeb, A. F. S. A., *Anal. Biochem.*, **1966**, *14*, 328.
14 Aso, K. *Agric. Biol. Chem.*, **1989**, *53*, 729-733.

RECEIVED January 21, 1993

Chapter 12

Molecular Design of Flavor Compounds Using O-Aminoacyl Sugars

Masahiro Tamura[1] and Hideo Okai[2]

[1]Biosignal Engineering Division, Kowa Research Institute, Kowa Company, Ltd., 1–25–5 Kannondai, Tsukuba, Ibaraki 305, Japan
[2]Department of Fermentation Technology, Faculty of Engineering, Hiroshima University, Higashihiroshima, Hiroshima 724, Japan

Design of flavor compounds was carried out using O-aminoacyl sugars. O-Aminoacyl sugar produced a bitterness when hydrophobic amino acids were introduced. A bitter potency of methyl 2,3-di-O-(L-phenylalanyl)-α-D-glucopyranoside was 20 times stronger than that of caffeine. O-Aminoacyl sugar produced a sweetness when neutral amino acids possessing low hydrophobic side chains were introduced. A sweet potency of methyl 2,3-di-O-(L-α-aminobutyryl)-α-D-glucopyranoside was 50 times stronger than that of sucrose. O-Aminoacyl sugar containing basic amino acids produced a saltiness and/or a *umami* taste. We also found that O-aminoacyl sugars containing basic amino acids have an enhancing effect of saltiness of sodium chloride.

In 1924, Henning, a psychologist, has proposed a famous "taste tetrahedron" which has four typical tastes (sweetness, bitterness, sourness, and saltiness) at each vertex (see Fig. 1). They are usually represented by sucrose, caffeine, acetic acid, and sodium chloride, respectively. His theory is that tastes of all compounds should be inside of the taste tetrahedron. Although the theory has been well accepted, it is difficult to elucidate the relationship between tastes and chemical structure of those four typical taste compounds using the Henning's taste tetrahedron. Those compounds have no common structure. We are not able to find the chemical structure relationships between sweetness and bitterness as long as we use two totally different compounds, a carbohydrate (sucrose) and heterocycle (caffeine). Sodium chloride, an inorganic compound, is also not preferable for the study. Therefore, it is of worth to design flavor compounds with a common structure which show four typical tastes.

In the present study, we selected O-aminoacyl sugars as a common structure. O-Aminoacyl sugars have been developed in our research group (1) and are composed of very simple essences (amino acids, sugars, and ester linkage). O-Aminoacyl sugars are very rare in natural products. We have confirmed their use as enzymatically removable protecting groups (1-b) and as substrate for measurements of enzyme activities (2). In this time, we developed the use of O-aminoacyl sugars as flavor compounds.

0097–6156/93/0528–0158$06.00/0

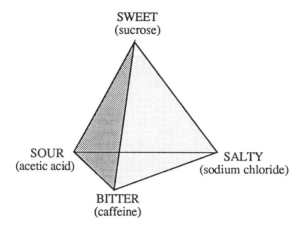

Fig. 1.   Henning'sTaste Tetrahedron

Design of Bitter Compounds

Design of bitter compounds seems to be useless. However, it should give us not only some idea when we study the taste tetrahedron, but much useful information when we study debittering of food materials.

Bitterness of *O*-Aminoacyl Sugars Containing Hydrophobic Amino Acids. Hydrophobic amino acids are known to produce a bitterness. Therefore, we introduced hydrophobic amino acids into the 2-*O*- and 3-*O*-positions of methyl α-D-glucopyranoside and carried out sensory analyses. Results of sensory analysis are listed in Table I.

It was surprising that a phenylalanyl derivative **1** produced a strong bitterness which was 20 times stronger than that of caffeine. By the systematic synthesis of bitter peptides (*3*), we know that a strong bitterness is produced when two "bitterness producing groups" locate close to each other. In the case of phenylalanyl derivative (**1**), two phenyl groups might be close enough to produce the strong bitterness.

The D-analog of phenylalanyl derivative (**2**) also produced a strong bitterness. The bitterness largely increased when two phenylalanyl-phenylalanyl residues were introduced (**3**) instead of phenylalanyl residues and the bitter intensity became more than 300 times stronger than that of caffeine. Aminoacyl sugars containing amino acids, which have lower hydrophobicity than that of phenylalanine (**4** and **5**), produced relatively weaker bitterness although all of their bitterness was far stronger than that of caffeine.

Bitterness of *O*-Aminoacyl Sugars Containing Peptides Composed of Valine or Phenylalanine. Since hydrophobicity of the amino acid moiety seemed to be an important key for bitterness of *O*-aminoacyl sugars, we then prepared *O*-

Table I. Tastes of O-Aminoacyl Sugars Containing Two Hydrophobic Amino Acids or Peptides

R^a	Taste	T.V. (mM)[b]	R_{caf}[c]
L-Phenylalanyl (**1**)	Bitter	0.05	20
D-Phenylalanyl (**2**)	Bitter	0.12	8.3
Phenylalanylphenylalanyl (**3**)	Bitter	0.003	333
Leucyl (**4**)	Bitter	0.08	12.5
Isoleucyl (**5**)	Bitter	0.2	5

SOURCE: Adapted from ref. 1-c.

[a]All aminoacyl groups were introduced into 2-O- and 3-O- positions of methyl α-D-glucopyranoside.

[b]Threshold value.

[c]R_{caf} is bitterness intensity. R_{caf} shows how the bitterness was weaker or stronger than that of caffeine. R_{caf} of caffeine is 1.

aminoacyl sugars containing some peptides. Results of sensory analysis are listed in Table II. O-Aminoacyl sugar (**6**) produced a sweetness when two valyl residues were introduced into 2-O- and 3-O-positions of methyl α-D-glucopyranoside. The sweetness changed to a bitterness when valyl-valyl, valyl-valyl-valyl or valyl-valyl-valyl-valyl residue replaced valyl residue. Bitter potency increased as amount of hydrophobicity of side chain of amino acids increased. The bitter potency of the valyl-valyl-valyl-valyl derivative (**9**) was 20 times stronger than that of caffeine. On the other hand, valine peptides did not produce strong bitterness. Valyl-valine (**11**) produced a *umami* taste (monosodium glutamate-like taste) instead of a bitterness. Valyl-valyl-valine (**12**) and valyl-valyl-valyl-valine (**13**) produced a bitterness, the potency of which was only 1/5 and half of that of caffeine, respectively.

O-Aminoacyl sugars containing phenylalanine peptides produced very strong bitterness. As listed in Table I, phenylalanyl-phenylalanyl derivative (**3**) produced the bitterness more than 300 times stronger than that of caffeine. Therefore, we expected that O-aminoacyl sugars might produce the strongest bitterness ever if two phenylalanyl-phenylalanyl-phenylalanyl residues were introduced into the sugar moiety. However, methyl 2,3-di-O-(phenylalanyl-phenylalanyl-phenylalanyl)-α-D-glucopyranoside, containing two phenylalanine trimers (**14**), produced almost the same bitterness (T.V. was 0.002 mM and R_{caf} was 500) as that of **3**. Probably, the bitterness receptor has a proper size and the O-aminoacyl sugar containing two phenylalanine trimers (**14**) was too big to enter the receptor site.

In order to confirm this idea, we prepared three O-aminoacyl sugars containing three different peptides which have the same hydrophobicity. All peptides were composed of two phenylalanine and glycine residues. Hydrophobicity of side chain of peptide moiety was all the same ($\Sigma\Delta f$=10.60 kcal/mol). The results of sensory analysis are listed in Table III. *O*-Aminoacyl sugars of this series produced the same bitterness no matter where phenylalanine locates. This result shows that the bitter potency is determined by hydrophobicity of the whole molecule.

Table II. Sensory Analysis of *O*-Aminoacyl Sugars and Peptides Composed of Valine Residues

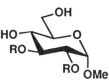

Compounds[a]	Taste	T.V. (mM)	Rcaf	Hydrophobicity of the Peptide Moiety $\Sigma\Delta f$ (kcal/mol)(4)
R=H-Val- (**6**)[b]	Sweet	0.41	----	3.38
R=H-Val-Val- (**7**)[b]	Bitter	0.25	4	6.76
R=H-Val-Val-Val- (**8**)[b]	Bitter	0.10	10	10.14
R=H-Val-Val-Val-Val- (**9**)[b]	Bitter	0.05	20	13.52
H-Val-OH (**10**)[c]	Sweet>Bitter	19	----	1.69
H-Val-Val-OH (**11**)[c]	*Umami*	25	----	3.38
H-Val-Val-Val-OH (**12**)[c]	Bitter	4.5	0.2	5.07
H-Val-Val-Val-Val-OH (**13**)[c]	Bitter	2	0.5	6.76

SOURCE: Adapted from ref. 5.

[a]H- refers to the amino terminus while -OH refers to the carboxyl terminus of peptides.

[b]Aminoacyl groups were introduced into 2-*O*- and 3-*O*-positions of methyl α-D-glucopyranoside.

[c]Free peptides.

Structure of the Bitterness Receptor. Ishibashi *et al.* studied the relationship between bitterness and chemical structure of peptides (*3*). Finally, they proposed the bitterness receptor model (*3-e*). According to them, bitterness is produced when two bitterness producing groups (B.U., hydrophobic group; S.U., hydrophobic or basic group) locate close to each other. They estimated the distance is about 5Å. However, it has been still unclear why different bitter potency was produced. Their model did not explain how those two receptor sites recognize the bitter potency. In order to explain how the receptor recognizes bitterness, we propose a new bitter receptor model. As shown in Fig. 2, if the bitterness receptor has a pocket, our model is well explained. The receptor recognizes both hydrophobicity of molecules and bitterness producing groups. We estimated the size of the pocket is about 15Å because the bitterness of a compound which is bigger than this size does not increase. Receptor sites corresponding to two active sites of bitter compounds locate the bottom of pocket.

Table III. Sensory Analysis of *O*-Aminoacyl Sugars and Peptides
Containing Phenylalanine and Glycine Residues

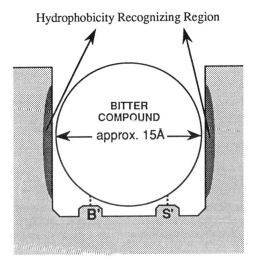

Compounds[a]	Taste	T.V. (mM)	Rcaf
R=H-Phe-Phe-Gly-Gly- (**15**)[b]	Bitter	0.0047	213
R=H-Phe-Gly-Phe-Gly- (**16**)[b]	Bitter	0.0058	172
R=H-Phe-Gly-Gly-Phe- (**17**)[b]	Bitter	0.0029	340
H-Phe-Phe-Gly-Gly-OH (**18**)[c]	Bitter	3.0	0.05
H-Phe-Gly-Phe-Gly-OH (**19**)[c]	Bitter	3.3	0.8
H-Phe-Gly-Gly-Phe-OH (**20**)[c]	Bitter	8.3	0.2

SOURCE: Adapted from ref. 5.

[a]H- refers to the amino terminus while -OH refers to the carboxyl terminus of peptides.

[b]Aminoacyl groups were introduced into 2-*O*- and 3-*O*-positions of methyl α-D-glucopyranoside.

[c]Free peptides.

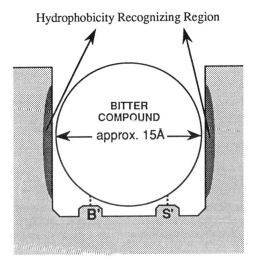

Fig. 2. Bitterness Receptor Model (Reproduced with permission from ref. 6. Copyright 1990 Japan Society for Bioscience, Biotechnology, and Agrochemistry.)

Design of Sweet Compounds

Tastes of O-Aminoacyl Sugars Containing Neutral Amino Acids with Low Hydrophobicity. We synthesized some O-aminoacyl sugars containing amino acids, the hydrophobicity of which is lower than that of bitter compounds, at 2-O- and 3-O-posidions of methyl α-D-glucopyranoside. The results of sensory analysis of these compounds are listed in Table IV. When hydrophobicity of amino acid was low, O-aminoacyl sugar produced a sweetness. The sweet intensity became stronger as hydrophobicity of amino acid increased. Sweetness of O-aminoacyl sugar (**22**) containing alanine was 16 times stronger than that of sucrose. α-Aminobutyryl derivative (**23**), which has one more methylene than the alanyl derivative, produced sweetness 50 times stronger than that of sucrose. We then prepared valyl or norvalyl derivative expecting that they would produce the sweetness of the same level of aspartame because hydrophobicity of the amino acids were higher than alanine or α-aminobutyric acid. However, the valyl derivative (**6**) produced the sweetness just 12 times stronger than that of sucrose while norvalyl derivative (**24**) produced a bitterness. This result shows hydrophobicity of amino acid moiety is also important in the case of sweetness although higher hydrophobicity makes the compounds bitter.

Table IV.   Relation between Tastes of O-Aminoacyl Sugars and Side Chain Length of Amino Acid Moiety

Amino Acid[a]	Side Chain Length	Taste	T.V. (mM)	R_{caf}, R_{suc}[a]
Glycyl (**20**)	CH_2	Sweet	2.1	2.4
Alanyl (**22**)	$CH-CH_3$	Sweet	0.32	16
Aba[b] (**23**)	$CH-CH_2-CH_3$	Sweet	0.09	53
Valiyl (**6**)	$CH-CH(CH_3)-CH_3$	Sweet	0.41	12
Norvalyl (**24**)	$CH-CH_2-CH_2-CH_3$	Bitter	0.35	2.9
Isoleucyl (**5**)	$CH-CH(CH_3)-CH_2-CH_2-CH_3$	Bitter	0.2	5
Leuciyl (**4**)	$CH-CH_2-CH_2-CH(CH_3)-CH_3$	Bitter	0.08	12.5

Source: Adapted from ref. 1-c.

[a]Aminoacyl groups were introduced into 2-O- and 3-O-positions of methyl α-D-glucopyranoside.

[b]R_{suc} is the sweet intensity. R_{suc} shows how the sweetness was weaker or stronger than that of sucrose. R_{suc} of sucrose is 1.

[c]α-Aminobutyryl.

Taste Variation of O-Aminoacyl Sugars Containing Alanyl Residues.
Because the alanyl derivative produced a good sweetness, we studied the relationship
between structure and sweetness using O-aminoacyl sugars composed of alanyl
residues. We introduced alanyl residues into various hydroxyl groups of methyl α-D-
glucopyranoside, methyl α-D-galactopyranoside and methyl α-D-mannopyranoside.
Results of sensory analysis are listed in Table V. O-Aminoacyl sugar containing
methyl α-D-glucopyranoside and two alanyl residues at the 2-O- and 3-O-positions
(22) produced a good sweetness as described previously. However, Compound 26,
two alanine residues were introduced into 4,6-position of methyl α-D-glucopyranoside
produced a bitterness instead of sweetness. O-Aminoacyl sugar containing methyl α-D-
glucopyranoside and four alanine residues (28) produced a sweetness. However, the
sweet potency was not very strong. We then introduced two alanine residues into 2-O-
and 3-O-position of methyl α-D-galactopyranoside to found that 30 produced a
bitterness.

   In conclusion, position of which amino acid introduced is limited when we
design sweet compounds using O-aminoacyl sugars. Not only orientation of alanine
introduced hydroxyl groups but orientation of free hydroxyl group is important. We
should introduce amino acids into 2,3-position of methyl α-D-glucopyranoside to
design a good sweetener.

Table V. Taste Variation of O-Aminoacyl Sugars Containing Alanine Residues

Seq.	Sugar Moiety[a]	Positions Alanyl Residues were Introduced	Taste	T.V. (mM)	R_{caf}, R_{suc}
25	MGlc	R_2	Sweet	2.3	2.1
22		R_1, R_2	Sweet	0.32	16
26		R_3, R_4	Bitter	0.68	1.47
27		R_1, R_2, R_3	Sweet>Bitter	0.55	-------
28		R_1, R_2, R_3, R_4	Sweet	2.3	2.1
29		R_4	Sweet	0.83	6
30	MGal	R_5, R_6	Bitter	0.31	3.2
31		R_5, R_8	Bitter	0.31	3.2
32		R_5, R_6, R_7, R_8	Sweet (aftertaste)	0.27	19
33	MMan	$R_9, R_{10}, R_{11}, R_{12}$	Sour>Sweet	0.49	-------

[a]MGlc, methyl α-D-glucopyranoside; MGal, methyl α-D-galactopyranoside;
MMan, methyl α-D-mannopyranoside.

Design of Salty Compounds

O-Aminoacyl Sugars Containing Basic Amino Acids. We have synthesized many salty peptide analogs and obtained some interesting results (*6, 7, 8*). We prepared methyl ester of linear amino acids and found that glycine methyl ester hydrochloride, the most simple methyl ester produced a saltiness. Glycine ethyl ester hydrochloride produced a saltiness stronger than that of the methyl ester (*8*). By this study, we knew ionic functions (*cf.* amino group) is very important for producing the saltiness. Therefore, we expected if amino groups were concentrated, the saltiness would be produced. We designed *O*-aminoacyl sugars composed of basic amino acids so that amino groups could locate close each others using sugars as skeletons.

Table VI.   Sensory Analysis of *O*-Aminoacyl Sugars Containing Basic Amino Acids

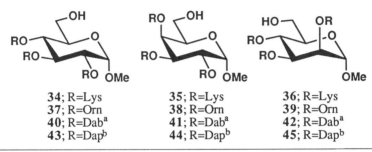

34; R=Lys	35; R=Lys	36; R=Lys
37; R=Orn	38; R=Orn	39; R=Orn
40; R=Dab[a]	41; R=Dab[a]	42; R=Dab[a]
43; R=Dap[b]	44; R=Dap[b]	45; R=Dap[b]

Compounds[c]	Taste	T.V. (mM)
34	Sweet (aftertaste)	0.47
35	Sour>Sweet (aftertaste)	0.23
36	Sour>Sweet (aftertaste)	0.47
37	Salty (unclear)	0.47
38	*Umami*	0.25
39	Sour>*Umami*	0.16
40	*Umami*>Sour>Salty	0.40
41	*Umami*>Sour>Salty	0.39
42	*Umami*>Sour>Salty	0.23
43	Sour>Sweet	0.73
44	Sour>Sweet	0.61
45	Sour>Sweet	0.36

[a]α,γ-Diaminobutyryl
[b]α,β-Diaminopropionyl
[c]Aminoacyl groups were introduced into 2-*O*-, 3-*O*-, and 4-*O*-positions of methyl α-D-glucopyranoside, methyl α-D-galactopyranoside, and methyl α-D-mannopyranoside.

Several *O*-aminoacyl sugars were prepared to study a relationship between taste and chemical structure. Methyl α-D-glucopyranoside, methyl α-D-galactopyranoside and methyl α-D-mannopyranoside were selected as sugar skeletons. As basic amino acids, esters of lysine, ornithine, α,γ-diaminobutyric acid, and α,β-diaminopropionic acid were introduced into 2-*O*-, 3-*O*-, and 4-*O*- positions of sugars leaving only 6-hydroxyl group free. The results of sensory analysis are listed in Table VI. *O*-

Aminoacyl sugars containing lysine esters produced a sweetness and a sourness with aftertastes. Ornithine ester derivatives produced *umami* basted tastes. *O*-Aminoacyl sugars containing α,γ-diaminobutyric acid esters clearly produced a saltiness with a *umami* taste and a strong sourness. We can see some rule from the result of sensory analysis. *O*-Aminoacyl sugars produce a saltiness as the side chain length of the amino acid moiety shortened. However, α,β-diaminopropionyl derivatives, of which side chain lengths of amino acid were shortest, produced only a sourness with a weak sweetness.

Table VII. Taste Variation of *O*-Aminoacyl Sugars Containing Ornithyl Residues

Seq.	Compounds	Taste	T.V. (mM)
47	3-*O*-ornithyl-D-glucose	Sweet>*Umami*	0.66
48	methyl 2,3-di-*O*-ornithyl-α-D-glucopyranoside	*Umami*	0.20
49	methyl 2,3-di-*O*-ornithyl-α-D-galactopyranoside	*Umami*	0.73
38	methyl 2,3,4-tri-*O*-ornithyl-α-D-glucopyranoside	Salty (not clear)	0.47
39	methyl 2,3,4-tri-*O*-ornithyl-α-D-galactopyranoside	*Umami*	0.25
40	methyl 2,3,4-tri-*O*-ornithyl-α-D-mannopyranoside	Sour>*Umami*	0.16
50	methyl tetra-*O*-ornithyl-α-D-galactopyranoside	Sour>*Umami*	0.44
51	methyl tetra-*O*-ornithyl-α-D-mannopyranoside	Sour>Astringency	0.13

When we compare taste of aminoacyl sugars which produced bitterness, the sweetness and the saltiness, we can see the difference between sensing the sweetness or the bitterness and the saltiness (see Tables I, IV, and VI). The bitterness can be easily enhanced greatly when the type of amino acid or position of amino acid ester introduced was changed. The sweetness also can be easily enhanced. Additionally, the sweetness can be changed to the bitterness easily by a small change of chemical structure. On the other hand, we could not see such a drastic change when we studied salty compounds. Taste intensity did not change very much even though we changed amino acids. In order to study this phenomenon more, we introduced ornithine into several positions of several sugars. Table VII shows the results of sensory analysis. The result of sensory analysis clearly shows the characteristics of *O*-aminoacyl sugars composed of basic amino acids. All compounds listed in Table VII produced the taste strength in the same level. The compound having the strongest taste intensity is methyl α-D-mannopyranoside with four ornithyl ester residues (**51**). The compound having the weakest taste intensity is methyl α-D-galactopyranoside with two ornithine residues at 2-*O*- and 3-*O*-position (**49**). The taste intensity of **51** was about six times more potent than that of **49**. Less than 10-fold differences between threshold values of compounds is not significant since the method for sensory analysis has that broad a deviation due to the experimental conditions. Therefore, we should report that O-aminoacyl sugars containing basic amino acids produced the similar taste at the same intensity level no matter how many ornithyl ester residues are contained. This result also supports the result of sensory analyses of O-aminoacyl sugars containing various basic amino acids shown in Table VI.

Enhancing Effect of O-Aminoacyl Sugars of Sodium Chloride. In order to study the interaction between O-aminoacyl sugars and sodium chloride, we carried out sensory analyses on a solution of the O-aminoacyl sugars, methyl 2,3,4-tri-O-(L-α,γ-diaminobutyryl)-α-D-glucopyranoside (**40**) and sodium chloride according to the method previously reported (7). As shown in Table VIII, **40** showed not only the saltiness, but an enhancing effect to the saltiness of sodium chloride. The most concentrated solutions (30 mM) of O-aminoacyl sugar produced saltiness which was stronger than that of 0.25 % sodium chloride solution alone. We evaluated the saltiness of 30 mM solutions of **40** as the same as that of 0.5 % sodium chloride solution. The only problem of these compounds was the strong sourness produced by presence of the hexahydrochloride. However, the sourness was weakened by adding sodium chloride. A saltiness became stronger than the sourness when 7.5 mM O-aminoacyl sugars and 0.188 % sodium chloride were mixed. The saltiness of this mixed solution was the same as that of 0.375 % sodium chloride solution, indicating that the saltiness of 0.188% sodium chloride solution was enhanced twice by adding O-aminoacyl sugars.

Table VIII. Saltiness of O-Aminoacyl Sugar 40 on Adding NaCl

Combinations				Score	Note
40[a]		NaCl			
-		0.0 %	(0mM)	0	tasteless
-		0.063 %	(10.7mM)	0.5	almost tasteless
-		0.125 %	(21.4mM)	2	
-		0.188 %	(32.1mM)	2.5	
-		0.25 %	(42.8mM)	3	
30 mM	(2.14%)	0.0 %	(0mM)	5	with strong sourness
22.5 mM	(1.61%)	0.063 %	(10.7mM)	5	with strong sourness
15 mM	(1.07%)	0.125 %	(21.4mM)	4	with weak sourness
7.5 mM	(0.53%)	0.188 %	(32.1mM)	4	with very weak sourness
0 mM	(0%)	0.25 %	(42.8mM)	3	

[a]Methyl 2,3,4-tri-O-(L-α,γ-diaminobutyryl)-α-D-glucopyranoside.

We calculated the weight percentages of sodium ions in this series of sample solutions in the same way as described in the previous paper (8). By mixing O-aminoacyl sugar with sodium chloride, the weight percentage of sodium ions was reduced to around 10 % of that conventionally used, while the saltiness was maintained at the same level. Sodium ion diet effects of ornithyl-β-alanine, glycine ethyl ester hydrochloride and basic amino acids were 75 %, 50 % and 25 %, respectively (8). O-

Aminoacyl sugars possess stronger sodium ion diet effect (90 %) than the compounds prepared previously.

Conclusions

We thus succeeded in designing molecules with the sweetness, the bitterness and the saltiness by using *O*-aminoacyl sugars. We designed an*O*-aminoacyl sugar from which bitter potency was more than 300 times stronger than that of caffeine. We also prepared an *O*-aminoacyl sugar from which sweet potency was more than 50 times stronger than that of sucrose. We then could design *O*-aminoacyl sugars from which saltiness was almost the same as that of sodium chloride. Now we can describe a taste tetrahedron using only *O*-aminoacyl sugars having common structure. Comparing aminoacyl sugars on the taste tetrahedron as shown in Fig. 3, all compounds locate at the vertex or edge of the tetrahedron.

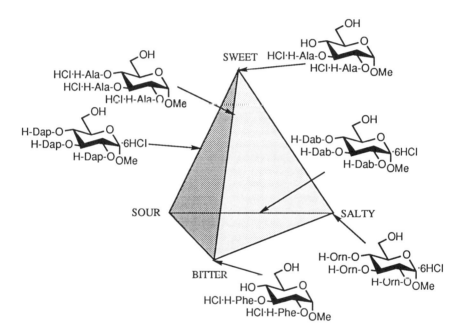

Fig. 3.   Taste Tetrahedron with *O*-Aminoacyl Sugars

Of course, we do not think that we have completely solved the puzzle of the taste tetrahedron. We are still studying saltiness of *O*-aminoacyl sugars although we have not found even a piece of the answer to a saltiness mechanism for these compounds. The most serious problem, which we have to solve first, is why we have been unable to design *O*-aminoacyl sugars from which salty potency is stronger than that of sodium chloride. However, we believe that we are very close to the entrance of the saltiness mechanism.

Literature Cited

(1)   (a) Kinomura, K.; Tamura, M.; Oga, T.; Okai, H. *J. Carbohydr. Chem.*, **1984**, *3*, 229. (b) Tamura, M.; Kinomura, K.; Tada, M.; Nakatsuka, T.; Okai, H.; Fukui, S. *Agric. Biol. Chem.*, **1985**, *49*, 891. (c) Tamura, M; Shoji, M; Nakatsuka, T; Kinomura, K; Okai, H; Fukui, S. *Agric. Biol. Chem.*, **1985**, *49*, 2579.

(2)   Azuse, I.; Tamura, M.; Kinomura, K.; Kouge, K.; Kawasaki, Y.; Okai, H. *Chem. Express*, **1989**, *4*, 539.

(3)   (a) Ishibashi, N.; Arita, Y.; Kanehisa, H.; Kouge, K.; Okai, H.; Fukui, S. *Agric. Biol. Chem.*, **1987**, *51*, 2389. (b) Ishibashi, N.; Sadamori, K.; Yamamoto, O.; Kanehisa, H.; Kouge, K.; Kikuchi, E.; Okai, H.; Fukui, S. *Agric. Biol. Chem.*, **1987**, *51*, 3309. (c) Ishibashi, N.; Ono, I.; Kato, K.; Shigenaga, T.; Shinoda, I.; Okai, H.; Fukui, S. *Agric. Biol. Chem.*, **1988**, *52*, 91. (d) Ishibashi, N.; Kubo, T.; Chino, M.; Fukui, H.; Shinoda, I.; Kikuchi. E.; Okai, H.; Fukui, S. *Agric. Biol. Chem.*, **1988**, *52*, 95. (e) Ishibashi, N.; Kouge, K.; Shinoda, I.; Kanehisa, H.; Okai, H. *Agric. Biol. Chem.*, **1988**, *52*, 819.

(4)   Barrolier, J, *Naturwissenschaften*, **1961**, *48*, 554.

(5)   Tamura, M; Miyoshi, M; Mori, N; Kawaguchi, M; Kinomura, K; Ishibashi, N; Okai, H. *Agric. Biol. Chem.*, **1990**, *54*, 1401.

(6)   Tada, M.; Shinoda, I.; Okai, H. *J. Agric. Food Chem.*, **1984**, *32*, 992.

(7)   Seki, T.; Kawasaki, Y.; Tamura, M.; Tada, M.; Okai, H. *J. Agric. Food Chem.*, **1990**, *38*, 25.

(8)   Tamura, M.; Seki, T.; Kawasaki, Y.; Tada, M.; Kikuchi, E.; Okai, H. *Agric, Biol. Chem.*, **1989**, *53*, 1625.

RECEIVED December 30, 1992

Chapter 13

Modification of Lipids by Modified Lipase

Mototake Murakami[1], Masami Kawanari[1], and Hideo Okai[2]

[1]Technical Research Institute, Snow Brand Milk Products Company, Ltd.,
1-1-2 Minamidai, Kawagoe, Saitama 350, Japan
[2]Department of Fermentation Technology, Faculty of Engineering,
Hiroshima University, Higashihiroshima, Hiroshima 724, Japan

In order to enhance the potential of synthetic reactions of lipids and the transesterification in organic solvents, a fungal lipase from *Phycomyces nites* was chemically modified. The promotion of dispersibility in organic solvents resulted in a much higher reactivity. Chemically modified lipases showed higher reactivity than unmodified lipase when they were utilized for the transesterification of triglycerides and other lipids. The initial rate of transesterification in organic solvents by modified lipase was 40 times faster than that of unmodified lipase. Chemically modified lipase was also found applicable for the synthesis of other fatty acids esters.

Natural fats and oils are mixtures of various triglyceride compounds. The physical and chemical properties of a triglyceride depend on its molecular structure; for example, chain length, unsaturation, combination, and binding position of constituent fatty acids on the glycerol backbone. The modification of lipids is an attempt to change the combination and the molecular structures of a triglyceride through certain specific procedures. The modification technologies for lipids which have been used thus far are: mixing, fractionation, hydrogenation and transesterification. However, these technologies have some critical problems in view of the diversified demand to freely modify the lipids. Recently, enzymatic processes have come into the limelight for their potential breakthrough in solving problems of the traditional modification technologies. This is because enzymatically catalyzed reactions can be very selective. The goal of this study was to develop, more fully, the potential of lipase as a catalyst. This paper will be presented the properties of lipases, why chemically modified lipase was used, the characteristics of the modified lipase, and some applications of its use.

Property of lipases

In general, a fungal lipase from *Phycomyces nites* catalyzed the hydrolysis of lipids

0097-6156/93/0528-0170$06.00/0

in water (Figure 1). Triglyceride is hydrolyzed to glycerol and fatty acids. However, in organic solvents, it is known that lipase catalyzes the reversible reaction, the synthesis of ester and transesterification (*1*). Enzymatic reactions generally exhibit positional specificity. All chemical catalysts release and bind the fatty acids from all three positions of the glycerol backbones of the triglycerides. On the other hand, some lipases release and bind the fatty acids only from the first and third position of the glycerol backbone. Using this positional specific lipase, it is possible to make positional specific triglycerides such as POP and SOP similar to cocoa butter.

Enzyme in organic solvents. In an aqueous environment, the enzymatic reaction shifts to hydrolysis. Thus, to perform esterification and transesterification, it is necessary to conduct the reaction in organic solvents. Unfortunately, the enzyme is thought to form aggregates and denature in organic solvents (Figure 2). Inada and co-workers have prepared some amphipathic enzymes by conjugating polyethylene glycol. These enzymes are soluble in organic solvents and can effectively catalyze the esterification of lipids in organic solvents (*2*). The method to introduce any hydrophobic group into the amino group (lysine residue) in the enzyme was developed using the water-soluble acylating reagent. This would permit the enzyme to become soluble and stable in the organic solvent.

Characteristic of modified lipase

Method of enzyme modification. The method of preparation of the organic solvent-soluble enzyme called modified lipase (*3*) is presented below:

Modifying group (Hydrophobic group) + DSP → DSP-ester

Lipase(Lys-) + DSP-ester → Modified lipase (Lys-Hydrophobic group) + DSP

The reagent used for this modification was p-dimethyl sulfonio phenol (DSP). First, DSP and the modifying group (hydrophobic group) were reacted to synthesize the active DSP-ester. The modifying groups were then reacted with amino groups, i.e., the lysine residues on the surface of lipase molecules. This coupling reaction was performed in water under mild conditions (4°C and neutral pH) to avoid enzyme denaturation.

    Several functional groups were used for modification: lauric acid, palmitic acid, arachidic acid, behenic acid, stearic acid, oleic acid, linoleic acid, linolenic acid, Z (Benzyloxycarbonyl) and Fmoc (9-Fluorenylmethoxycarbonyl). Using these modifying groups, the dispersibility become higher with increasing modification ratio in organic solvents.

Hydrolysis and synthesis of triglyceride using the modified lipase. The hydrolytic activity of lipase was examined using a two-phase reaction mixture of olive oil and phosphate buffer. As shown in Figure 3a, there was a significant

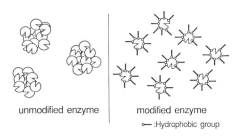

unmodified enzyme   modified enzyme
                    ○— :Hydrophobic group

Fig. 1 Schematic diagrams of lipase catalyzed reaction.

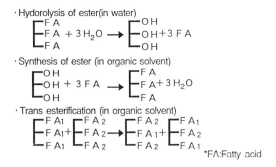

· Hydorolysis of ester(in water)

$$\begin{bmatrix} FA \\ FA \\ FA \end{bmatrix} + 3H_2O \longrightarrow \begin{bmatrix} OH \\ OH \\ OH \end{bmatrix} + 3FA$$

· Synthesis of ester (in organic solvent)

$$\begin{bmatrix} OH \\ OH \\ OH \end{bmatrix} + 3FA \longrightarrow \begin{bmatrix} FA \\ FA \\ FA \end{bmatrix} + 3H_2O$$

· Trans esterification (in organic solvent)

$$\begin{bmatrix} FA_1 \\ FA_1 \\ FA_1 \end{bmatrix} + \begin{bmatrix} FA_2 \\ FA_2 \\ FA_2 \end{bmatrix} \longrightarrow \begin{bmatrix} FA_2 \\ FA_1 \\ FA_2 \end{bmatrix} + \begin{bmatrix} FA_1 \\ FA_2 \\ FA_1 \end{bmatrix}$$

*FA:Fatty acid

Fig. 2 Enzyme in organic solvent.

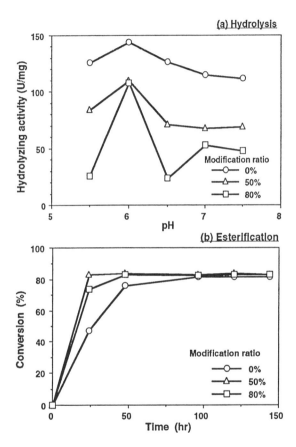

Fig. 3 Hydrolysis and esterification by modified lipase (1mg) at 37°C. (a) was obtained at triolein (3mL) and (7mL). (b) was obtained at oleic acid (0.2mL) and glycerol (5mL). Conversion was defined as the decrement of oleic acid concentration.

seen at pH 6.0. On the other hand, modification of the lipase lowered the dispersibility caused by the combined hydrophobic groups in water, resulting in lower hydrolyzing activity.

The modified lipase catalyzed the esterification of oleic acid and glycerol without solvent. It showed higher activity than the unmodified lipase within the first 50 hr of reaction (Figure 3b). While conversion reached 90% for all lipases, the initial rate of reaction using the modified lipase was about twice that of the unmodified lipase.

Transesterification of triglyceride using the modified lipase. Transesterification using the modified lipase in organic solvents was investigated. The activity of transesterification in organic solvent (n-hexane) was examined with triolein and linoleic acid as the substrate. The enzyme for transesterification was a stearic acid-modified lipase. Its activity was much higher than that of the unmodified form. The initial rate of transesterification with the stearic acid-modified lipase was about 40 times higher than the unmodified form (Figure 4). Dispersibility of the modified lipase increased linearly with increasing stearic acid modifying ratio (4), i.e., the initial rate increased exponentially with increasing dispersibility.

Effect of modifying groups on characteristics of modified lipase. The transesterification activity of the modified lipase in organic solvents was presumed to be dependent upon hydrophobic interactions. Arachidic and behenic acid-modified lipase were prepared, which had a much longer chain than that of the stearic acid-modified lipase. The modification ratio of the modified lipases was adjusted to about 50% in all cases. The reaction mixture and solvent were the same as describe above for the stearic acid modified lipase. However, the activity of these lipases was not as good as that of the stearic acid-modified lipase (Figure 5a).

The effect of unsaturation on modified lipase activity was examined. A modified lipase was prepared using stearic (C18:0), oleic (C18:1), linoleic (C18:2) and linolenic acids (C18:3), i.e., these lipases used fatty acids that had the same chain length but different unsaturation degrees. Oleic (C18:1) and linoleic (C18:2) acids-modified lipase behaved similarly and had unexpectedly higher activities than the stearic and linolenic acid-modified lipase (Figure 5b). Although conversion using all modified lipases eventuality reached finally 70%, it was apparent that the initial rate of the oleic and linoleic acid-modified lipase was greater than the others.

Effect of modifying groups on selectivity. In order to insure the selectivity of stearic, oleic, linoleic and linolenic acid-modified lipases, tripalmitin and equal weight mixtures of stearic, oleic, linoleic and linolenic acids were used as substrates in the transesterification. Figure 6 illustrates the effects of modifying groups on the transesterification selectivity. The stearic acid-modified lipase selectively transesterified stearic acid. On the contrary, linolenic acid was transesterified by only the linoleic and linolenic acid-modified lipase. From these reaction behavior, it appears that modifying groups with a saturated fatty acid tended to selectively react with the saturated fatty acid. Similarly, groups with unsaturated fatty acids

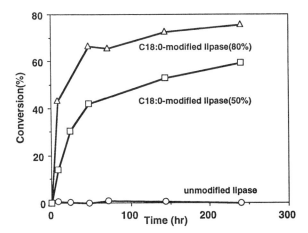

Fig. 4 Transesterification with triolein (5mg) and linoleic acid (50mg) by C18:0-modified lipase (1mg) in hexane (1mL) at 37°C. Conversion was defined as the ratio of linoleic acid in triglyceride. (Adapted from ref. 4.)

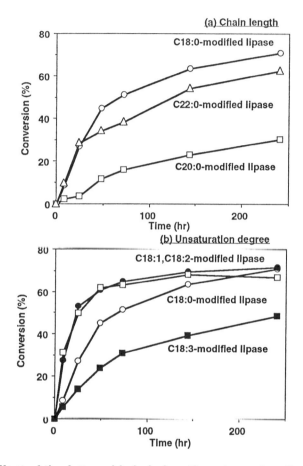

Fig. 5 Effect of the fatty acid chain length and unsaturation degree of lipase modifying group on transesterification. Triolein (5mg) and linoleic acid (50mg) and modified lipase (1mg) reacted at 37°C in hexane (1mL). (Adapted from ref. 4.)

transesterification by a modified lipase with a similar fatty acid is a suggested out some of these data.

Applications using modified lipase.

Modification of phospholipids by modified lipase. Phospholipids are natural emulsifiers which have many applications in foods, cosmetics, pharmaceutical and other industries. Yagi *et al.* (5) reported the transesterification of phospholipids with native lipase.   Yoshimoto *et al.* (6) reported the transesterification of phospholipids with polyethylene glycol modified lipase.

Since phospholipids were not soluble in water, it was necessary to emulsify them or use organic solvents for the enzymatic phospholipid hydrolyzing. Consequently, the phospholipid (phosphatidylcholine) was hydrolyzed in a two phase system of diethylether and water by a Z-modified lipase.   Table I shows the hydrolysis of phospholipid by a modified lipase.

Table I. Hydrolysis of phospholipid by modified lipase

Enzyme (lipase)	Modification ratio (%)	Conversion (%)
Unmodified lipase	0	8.3
Modified lipase	50	16.2
	80	24.0

Phosphatidyl choline was hydrolyzed to lyso-phosphatidyl choline; the activity of the modified lipase was much higher than that of unmodified lipase.

It was necessary to add over 10% buffer for the transesterification of phosphatidyl choline by native lipase (5).   Hydrolysis occurred as a side reaction in the hydrophobic solvent-water system.   The transesterification of phosphatidyl choline and eicosapentaenoic acid (EPA) was carried out in water-saturated n-hexane using palmitic acid-modified lipase.   Table II shows the transesterification of phosphatidyl choline and EPA.   Modified lipase made it possible for the transesterification of phospholipids in organic solvents.

Table II. Transesterification of phosphatidyl choline and EPA

Enzyme(lipase)	Fatty acid composition of phosphatidyl choline (%)		
	Palmitic acid	Stearic acid	Eicosapentaenoic acid
Control	50.0	50.0	0.0
Unmodified	50.0	50.0	0.0
Modified	50.0	31.9	8.1

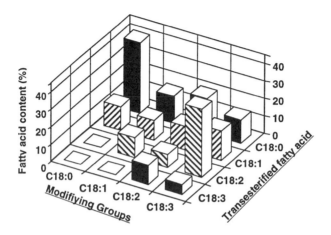

Fig. 6   Effect of modifying groups on the selectivity of transesterification.   Modified lipase (1mg), tripalmitin (5mg) and 12.5mg each of stearic, oleic, linoleic and linolenic acid reacted at 37°C in hexane (1mL). (Adapted from ref. 4.)

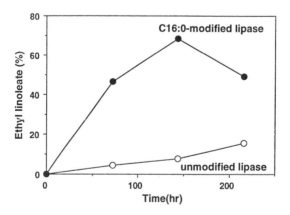

Fig. 7 Esterification of linoleic acid (5mg) with ethanol (1mL) by modified lipase (1mg) at 37°C.

Ester synthesis of cholesterol linoleate. Cholesterol fatty acid ester is an important cell membrane lipids and has many applications in cosmetics, pharmaceutical and other industries. Akehoshi *et al.*(7) reported the ester synthesis of the cholesterol fatty acid ester with native lipase. Synthesis of the cholesterol fatty acid ester was also carried out in water-saturated n-hexane by palmitic acid-modified lipase. As shown in Table III, this system made it possible for the synthesis of the cholesterol fatty acid ester in organic solvents using the modified lipase.

Table III. Ester synthesis of cholesterol linoleate by modified lipase

Enzyme (lipase)	Cholesterol	Cholesterol linoleate
Control	100.0	0.0
Unmodified lipase	100.0	0.0
Modified lipase	91.9	8.1

Ester synthesis of fatty acid ethyl ester. The lipase-catalyzed esterification of fatty acid and alcohol is well-known. It was also favorable for the esterification of poly unsaturated fatty acids under mild conditions with the enzyme. However, the activity of native lipase is lower in polar organic solvents, i.e. ethanol and methanol. The synthesis of the fatty acid ethyl ester was carried out in ethanol using the palmitic acid-modified lipase. As shown in Figure 7, the reactivity of the modified lipase in this system was much higher than that of the unmodified lipase.

Discussion

Lipase has been used in organic solvents to produce useful compounds. For example, Zark and Klibanov (8) reported wide applications of enzymes to esterification in preparing optically active alcohols and acids. Inada *et al.* (9) synthesized polyethylene glycol-modified lipase, which was soluble in organic solvent and active for ester formation. These data reveal that lipases are very useful enzymes for the catalysis different types of reactions with rather wide substrate specificities. In this study, it was found that modified lipase could also synthesize esters and various lipids in organic solvents. Chemically modified lipases can help to solve today's problems in esterification and hopefully make broader use of enzymatic reactions that are attractive to the industry.

Literature Cited

1. Iwai, M.,*Lipase*; Saiwaisyobo: Tokyo, Japan, 1991; Vol.1.
2. Inada, Y.;Matushima, A.;Takahashi, K.;Saito Y. *Biocatalysis*, 1990, Vol.3, pp.317

wasaki, Y.;Murakami, M.;Dosako, S.;Azuse, I.;Nakamura, T.;Okai, H. Biosci. h. Biochem. 1992, Vol.4, pp.441

rakami M.;Kawasaki Y.;Kawanari M.;Okai H. *J. Am. Oil Chem. Soc. in review*

gi, T.;Nakanishi, T.;Yoshizawa, Y.;Fukui, F. *J. Ferment. Bioeng.* 1990, Vol.69, p.5

6. oshimoto, T.;Nakata, M.;Yamaguchi, S.;Funada, T.;Saito, Y.;Inada, Y. *chnol. Lett.* 1986, Vol.8, pp.771

7. hoshi ;Matufune; Yoshikawa; *Kagakukougyou* 1987, Vol.34, pp.410

8. ak A.;Klibanov A. M. *Proc. Natl. Acad.Sci USA* 1985. Vol.82, pp.3192

9. nda, Y.;Nishimura, H.;Takahashi, K.;Yoshimoto T.;Saha, A.R.;Saito, Y. *B iochem. Biophys. Res. Comm.* 1984, Vol.122, pp.845

R ECEIVED December 30, 1992

Chapter 14

Enzymatic Modification of Soy Proteins To Improve Their Functional Properties for Food Use

F. F. Shih and N. F. Campbell

Southern Regional Research Center, Agricultural Research Service, U.S. Department of Agriculture, 1100 Robert E. Lee Boulevard, New Orleans, LA 70124

Soy proteins modified by proteolysis, deamidation, and phosphorylation were characterized to determine the extent of modification and changes in molecular structure. Effects of modification on solubility and emulsifying activity were investigated. Limited proteolysis in treatments with pronase E produced hydrolysates with up to 64 mmoles α-amino groups per 100 g protein or 10.2% peptide bond hydrolysis. Also investigated were enzymatic deamidation which occurred with limited proteolysis during germination of soybean seeds. Soy proteins were also phosphorylated by a protein kinase from bovine cardiac muscle. Both deamidation and phosphorylation caused increase in negative charge of the protein and generally resulted in improved functional properties that are essential for use of the proteins in food products. For example, soy proteins with 5-10% peptide bond hydrolysis or 12% deamidation and less than 3% peptide bond hydrolysis could be used as a sodium caseinate substitute in coffee whiteners.

Soy protein is a low-cost food protein with good nutritional value, but its uses in foods are limited because of inferior functional properties as compared to those of commonly used animal proteins such as casein and albumin (1,2). Therefore, modifications are often required to make soy protein more suitable for food use. Improved functional properties, particularly in the pH range of 3 to 7 where most food systems belong, have been achieved by non-enzymatic methods, including succinylation (3-5), deamidation (6,7), and phosphorylation (8,9).

However, effective enzymatic modifications, if achievable, will be a viable alternative because the methods are more specific and, most importantly, more likely to clear regulatory safety guidelines. Recently, functional changes or improvements in soy proteins have been reported by limited proteolysis with various proteases (10-12). Hamada et al. (13) investigated the use of peptidoglutaminase, a deamidating enzyme, for soy protein modification. They later reported that deamidation was enhanced by prior proteolysis or heating of the protein substrate and the deamidated products showed improved functional properties (14). Soy proteins have also been modified by enzymatic phosphorylation using a protein kinase from bovine cardiac muscle (15,16).

The present study was conducted to obtain additional information on changes in soy protein subunits during limited proteolysis. Enzymatic soy protein deamidation that occurred, in addition to limited proteolysis, during germination of soybean seeds was investigated. The effects of proteolysis and deamidation on solubility and emulsifying activity were compared. Phosphorylation of soy protein with a commercially available protein kinase and its effects on subsequent changes in functional properties of the protein were also studied.

Materials and Methods

Materials. Soy isolate, Mira Pro 111, was obtained from A.E. Staley MFG. Co. (Decatur, IL). BCA protein assay reagent was purchased from Pierce Chemicals (Rockford, IL). Reagents for ammonia analysis which included glutamine dehydrogenase (GIDH), 2-oxoglutarate, and reduced nicotinamide adenine dinucleotide (NADH) were from Bowhinger Mannheim Biochemicals (Indianapolis, IN). Pronase E and the protein kinase from bovine heart muscle were from Sigma Chemical Co. (St. Louis, MO). [γ-^{32}P]ATP was obtained from DuPont Chemical Co. (Boston, MA).

Preparation of Protease-Treated Proteins. The proteolysis of soy isolate was carried out by introducing 6 mL pronase E (mg/mL) to a well dispersed mixture of 12 g Mira Pro 111 in 760 mL water. Different levels of proteolysis were achieved by reaction at 50 °C for various periods of time. The reaction was stopped by heating at 100 °C for 3 min, and the modified protein was then recovered by lyophilization.

Preparation of Modified Soy Proteins from Germinating Soybeans. Soybean seeds (12 g dry weight) were germinated on wet Vermiculite in a metal pan covered with aluminum foil. Water was added to the tray as needed during the course of germination. To collect products, the germinated seeds were washed and then ground with mortar and pestle. Extraction of ground seeds was carried out with 200 mL 0.2 M sodium bicarbonate buffer (pH 8.0). The extract was collected by filtration through 5 layers of cheesecloth. Enzymatic reactions in the extract were stopped by protein precipitation at pH 4.7 with the addition of HCl. The residues were redissolved in water at pH 7 and filtered, and the proteins in the filtrate were recovered by lyophilization.

Protein phosphorylation. Soy proteins were phosphorylated at 37 °C in a reaction mixture containing 0.07% Miro Pro 111, 60 μM ATP (100-250 cpm/pmole), 2 mM $MgCl_2$, and 2.5 U of protein kinase in a 100 μL reaction volume. The amount of ^{32}P incorporation was determined by liquid scintillation counting as described previously (15). When larger amounts (up to 350 mg) of phosphorylated protein were needed for functionality tests, all reactants were increased by a linear scale-up, with nonradioactive ATP substituting for radioactive ATP.

Analysis of Proteolysis and Deamidation. Degree of proteolysis for the modified proteins was measured in terms of peptide bonds hydrolyzed or increase of amino groups. Total amino groups was analyzed by the trinitrobezene sulfonic acid (TNBS) method (17). Alpha-amino and ϵ-amino groups were distinguished according to Kakade and Liener (18). Complete peptide bond hydrolysis (100%) was achieved by hydrolysis with 6N HCl at 110 °C for 24 hr. Degree of deamidation was analyzed by measuring the ammonia generated using the enzymatic method of Bergmeyer and Beutler (19). Complete protein deamidation (100%) was achieved by hydrolysis with 2N HCl at 100 °C for 3 hr.

Electrophoresis. SDS polyacrylamide gel electrophoresis (4-20% gradient) was performed using the method of Laemmli (20). Autoradiographic analysis was performed on gels stained with Coomassie Brilliant Blue (R-250), dried, and stored at -70 °C with a sheet of X-ray film (Kodak X-OMAT RP) in direct contact with the gel.

Functionality Analysis. Solubility was determined on 1% (w/v) dispersion of protein in 0.2 M phosphate buffers at a pH of 3.0 to 8.0. After stirring for 0.5 hr, the dispersion was filtered (0.45 μm. Millipore), and the filtrate was analyzed for protein by the BCA method (21). Emulsifying activity index (EAI), expressed as interfacial area/unit weight protein (m^2/g), was assessed by the turbidometric method of Pearce and Kinsella (22).

Feathering Test. Modified soy proteins were investigated as a substitute for sodium caseinate in coffee whitener. A 15 mg/mL protein solution was heated at 70 °C for one hour with intermittent stirring. To 10 mL of the solution, 1.5 g of Louana vegetable oil was added. The

mixture, heated at 70 °C for 15 min, was then homogenized
for 2 min. After cooling to room temperature, 3 mL of
this liquid coffee whitener was added to 30 mL 2% coffee
solution at 90-95 °C. The whitening capacity and the
extent of feathering (coagulation) were recorded. A
control was conducted using sodium caseinate.

Results and Discussion

Protease-Treated Soy Proteins. Proteolysis of soy
proteins with pronase E proceeded at a consistent rate
during the time of investigation (Figure 1). Alpha-amino
groups increased from 3.0 mmoles/100 g protein at 0 hr to
about 64 mmoles/100 g at 3 hr. The extent of peptide
bonds hydrolyzed was calculated to be from 0 to 10.2% in
the same period of time.
    Electrophoresis analysis showed more details about the
proteolysis of soy proteins (Figure 2). The 7S subunits
disappeared from the profile faster than the 11S subunits
as proteolysis proceeded. At the end of 3 hr, 7S subunits
were completely converted to smaller polypeptides, whereas
both acidic and basic 11S subunits remained mostly
unchanged. Similar findings that the smaller 7S
conglycinin of soy proteins, more than the larger 11S
glycinin, was susceptible to reaction with proteolytic
enzymes such as trypsin and chymotrypsin have been
reported before (10). As soy proteins consist of about
one third each of 7S conglycinin and 11S glycinin, the
effect of 7S proteolysis on the functional properties of
soy protein, such as solubility, may be limited.

Proteins Modified by Deamidation. The presence of
proteases in germinating seeds is well known, but these
enzymes may or may not be activated depending on the
germinating conditions. According to α-amino group
analysis, relatively low levels of proteolysis were
observed during germination. Instead, we found
significant amount of ammonia, indicating deamidation. As
can be seen in Figure 3, both deamidation and peptide bond
hydrolysis of soybean storage protein increased with
increased germination. However, at the end of 4 days,
about 12% of the amide bonds were hydrolyzed, with only
2.6% peptide bond hydrolysis. The limited peptide bond
hydrolysis was confirmed by both HPLC analysis (not shown)
and electrophoresis. The latter showed that, for the
hydrolysate with 12% deamidation, 7S subunits were
partially hydrolyzed whereas the 11S subunits remained
practically unchanged and very little small peptides were
generated (Figure 2). For this modified protein,
therefore, amide bond hydrolysis may be responsible, more
so than peptide bond hydrolysis, for the subsequent
changes in its functionality.

Proteins Modified by Phosphorylation. The reaction of
protein kinase with soy protein was very specific. Only

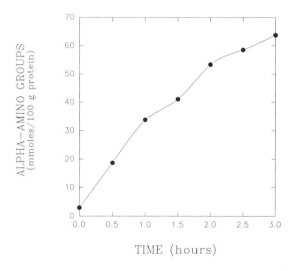

Figure 1. Proteolysis of soy protein with pronase E in termas of α-amino groups generated as a function of time.

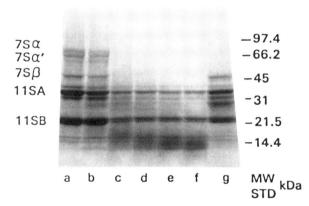

Figure 2. Electrophoretic profiles for intact and pronase E-modified soy proteins. Intact soy protein (a); soy protein proteolyzed for 0.5 h (b), 1.0 h (c), 1.5 h (d), 2.0 h (e), and 3.0 h (f); deamidated soy isolate (g).

13 μmoles of phosphorus per g protein was incorporated in soy protein after reaction with the enzyme for 4 hr at 37 °C. Autoradiography of the gels showed $^{32}P_i$ incorporated into the glycinin acidic polypeptides first, then the glycinin basic polypeptides, and finally the β and α/α' subunits of β-conglycinin (Figure 4). The order in which the subunits were phosphorylated corresponded to the number of potential phosphorylation sites in the known protein primary structures (23-26).

Effects of Protein Modification on Functionality. Figure 5 shows the solubility-pH profiles of soy proteins modified by pronase E. Solubility in the pH range of 3 to 8 increased initially with increased proteolysis. The solubility increase was caused mostly by the decrease in size of the 7S subunits. After 2 hr, almost all the down-sized 7S subunits became soluble and the insoluble residues were the remains of slightly hydrolyzed 11S subunits. The solubility ceased to increase and even decreased slightly for hydrolysates with more than 2 hr proteolysis indicating crosslinking among the hydrolysates. Minimum solubility occurred at pH ranging form 4.5 to 5.0, depending on proteolysis. The effects of pH on the solubility became insignificant after 2 hr.

For comparison, the solubility-pH profile of the deamidated protein was added to the plot containing the profiles for the pronase E-treated proteins (Figure 5). The deamidated protein, with 2.6% peptide bond hydrolysis, showed improved minimum solubility, comparable to the protein with 5.7% peptide bond hydrolysis and no deamidation. The shape of solubility-pH profile for the deamidated sample resembled that of the intact protein more than those of the pronase E-treated samples. For the deamidated sample, both the increase in solubility and the slight shift of minimum solubility to the acid side were the result of the increase in negative charges from deamidation. Obviously, deamidation was more capable of maintaining the original protein structure than proteolysis, which is essential for the development of desirable functional properties.

Emulsifying activity index (EAI) is a measure of the ability of protein to emulsify oil, which depends on solubility, size, charge, and surface activity of the protein molecules. The effect of proteolysis with pronase E on EAI of the modified protein was relatively insignificant (Figure 6). However, deamidation appeared to enhance EAI, especially at pH values more basic than the isoelectric point (pH 4.7).

The effect of phosphorylation at 13 μmoles/g protein on functional properties was minimal. Both solubility and EAI of the phosphorylated protein were slightly higher, compared to those of intact protein (not shown). Phosphorylation with protein kinase from bovine cardiac muscle is restricted by the limited number of potential phosphorylation sites in soy proteins. Experiments are

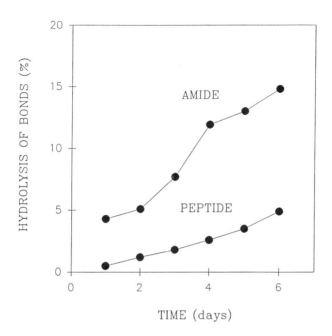

Figure 3.   Deamidation and peptide bond hydrolysis of soy protein during germination of soybean seeds.

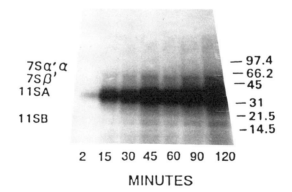

Figure 4.   Autoradiogram of SDS-PAGE gel for the phosphorylation of soy proteins.

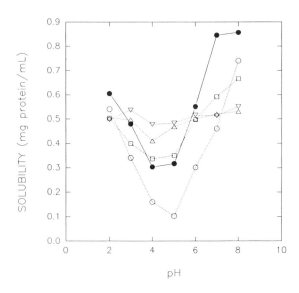

Figure 5. Solubility of soy proteins as a function of pH. Intact soy protein (○); soy protein proteolyzed with pronase E for 1 h (□), 2 h (▽), and 3 h (△); deamidated soy protein (●).

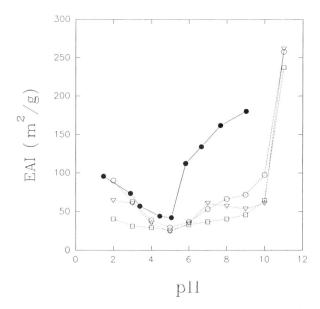

Figure 6. Emulsifying activity index of soy proteins as a function of pH. Intact soy protein (○); soy protein proteolyzed with pronase E for 1 h (▽) and 3 h (□); deamidated soy protein (●).

being conducted in our laboratory to prepare new protein kinases with different and less specific phosphorylation site requirements. For instance, protein kinase from germinating soybeans, if available, will most likely be more effective in soy protein phosphorylation, because both enzyme and substrate are from the same source.

Potential Uses of Modified Proteins. Deamidation and limited proteolysis are effective in the development of desirable functional properties, because they cause increase in negative charges without significantly changing the basic protein structure. This can be demonstrated by using the modified soy proteins as casein substitute in coffee whiteners. The coffee whitener was formulated using only the key ingredients, protein and oil, without other emulsifying additives that are normally presence in commercial coffee whiteners. The method was chosen to provide the most severe test for the protein ingredient of its whitening capacity and the ability to resist coagulation (feathering). The results showed that proteolyzed soy proteins, at 10-60 mmoles α-amino groups/100 g protein, and the deamidated protein, at 12% deamidation, were most effective as sodium caseinate substitute in coffee whiteners (Table 1).

Table 1. Relative Effectiveness of Proteins as an Ingredient in Coffee Whitener

Proteins	Whitening	Feathering
Sodium Caseinate	+ + + + +	non-feathering
Intact Soy Protein	+	almost instantly
Soy Protein		
Proteolyzed 0.5 hr	+ +	extensively within 1 min
Proteolyzed 1.0 hr	+ + +	extensively within 1 min
Proteolyzed 2.0 hr	+ + + +	slightly after 1 min
Proteolyzed 3.0 hr	+ + + +	slightly after 1 min
Deamidated	+ + + +	slightly after 1 min

Conclusions

In limited proteolysis, proteases such as pronase E hydrolyzed the 7S subunits of soy proteins more than the 11S subunits, resulting in enhanced protein solubility. Deamidation with relatively insignificant peptide bond hydrolysis that occurred during the germination of soybeans imparted to the storage protein improved solubility and emulsifying activity. On the other hand, the incorporation of phosphorus in soy proteins by the protein kinase cAMPdPK was too low to effect significant improvements in the functional properties for the modified proteins. Consequently, soy proteins modified by deamidation and/or limited proteolysis were useful as a sodium caseinate substitute in food systems such as coffee whiteners. However, phosphorylating enzymes more effective than cAMPdPK are needed to raise the level of phosphorylation before it can be considered for food use.

Acknowledgment. We thank Kim Daigle for assistance in conducting experiments for the investigation and in the preparation of the manuscript.

Literature Cited

(1) Kinsella, J.E. J. Am. Oil Chem. Soc. **1979,** 56, 542.
(2) Morr, C.V. J. Am. Oil Chem. Soc. **1979,** 56, 383.
(3) Franzen, K.L.; Kinsella, J.E. J. Agric. Food Chem. **1976,** 24, 788.
(4) Kim, K.S.; Kinsella, J.E. Cereal Chem. **1986,** 63, 342.
(5) Kim, K.S.; Rhee, J.S. J. Agric. Food Chem. **1990,** 38, 669.
(6) Shih, F.F. J. Food Sci. **1987,** 52, 1529.
(7) Shih, F.F. J. Food Sci. **1991,** 56, 452.
(8) Sung, H.; Chen, H.; Liu, T.; Su, J. J. Food Sci. **1983,** 48, 716.
(9) Hirotsuka, M.; Taniguchi, H.; Narita, H.; Kito, M. Agric. Biol. Chem. **1984,** 48, 93.
(10) Kim, S.Y.; Park, P.S.-W.; Rhee, K.C. J. Agric. Food Chem. **1990,** 38, 651.
(11) Mohri, M.; Matsushita, S. J. Agric. Food Chem. **1984,** 32, 486.
(12) Lehnhardt, W.F.; Orthoefer, F.T. Eur. Patent Appl. 0,072,617, **1982.**
(13) Hamada, J.S.; Shih, F.F.; Frank, A.W.; Marshall, W.E. J. Food Sci. **1988, 53,** 671.
(14) Hamada, J.S.; Marshall, W.E. J. Food Sci. **1988,** 53, 1132.
(15) Ross, L.F.; Bhatnagar, D. J. Agric. Food Chem. **1989,** 37, 841.
(16) Ross, L.F. J. Agric. Food Chem. **1989,** 37, 1257.
(17) Adler-Nissen, J. J. Agric. Food Chem. **1979,** 27, 1256.
(18) Kakade, M.L.; Liener, I.E. Anal. Biochem. **1969,** 27, 273.
(19) Bergmeyer, H.U.; Beutler, H.O. In Methods of Enzymatic Analysis; Bergmeyer, H.U., Ed.; 3rd ed.; Verlag Chemie Weinheim: Deerfield Beach, FL, 1985; Vol. VIII.
(20) Laemmli, U.K. Nature **1970,** 227, 680..
(21) Smith, P.K.; Krohn, R.I.; Hermanson, G.T.; Mallia, A.K.; Gartner, F.H.; Provenzano, M.D.; Fujimoto, E.K.; Goeke, N.M.; Olson, B.J.; Klenk, D.C. Anal. Biochem. **1985,** 150, 76.
(22) Pearce, K.N.; Kinsella, J.E. J. Agric. Food Chem. **1978,** 26, 716.
(23) Moreira, M.A.; Hermodson, M.A.; Larkins, B.A.;

Nielsen, N.C.   Arch. Biochem. Biophys. **1981,** 210, 633.

(24) Staswick, P.E.; Hermodson, M.A.; Nielsen, N.C.   J. Biol. Chem. **1984, 259,** 13424.

(25) Coates, J.B.; Medeiros, J.B.; Thanh, V.H.; Nielsen, N.C.   Arch. Biochem. Biophys. **1985,** 243, 184.

(26) Utsumi, S.; Kohno, M.; Mori, T.; Kito, M. J. Agric. Food Chem. **1987, 35,** 210.

RECEIVED October 28, 1992

Chapter 15

Use of Enzyme-Active Soya Flour in Making White Bread

J. P. Roozen, P. A. Luning, and S. M. van Ruth

Department of Food Science, Wageningen Agricultural University, P.O. Box 8129, 6700 EV Wageningen, Netherlands

Native soya flour contains at least 3 lipoxygenase isoenzymes, which improve dough characteristics by peroxidizing unsaturated fatty acids followed by oxidation of proteins (rheology) and carotenoids (bleaching). The fatty acid peroxides formed can also degrade into volatile compounds interfering with the flavor of white bread. Dynamic headspace samples of this bread made with and without soya flour were analysed by gas chromatography (and mass spectrometry). Addition of soya flour increased the concentrations of several secondary lipid oxidation products. Storage of bread improvers containing enzyme active soya flour caused more loss of lipoxygenase activity in a paste than in a powder type of improver. As a consequence their bleaching action and formation of volatile lipid oxidation products were diminished to the same extent.

Haas and Bohn patented in 1934 (*1*) the use of enzyme active soya flour as a dough bleaching agent. The bleaching activity of soya bean lipoxygenase L2 is much higher than that of wheat lipoxygenase (*2*). Lipoxygenase catalyzes polyunsaturated fatty acid oxidation to yield hydroperoxides with intermediate free radicals able to cooxidize lipophilic pigments (*3*), fat soluble vitamins, and thiol groups of proteins. Lipoxygenase is entangled in the gluten fraction, when it is added to a flour-water suspension (*4*).

To date the aroma of bread consists of a total of 296 volatile compounds (*5*), which can originate from different stages of bread making (Table I). The

0097–6156/93/0528–0192$06.00/0

volatile compounds of wheat kernels normally increase on milling and storage of flours (maturation). New flavor compounds are probably formed by the disturbed metabolism of the grain after grinding. Maturation improves the performance of the flour in bread making, which is probably caused by oxidative changes in the flour components (6). These changes are propagated by the activity of lipase because it acts as a supplier of substrate for lipoxygenase.

During dough making and development yeast fermentation induces enzyme activity responsible for additional flavor compounds or precursors (Table I). These compounds relate strongly to the ingredients used for the dough mixture, e.g. type of bread improver. The aroma of the finished bread depends to a large extent on the way of baking. In general the amine-carbonyl reactions play a dominant role in flavor formation in the crust, while the crumb retains mainly fermentation volatile compounds (7).

For estimating the contribution of volatile compounds to bread aroma Rothe and coworkers (8) defined "aroma value" as the ratio of the concentration of some volatile compounds to the taste threshold value of the aroma. This concept was further developed by Weurman and coworkers (9) by introducing "odor value", in which aroma solutions were replaced by synthetic mixtures of volatile compounds in water. These mixtures showed the complexity of the volatile fractions of wheat bread, because none of them resembled the aroma of bread. Recently two variations of GC-sniffing were presented (10-11), in which the aroma extract is stepwise diluted with a solvent until no odor is perceived for each volatile compound separately in the GC effluent. The dilution factors obtained indicate the potency of a compound as a contributor to the total aroma.

Bread improvers are utilized for enhancing the processing capabilities of raw materials. The bread made should have desirable physical and chemical properties and kept for some time during storage. The improvers are made up to fulfil specific dough and bread characteristics like texture, whiteness and aroma. The aroma of bread can be modified by oxidative degradation products of polyunsaturated fatty acids, and by physical absorbtion (selective) of flavor compounds in the matrix (e.g. starch). However, volatile products from hydroperoxide breakdown are considered beneficial in German bread (12) and detrimental in French bread (13) depending on the opinions of the consumers involved.

In many countries additives are forbidden in bread, except aromas in, for instance, fruit-flavored bread. Natural additives like enzyme active soya flour get only little attention for application in bread improvers because:

- the active substances like enzymes are not fully understood and are present in variable amounts;
- beany and rancid flavors can negatively affect the aroma of bread;
- the risk of unknown volatile compounds formed during bread making, e.g. by enzymic reactions.

Experimental

Bread Sample Preparation. The recipes for making white bread consisted of wheat flour (regular, 1 kg), water (520 ml), baker's yeast (25 g), improver mix (24 g), salt (20 g) and soya flour (0 or 30 g). For studying the shelf-life of soya containing improver mixtures (20 °C, 80 % rH) an improver paste with 0.5 % and an improver powder with 2.5 % soya flour were used in the recipe without further soya flour addition.

The dough was mixed for about 8 min in a spiral mixer (Kemper) at 26 °C and 80 % rH. Four loaves were formed, allowed for two fermentation periods (40 resp. 70 min) at 34 °C and 80 % rH, and then baked (30 min at 225 °C) in a rotary oven (Siemens). After one hour of cooling about 90 g slices were packed in baker's paper and aluminum foil, and afterwards kept in cold storage for max fourteen days until sampling.

Sampling and Analysis. A frozen slice of bread was cut in pieces and stacked in an enlarged sample flask of an aroma isolation apparatus according to MacLeod and Ames (*14*). Volatile compounds were trapped on Tenax TA and afterwards thermally desorbed and cold trap injected in a Carlo Erba GC 6000 vega equipped with a Supelcowax 10 capillary column (60 m x 0.25 mm i.d.) and a flame ionisation detector. Similar GC conditions were used for GC-MS identification of volatile compounds by dr. M.A. Posthumus (Dept. Organic Chemistry, VG MM7070F mass spectrometer at 70 eV EI, *15*).

Color evaluation was carried out with a panel of 10 assessors, who judged coded slices of white bread in a Kramer ranking test design (*16*).

A spectrophotometric assay procedure was used to measure the lipoxygenase isoenzyme activities of fresh and stored soya containing bread improver pastes and powders (*17*).

Results and Discussion
The number of volatile compounds collected from white bread by the dynamic headspace method are presented per class in Table II. The values give only a partial impression, because they depend strongly on the conditions used for isolation and detection of volatile compounds. Overall this method seems to release 1/3 of the total number of compounds published for white bread (*5*). So, the data obtained can only be used for studying differences compared to control samples.

Table III shows that 1-octen-3-ol and 2-heptenal were only detected in soya containing bread and that the latter has significant larger peak areas of 1-pentanol, 1-hexanol, 1-penten-3-ol, hexanal and 2- heptanone compared to the control ($p < 0.05$). These differences could be caused either by addition of volatile compounds present in soya flour or by its lipoxygenase activity. The main volatile compound found in soya flour was limonene (*15*), of which the peak areas were similar in the soya flour containing bread and its control sample. Moreover minor volatile compounds of soya flour, like 1-pentanol, 1-hexanol and hexanal, increased steadily in soya samples, as can be seen in Table

Table I. Kind of volatile compounds formed in different stages of bread making

Grain	Flour	Dough making	Baking
(disturbed) metabolism of grain		Enzymes/yeast metabolism	Amine-carbonyl reactions
alcohols	esters	alcohols	alkyl pyrazines
aldehydes	lactones	acetoïn	pyrroles
ketones	hetero-	organic acids	aldehydes
hydro-	cyclic		ketones
carbons	compounds		

Table II. Number of volatile compounds detected in white bread samples with the dynamic headspace method

Class of compound	Number found	Proportion*
Acids	6	.3
Alcohols	13	.7
Aldehydes	11	.3
Bases	6	.1
Esters	9	.5
Furans	5	.3
Hydrocarbons	2	.2
Ketones	10	.3
Sulfur Compounds	2	.1

*Proportion to the total number of these compounds presented by Maarse and Visscher (5)

III. These results indicate that the composition of volatile compounds of soya containing white bread is hardly influenced by the volatile compounds of the soya flour itself and that lipoxygenase activity plays a major role.

It is interesting to note that the two most potent detrimental volatile compounds of autoxidized white bread (2-nonenal and 2,4-decadienal; *18*) were not detected in our soya containing white bread samples. But, these samples had additionally 2-heptenal and 1-octen-3-ol, which belong to the flavor significant volatile compounds of oxidized soya bean oil (*19*). Apparently, soya lipoxygenase peroxidized available soya oil constituents more easily than wheat lipids. The enzymes of legume flour have been shown to peroxidize unsaturated fatty acids in all major lipids occurring in wheat dough (*20*). It has been noticed, that these enzymes were responsible for a "green" and bitter off-flavor of bread (*21*).

In todays practice of making bread enzyme active soya flour will not be used as an additional ingredient, but as a component of a bread improver mixture. In that case the storage stability of soya lipoxygenase in bread improvers is an important factor for their bleaching action and formation of volatile lipid oxidation products. The latter effect is demonstrated in Table IV. Bread improver paste products declined fully during 5 months of storage, while powder products decrease only about 50 %. In average the difference between paste and powder is 32 % and significant at $p < 0.01$ in the Student-t test.

The bleaching capacity of the soya flour added to bread improver paste and powder is shown in Figure 1. The sums of ranks of the slices of control bread are much higher (= less white) than of the slices of the soya added ones. The difference in soya flour content of the improvers is also reflected here: sum of ranks for whiteness is significant lower (= more white) for bread with fresh and stored bread improver powder ($p < 0.05$). The effect of storage of the soya containing bread improvers gives a not significant increase in the sum of ranks. So, within the studied shelf-life of 5 months, these bread improvers can be considered as satisfactory for the end-users.

The bleaching of plant pigments responsible for greyness of bread crumb, is related to lipoxygenase isoenzyme activities of native soya flour. These activities were measured in the soya containing bread improvers (Figure 2). The activities of the enzymes decreased much more in the paste than in the powder during storage. The higher activities of L2 and L3 isoenzymes of the powder are likely to be responsible for its larger bleaching capacity presented in Figure 1. Weber and coworkers (*23*) found that the isoenzymes mentioned are able to cooxidize carotenoïds and that isoenzyme L1 exhibits slightly carotene oxidation. The activity of isoenzyme L1 decreased much more in the paste than in the powder improver (Figure 2). This might explain the differences in yield of volatile compounds in Table IV, assuming that these compounds originated mainly from lipids peroxidized by isoenzyme L1.

As discussed, addition of enzyme active soya flour changes the composition of volatile compounds of white bread. In its practical application as a bread improver component, the soya lipoxygenase isoenzymes are sufficient stable for 5 months to meet the bleaching requirements.

Table III. Amounts of volatile compounds (ng/kg) detected in white bread by dynamic headspace analysis; CONTROL is without and SOYA is with 30 g soya flour addition (see recipe; *15*)

Compound	CONTROL	SOYA
1-Pentanol	6 ± 1.5	14 ± 2
1-Hexanol	9 ± 2	61 ± 11
1-Penten-3-ol	4 ± 1.5	26 ± 2
1-Octen-3-ol	nd	3 ± 0.6
Hexanal	78 ± 14	173 ± 23
2-Heptenal	nd	14 ± 1.9
2-Heptanone	6 ± 1.4	10 ± 1.9

nd = not detected.

Table IV. Influence of 16-23 weeks of storage of soya containing bread improver PASTE and POWDER on differences (d) in relative decreases (%) of GC peak areas of selected volatile compounds of white bread (*22*)

Compound	PASTE	POWDER	d*
1-Butanol	53	-7	60
1-Pentanol	112	50	62
1-Penten-3-ol	106	80	26
1-Hexanol	100	36	64
1-Octen-3-ol	87	62	25
Hexanal	25	43	-18
Heptanal	25	50	-25
Octanal	100	50	50
Nonanal	100	46	54
Decanal	90	64	26

* mean of d is 32 % and Student-t: $p < 0.01$.

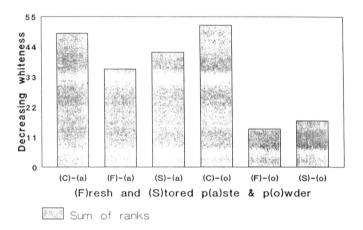

Figure 1. Influence of 16-23 weeks of storage of soya containing bread improver paste and powder on visual whiteness of slices of bread. Kramer ranking test design (1 = most white sample, etc.); (C) = control, bread improver without soya flour has been used (based on data mentioned in text of 22).

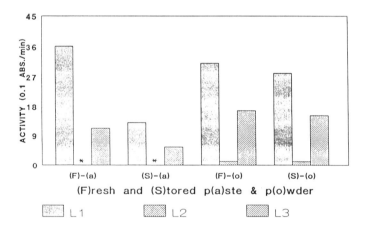

Figure 2. Influence of 16-23 weeks of storage of soya containing bread improver paste and powder on lipoxygenase isoenzyme L1, L2 and L3 activity. * = no activity detected (based on data mentioned in text of 22)

Literature Cited

1. Haas, L.W.; Bohn, R.M. *Chem. Abstr.* **1934**, 28, 4137.
2. Grosch, W.; Laskawy, G.; Kaiser, P. *Z. Lebensm. Unters. Forsch.* **1977**, 165, 77-81.
3. Nicolas, J. *Ann. Technol. Agric.* **1978**, 27, 695-713.
4. Graveland, A. *Biochem. Biophys. Res. Commun.* **1970**, 41, 427-434.
5. Maarse, H.; Visscher, C.A. *Volatile Compounds in Food. Qualitative Data;* TNO-CIVO Food Analysis Institute: Zeist, The Netherlands, 1987.
6. Bellenger, P.; Godon, B. *Ann. Technol. Agric.* **1972**, 21, 145-161.
7. Schieberle, P.; Grosch, W. *Z. Lebens. Unters. Forsch.* **1991**, 192, 130-135.
8. Rothe, M.; Thomas, B. *Z. Lebensm. Unters. Forsch.* **1963**, 119, 302-310.
9. Mulders, E.J.; Maarse, H.; Weurman, C. *Z. Lebens. Unters. Forsch.* **1972**, 150, 68-74.
10. Acree, T.E.; Barnard, J.; Cunningham, D.G. *Food Chem.* **1984**, 14, 273-286.
11. Schieberle, P.; Grosch, W. *Z. Lebensm. Unters. Forsch.* **1987**, 185, 111-113.
12. Kleinschmidt, A.W.; Higashiuchi, K.; Anderson, R.; Ferrari, C.G. *Bakers Dig.* **1963**, 37(5), 44-47.
13. Drapron, R.; Beaux, Y.; Cormier, M.; Geffroy, J.; Adrian, J. *Ann. Technol. Agric.* **1974**, 23, 353-365.
14. MacLeod, G.; Ames, J. *Chem. Ind.* **1986**, 175-177.
15. Luning, P.A.; Roozen, J.P.; Moëst, R.A.F.J.; Posthumus, M.A. *Food Chem.* **1991**, 41, 81-91.
16. O'Mahony, M. *Sensory evaluation of food: statistical methods and procedures;* Marcel Dekker Inc.: New York, NY, 1986.
17. Engeseth, N.J.; Klein, B.P.; Warner, K. *J. Food Sci.* **1987**, 52, 1015-1019.
18. Grosch, W. In *Autoxidation of unsaturated lipids;* Chan, H.W.-S., Ed.;Academic Press: London, 1987; pp 95-139.
19. Frankel, E.N. *J. Sci. Food Agric.* **1991**, 54, 495-511.
20. Morrison, W.R.; Panpaprai, R. *J. Sci. Food Agric.* **1975**, 26, 1225-1236.
21. Drapron, R.; Beaux, Y. *C. R. Acad. Sci., Ser. D* **1969**, 268, 2598-2601.
22. Van Ruth, S.M.; Roozen, J.P.; Moëst, R.A.F.J. *Lebensm. Wiss. Technol.* **1992**, 25, 1-5.
23. Weber, F.; Laskawy G.; Grosch, W. *Z. Lebensm. Unters. Forsch.* **1974**, 155, 142-150.

RECEIVED January 29, 1993

Chapter 16

Rapid Fermentation of Soy Sauce

Application to Preparation of Soy Sauce Low in Sodium Chloride

Shunsuke Muramatsu[1], Yoshihito Sano[1], and Yasuyuki Uzuka[2]

[1]Takeda Shokuryo Company, Ltd., 9–30 Saiwai-cho, Kofu,
Yamanashi 400, Japan
[2]Department of Applied Chemistry and Biotechnology, Faculty
of Engineering, Yamanashi University, Kofu, Yamanashi 400, Japan

Because of the importance of soy sauce as a food flavorant, a new method for its production was developed. The importance of this novel processing procedure is that it significantly shortens the fermentation processing time by half or to about three months. Attempts at further modification of the process to prepare low-sodium-chloride-containing soy sauce are also described.

Soy sauce is a uniquely Japanese seasoning prepared from soy beans and wheat. Although soy sauce is made from soy beans, it has a highly desirable flavor without the beany aroma. Soy sauce is currently used not only in the preparation of many Japanese cuisines but also in the preparation of many popular western dishes such as beef steaks, poultry, and stews. Japan has more than 2,300 soy sauce producing firms. Soy sauce production in Japan is 1,200,000kL per year at a cost of about 239.4 billion yen ($1,840 million U.S.). The per capita consumption of soy sauce in Japan is 2,947 yen ($22.70 U.S.) per year. By comparison, the per capita yearly consumption of sugar in Japan is 2,318 yen ($17.80 U.S.). This indicates the relative impact of soy sauce as a food adjuvant for the Japanese consumer.

Conventional production of Soy Sauce

Koji making: The conventional process of soy sauce production is summarized in Figure 1. Soy beans are boiled for 10 minutes at 120°C or higher. Wheat grains, the second extremely important component in soy sauce production, are roasted and crushed. The pulverized wheat and the boiled soy beans are mixed with the spores of a food grade strain of *Aspargillus oryzae* (AO). The mixture is placed in a well-ventilated room with high humidity. The mold, AO, utilizes the starch of the wheat and the humidity of the storage area as a source of substrate for its growth. At the same time the mold is growing it is hydrolytically converting the starch rich wheat to glucose by amylase. The glucose will serve as substrate for lactic acid bacteria and yeast described below. The mold and its associated mycelia at this point are called "*koji*." The *koji* is a rich source of other hydrolytic enzymes such as proteases, peptidases, and amylase (*1, 2*). Protein rich soy beans and wheat grains are proteolytically converted to peptides and amino acids by enzymes from the *koji*. The distinctive *umami* taste of soy

0097–6156/93/0528–0200$06.00/0

sauce comes from the degradation of glutamic acid rich proteins from the wheat and soy bean.

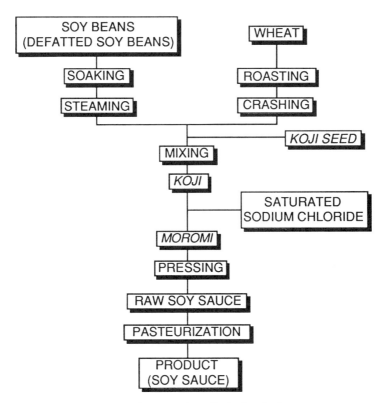

**Fig. 1. Conventional Method of
Soy Sauce Production**

***Moromi* making:** When *koji* mixed with saturated sodium chloride solution and allowed to stand, the mixture is called "*moromi.*" The mixing of the *koji* with the salt solution makes the internal region of the *moromi* mix anaerobic, a condition that kills *Aspergillus oryzae.* On the other hand the high salt, anaerobic conditions are quite suitable for the growth of other microorganisms. Halotolerant microorganisms such as *Pediococcus* halophilus are the first to increase in population densities. *Pediococcus halophilus*, a lactic bacteria, produces lactic acid to sour the soy sauce and makes the saltiness milder. As the lactic acid concentration in the *moromi* increases, the pH decreases, leading to the cellular densities of *Pediococcus halophilus* to reach maximal levels. At the same time, as the pH approaches 5.0, halophilic yeast species such as *Zygosaccharomyces rouxii* begin to grow. By 60 days these yeast have produced large amounts of ethanol (Figure 2 & 3) (*3*). The switch from lactic acid fermentation to alcohol fermentation in the *moromi* is mainly due to the difference in growth characteristics of the bacteria and yeast based on the pH. The hydrolysis of starch, lactic acid fermentation, and alcohol fermentation, proceed very slowly in the *moromi* vessel.

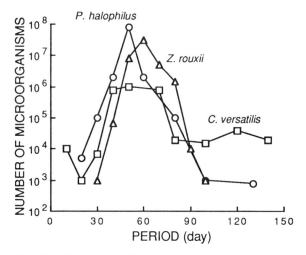

Fig. 2.   Changes of *P. halophilus*, *Z. rouxii* and *C. versatilis* in *Moromi* Mash (Reproduced with permission from ref. 3.   Copyright 1988 Brewing Society of Japan)

Once the lactic acid and alcohol fermentation is complete, the *moromi* is aged until its moldy odor has disappeared. During this portion of the aging period other flavor molecules, particularly 4-ethylguaiacol and 4-ethylphenol are added to the *moromi* by *Candida versatilis* (*4*). These compounds are the major components contributing to the flavor of soy sauce. The entire *moromi* process is completed in about six months.

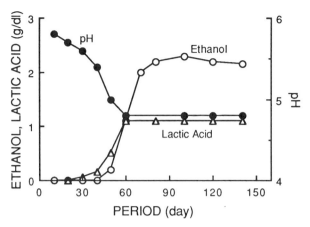

Fig. 3.   Changes of pH, Lactic Acid and Ethanol in *Moromi* Mash (Reproduced with permission from ref. 3.   Copyright 1988 Brewing Society of Japan)

Final soy sauce product: When the *moromi* aging process is complete, the material is pressed yielding a raw soy sauce.   The raw pressed material is pasteurized and filtered prior to sending it to the market.

Rapid Fermentation of Soy Sauce

Analysis of each process in the conventional method of soy sauce production indicated that *moromi* required three processes: hydrolysis, fermentation, and aging.   It was hypothesized that the time required for the *moromi* process could be reduced into two process, i.e. autolysis and fermentation. These are discussed below.

Effect of Salt on Autolysis of *koji*. In addition to causing the interior of the *koji* mixture to become anaerobic the high salt used in the conventional method of soy sauce production uses high salt to avoid putrefaction of the autolysate (5).   Since high dietary levels of sodium chloride can exacerbate pre-existing medical conditions such as hypertension, it would be desirable to prepare soy sauce preparations low in sodium chloride.   Since high temperatures are used in the beer brewing industry to assist in the saccharification of malt, it seemed to be a likely choice to apply to the autolysis of *koji*.

   Figure 4 shows the autolysis of *koji* in the absence of sodium chloride at several different temperatures over a three day period (6, 7).   Maximal levels of total soluble nitrogen were obtained at temperatures of 60°C.   The lower the temperature the less total soluble nitrogen was observed with similar amounts of total nitrogen being found at 45°, 50°, and 55°C.   On the other hand, the maximum levels of glutamic acid were obtained at temperatures of 45°C anc 50°C. Higher temperatures produced lower yields of glutamic acid. A similar temperature response was observed for formol nitrogen.. The pH of the *koji* was significantly affected by the temperature at which the autolysis was carried out. At 45° and 50° the pH dropped below 4.5 within one day of autolysis suggesting that the lower pH might be due to lactic acid or other acidifying materials produced by unexpected microorganisms. Since the acidified autolysis is not suitable as the starting material for the next fermentation step, it was determined that autolysis should be carried out at 55°C.

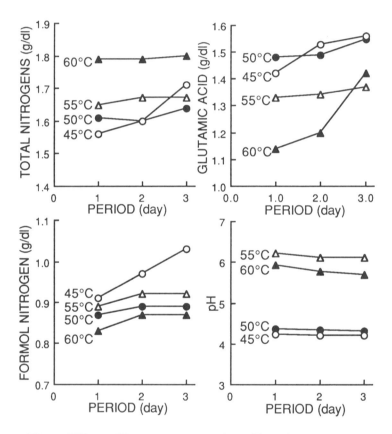

Fig. 4. Effect of Temperature on Autodigestion of Soy Sauce
***Koji* Materials in the Absence of Sodium Chloride**
(Reproduced with permission from ref. 6.
Copyright 1991 Brewing Society of Japan)

The effect of sodium chloride (NaCl) on autolysis of *koji* was also examined. NaCl concentrations were varied while the temperature of autolysis was maintained at 55°C. Maximum total nitrogen was obtained when NaCl was absent. The yield of total nitrogen decreased with increasing concentrations of NaCl. Similar responses were observed for glutamic acid and formol nitrogen. As opposed to the traditional means of soy sauce production the require high levels of NaCl, these data suggested that maximal levels of total nitrogen, glutamic acid and formol nitrogen could be obtained in the absence this salt (*8, 9*).

Having determined the effect of temperature and NaCl on *koji* autolysis, the next step in finding an optimal protocol for the rapid production of soy sauce with high flavor quality was to examine the effect of independent and combined effect of lactic acid fermentation and alcohol on *koji* autolysis.

Effect of Lactic Acid Fermentation on *Koji* Autolysis. The effect of NaCl on lactic acid fermentation (*10*) is seen in Figure 5. Four autolysates were prepared

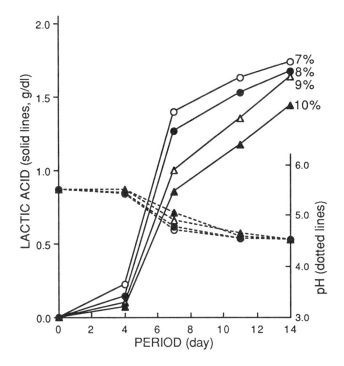

Fig. 5. Effect of Sodium Chloride Concentration on Lactic
Acid Fermentation by *P. halophilus* (Reproduced with
permission from ref. 10. Copyright 1992
Brewing Society of Japan)

with 7, 8, 9, and 10% NaCl. The autolysates were inoculated with 10^5 cells/mL of *P. halophilus*. The mixture was incubated at 30°C with the initial pH adjusted to 5.8. *P. halophilus* was observed to have a more rapid growth rate at lower concentrations of NaCl. As the fermentation proceeded, the pH of the autolysate would drop to 4.5 regardless of the NaCl concentration. The yield of lactic acid in this system was more than 15g/L, which is about twice that of regular soy sauce.

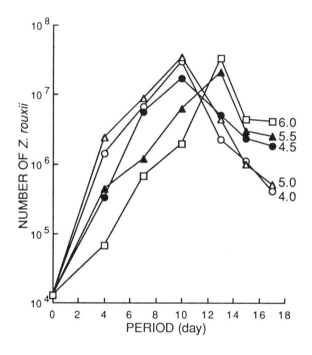

Fig. 6. Effect of Initial pH on the Growth of *Z. rouxii*
(Reproduced with permission from ref. 10.
Copyright 1992 Brewing
Society of Japan)

Effect of Alcohol Fermentation on *Koji* Autolysis. *Z. rouxii* was used as an inoculum in the study of alcohol fermentation. Reports have suggested that the initial pH of the inoculum is very important to the proliferation of yeast such as *Z. rouxii* when grown under high concentrations of NaCl. Five autolysates were prepared containing 13.8% NaCl with the pH of the inoculum adjusted in 0.5 pH increments from 4.0 to 6.0 (Figure 6). *Z. rouxii* was inoculated at 10^4 cells/mL and the inoculum incubated at 30°C. The maximal density of cells was reached by 10 days when pH levels were below 5.0. On the other hand, maximum cell levels were reached at 13 days if the initial pH was 5.5 or greater (*10*). Ethanol concentrations at the end of incubation was more the 20g/L, which is enough for soy sauce.

Scale-up of Soy Sauce Preparation. A potentially useful method for the large scale production of soy sauce was developed based on a combination of the data and optimal protocols determined in the experiments described above. The pH change, reducing sugar, ethanol and lactic acid concentration of a 1 ton scale incubation in which lactic acid and alcohol fermentation are set to proceed sequentially is shown in Figure 7. *P. halophilus* inoculations of 10^3 cells/mL were used for lactic acid fermentation with the mixture incubated at 30° in the presence of 10% NaCl. Lactic acid levels reached 10g/L by 14 days of incubation in the 1 ton vessel. After 14 days the autolysate was mixed with an equal amount of newly prepared autolyzate and was inoculated with *Z. rouxii* and *C. versatillis* and the inoculum incubated. Alcohol fermentation proceeded with ethanol levels reaching their maximum plateau level of 23g/L. The pH in this 1 ton vessel slowly dropped during the first 23 days and reached a low of 4.8 (*11*).

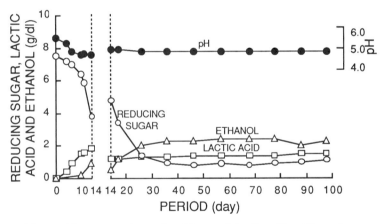

Fig. 7. Fermentation of *Moromi* **in One Ton Vessel**
(Reproduced with permission from ref. 11.
Copyright 1992 Brewing Society of Japan)

Further scaled-up of the soy sauce production to a 40 ton reaction vessel yielded result very similar to that obtained for the 1 ton vessel and comparable to the of the conventional method. Data on the composition of these raw soy sauces are seen in

Table I. The soy sauce produced by the scaled-up methods yielded a soy sauce with lower NaCl concentrations than the soy sauce produced by conventional means. Glutamic acid concentrations, while somewhat lower than that of the soy sauce produced by conventional method, were still quite high. While the lengthy *moromi* aging process of the conventional method produces pyroglutamic acid, a tasteless product of glutamic acid, the scale-up procedure reduces the pyroglutamic acid conversion to half that found in the conventional method. Consumer evaluating soy sauce produced by the scale-up method were unable to distinguish taste differences between the new (scale-up) soy sauce and soy sauce produced by the conventional method. This led to the opening of a new production plant that is currently producing the new soy sauce for marketing.

Table I.   Comparison of Autolyzate Compositions* of Raw Soy Sauce

Contents	Autolyzate (Unfermented)	Raw Soy Sauce Obtained from 1 Ton Vessel	Raw Soy Sauce Obtained from 40 Ton Vessel	Raw Soy Sauce Obtained by Conventional Method
NaCl	0	14.5	14.5	16.9
Total Nitrogen	1.99	1.90	1.83	1.80
Glutamic Acid	1.98	1.63	1.75	1.49
Pyroglutamic Acid	0.11	0.16	0.15	0.29
Reducing Sugars	6.68	1.20	2.05	3.13
Alcohol	0	2.20	1.97	1.83
Lactic Acid	0	1.00	1.02	1.20
pH	5.92	4.80	4.77	4.77

*All values, exsept pH, are given as g/dL

Preparation of Low-Sodium-Chloride-Containing Soy Sauce. While soy sauce is a highly flavorful seasoning its continuous use by consumers suffers from a major drawback, i.e., high NaCl level are potentially deleterious to the individuals with hypertension, hypertensive tendencies, and renal problems, to name just a few. Thus, it would be advantageous to a soy sauce producing company to develop a method that would produce a low-sodium-chloride-containig soy sauce that still maintains the high flavor quality of the sauce. Conventional methods have attempted to use potassium chloride (KCl) as a replacement for NaCl. Unfortunately this yields a soy sauce with an unsatisfactory taste for most consumers. Since the new scaled-up method described above yields a soy sauce with lower than conventional NaCl levels attempts were made to redesign the protocol with even lower levels of NaCl.

Soy sauce containing a final concentration of 4.6% NaCl were prepared by the scale-up method described above. Since traditional (conventional) soy sauce contains approximately 16.2% NaCl, experiments were designed to use the addition of several salt substitutes to make up the difference in salt concentration between the scale-up soy sauce (4.6%) and the traditional soy sauce (16.2%). Salt substitutes used included KCl, glycine ethyl ester hydrochloride, lysine hydrochloride, taurine and glutamic acid. According to Tamura et al, (12), glycine ethyl ester hydrochloride, lysine hydrochloride and glutamic acid have the effect of enhancing the saltiness of NaCl. Taurine was expected to have an affect on the final salty taste since the presence of a salty peptide, ornithyltaurine, had been previously reported (12). Sensory evaluation of the various samples by five panelists included a dilution of the samples with distilled water such that the final concentration of the additives were adjusted to the following percentages (%): 2.0, 1.0, 0.5, 0.125, and 0.063%. Threshold values were determined for each salt substitute.

The results of the sensory evaluation of the salt substitute investigation are shown on Table II. The threshold concentration for the perception of the "salty" taste remained the same for all groups. This suggested that the perception mechanism of each of the compounds was similar, if not identical, i.e., they enhanced the salty taste perception. The samples containing taurine yielded desirable flavor result, but the mechanism of how taurine effected this response remains unclear at this time. While glutamic acid and glycine ethyl ester hydrochloride enhanced the saltiness of the soy sauce, they also imparted a sour note. On the other hand, the panelists did not find the sourness unfavorable and reported that the net sensory response of both glutamic acid and glycine ethyl ester hydrochloride was to give a better flavor to the soy sauce. Lysine hydrochloride gave the soy sauce an unfavorable aftertaste which disappeared when diluted samples were tasted. Acceptable flavors and saltiness were unattainable even in diluted samples when using KCl as the salt replacement suggesting that KCl could not serve as a commercially marketable NaCl substitute in soy sauce.

Table II. Results of Sensory Analysis

Samples	Original Concentration of NaCl (%)	Threshold Value (%) of Additives	Comments
Conventional Soy Sauce	16.2	0.25	Tastes like soy sauce.
Soy Sauce Prepared by Our Method	4.6	0.25	Contains a little roasty flavor.
Soy Sauce Prepared by a New Method + 11.6% KCl	4.6	0.25	Unfavorable taste of KCl was detected even at lower concentration where saltiness could not sensed.
Soy Sauce Prepared by a New Method + 11.6 % Glycine Ethyl Ester Hydrochloride	4.6	0.25	Contains a sourness.
Soy Sauce Prepared by a New Method + 11.6% Lysine Hydrochloride	4.6	0.5	Unfavorable taste of Lys·HCl was detected at the higher concentration.
Soy Sauce Prepared by a New Method + 11.6% Taurine	4.6	0.5	Delicious.
Soy Sauce Prepared by a New Method + 11.6% Glutamic Acid	4.6	0.25	Contains a sourness.

An interesting outcome of this investigation is the birth of a potentially new seasoning. By dividing the *moromi* process one can produce a *koji* autolysate with the same chemical composition and flavors as soy sauce expect for the salt and saltiness, i.e. the *koji* autolysate contains the same amount of total nitrogen and glutamic acid as raw soy sauce but no salt. The new type of NaCl-free seasoning has the potential of being applied to new food materials.

Conclusion

A new method for the preparation of soy sauce has been developed. The new scaled-up method divides the *moromi* process into two processes: autolysis and fermentation. Because of the utilization of high temperatures, the new process permits the production of a NaCl free autolyzate from *koji*. Division of the fermentation process into two separated processes permit better control of lactic acid fermentation and alcohol fermentation processed which used to require great skill. The new scale-up procedure for soy sauce production yields a product in half the time required by the traditional (conventional) method and still produces a soy sauce with high levels of the desirable flavor component, glutamic acid. Utilization of this protocol by the soy sauce producing industry should have significant economic impact to both producers and consumers.

Literature Cited

(1)   Yokosuka, Y; Iwasa, T; Fuji, S. *J. Jpn. Soy Sauce Res. Inst.*, **1987**, *13*, 18.
(2)   Terada, M; Hayashi, K; Mizunuma, T. *J. Jpn. Soy Sauce Res. Inst.*, **1981**, 7, 158.
(3)   Kadowaki, K. in *Shoyu no Kagaku to Gijyutsu*,Tochikura, T., Ed; Nippon Jozo Kyokai; Tokyo; 1988.
(4)   Uchida, D. In *Shoyu no Kagaku to Gijyutsu*; Tochikura, T., Ed.; Nihon Jyozou Kyokai: Tokyo (1988).
(5)   Terada, M.; Hayashi, K.; Mizunuma, T.; Mogi, K. *Seasoning Sci.*, **1973**, *2*, 23.
(6)   Muramatsu, S; Sano, Y; Takeda, T; Uzuka, Y. *J. Brew. Soc. Jpn.*, **1991**, *86*, 610.
(7)   Muramatsu, S; Sano, Y; Uzuka, Y. *J. Brew. Soc. Jpn.*, **1992**, 87, 219.
(8)   Muramatsu, S; Sano, Y; Uzuka, Y. *J. Brew. Soc. Jpn.*, **1992**, 87, 150.
(9)   Muramatsu, S; Sano, Y; Uzuka, Y. *J. Brew. Soc. Jpn.*, **1992**, 87, 295.
(10)   Muramatsu, S; Ito, N; Sano, Y; Uzuka, Y. *J. Brew. Soc. Jpn.*, **1992**, *87*, 378.
(11)   Muramatsu, S; Ito, N; Sano, Y; Uzuka, Y. *J. Brew. Soc. Jpn.*, **1992**, *87*, 538.
(12)   Tamura, M.; Seki, T.; Kawasaki, Y.; Tada, M.; Kikuchi, E.; Okai, H. *Agric, Biol. Chem.*, **1989**, *53*, 1625.
(13)   Tada, M.; Shinoda, I.; Okai, H. *J. Agric. Food Chem.*, **1984**, *32*, 992.

RECEIVED October 28, 1992

Chapter 17

Large-Scale Preparation and Application of Caseinomacropeptide

Y. Kawasaki, H. Kawakami, M. Tanimoto, and S. Dosako

Technical Research Institute, Snow Brand Milk Products Company, Ltd., 1–1–2 Minamidai, Kawagoe, Saitama 350, Japan

Caseinomacropeptide (CMP) is a biologically active peptide derived from κ-casein during cheese making. The apparent molecular weight of CMP depends on pH; it is much higher at neutral pH than its theoretical value (9 kDa). A simple procedure for isolating CMP through ultrafiltration was developed, using the pH-dependent behavior of its molecular weight. CMP competitively inhibits the binding of cholera toxin to its receptor, ganglioside G_{M1}, and reduces the characteristic cholera toxin-derived morphological change which normally occurs in Chinese hamster ovary cells *in vitro*.

Caseinomacropeptide (CMP), also called glycomacropeptide (GMP), is a hydrophilic glycopeptide with molecular weight (MW) of approximately 9 kDa (*1*), released from bovine κ-casein by the action of chymosin at [105]Phe-[106]Met (*1*). Casein coagulates to form curd in the presence of calcium when CMP is released. CMP remains in the supernatant of the cheese whey together with whey components such as β-lactoglobulin (β-Lg), α-lactalbumin (α-La), lactose and minerals.

Some biological functions of CMP that have been reported recently include: inhibiting action for gastric secretion (*2*), periodic contractions of stomach and duodenum (*3*), aggregation of ADP-treated platelets (*4*), and adhesion of oral *Actinomyces* and *Streptococci* to erythrocytes or polystyrene tubes which simulate tooth structure (*5*). Previous studies (*6*) indicated that CMP inhibited the binding of cholera toxin to its receptor. These functions suggested that CMP might be a useful ingredient in dietetic foods or pharmaceuticals. This paper will discuss and review experimental data on the interaction with cholera toxin (*6*). Finally it will discuss a novel procedure to isolate CMP on a large scale (*7*).

Application: Inhibitory Effect of CMP against Binding of Cholera Toxin to Its Receptor

Cholera toxin (CT) and heat-labile enterotoxin (LT) cause gastroenteric disorders such as stomach ache and diarrhea (*8*). The toxins consist of one A subunit and five B

0097–6156/93/0528–0211$06.00/0

subunits; the A subunit activates adenylate cyclase in cells, leading to exclusion of cellular water. The B subunits are responsible for the attachment of the toxins to cells through the oligosaccharide portion of the receptors (9).

A well-known receptor of CT is ganglioside G_{M1}; its chemical structure is Gal-β-(1-3)GalNAc-β-(1-4) [NeuNAc-α-(2-3)]Gal-β-(1-4)Glc-β-(1-1) Cer (10). It is widely accepted that sialic acid at the terminal of the sugar moiety plays a key role in the binding of B subunits (11). CMP also has heterogeneous sugar chains containing sialic acids; its structures are NeuNAc-α-(2-3)Gal-β-(1-3)[NeuNAc-α-(2-6)] GalNAc, NeuNAc-α-(2-3)Gal β (1-3)GalNAc, and Gal-β-(1-3) [NeuNAc-α-(2-6)]GalNAc (12). A similarity in oligosaccharide structure exists between CMP and G_{M1}. These similar structures are Gal-β-(1-3)GalNAc and NeuNAc-α-(2-3)Gal. If CMP binds CT, then it could be hypothesized that the CT-CMP complex may no longer bind to the receptor on a target cell, that is, the toxicity of CT is eliminated. Thus, CMP may function as an inhibitor of CT and prevent the gastrointestinal disorders attributed to CT.

Inhibition by CMP of CT-Derived Morphological Change of Chinese Hamster Ovary-K1 Cell. Chinese hamster ovary (CHO)-K1 cells are frequently used to detect the toxicity of CT (13), since the toxin induces a morphological change in the CHO-K1 cell. As seen in Figure 1, the cell has originally spherical or ellipsoidal form (Figure 1a), but once infected with the toxin the cells shape changes to a spindle-like form (Figure 1b). When the mixture of CMP and CT was added to the cells, the morphological changes were effectively suppressed (Figure 1c,d). Using this bioassay system, the inhibitory activity of CMP for CT was evaluated (6). The assay was carried out according to Honda et al. (13). The extent of inhibition was estimated with the following equation,

Inhibition (%) = (1-A/B) x 100
where A is the number of cells morphologically changed in the presence of CMP, and B is the number of cells morphologically changed in the absence of CMP.

Table I summarizes the inhibitory effect of CMP and enzymatically digested CMP on the morphological change of CT-treated CHO-K1 cells. At 20 µg/ml, CMP effectively suppressed the CT-induced morphological change of CHO-K1 cells. Removal of sialic acid resulted in the complete loss of inhibitory activity. The results imply that the sialic acid moieties of CMP are essential for inhibition. The inhibitory activity remained to a small extent even after hydrolysis of CMP with pronase. Therefore, it seems that not only sialic acids residues but also the peptide portion of CMP contribute to the inhibition of morphological changes of CHO-K1 cells induced by CT.

Competitive Binding Assay. To evaluate the interaction between CMP and CT, a competitive binding assay using peroxidase-labeled cholera toxin B subunit (P-CT-B) was carried out. A modified the G_{M1}-ELISA method described by Svennerholm et al. (14) for the competitive binding assay was used. Figure 2 shows the concentration-dependent inhibition of CMP, its enzymatic (pronase and sialidase) digestion products and G_{M1}. G_{M1} showed nearly complete inhibition at concentrations higher than 0.2 nM. CMP inhibited the binding of P-CT-B to G_{M1} in a dose dependent manner; the maximum inhibition of CMP was rather low (about 50%) compared to G_{M1}. The lower affinity of CMP for CT could be attributable to its oligosaccharide structure.

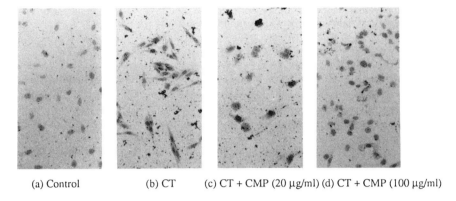

(a) Control   (b) CT   (c) CT + CMP (20 µg/ml) (d) CT + CMP (100 µg/ml)

Figure 1. Inhibitory Effect of CMP on CT-derived Morphological Change of CHO-K1 Cell
Forty µl of CT solution with CT concentration leading to 70 - 100 % morphological change in CHO-K1 cells was mixed with 50 µl of a sample solution. After preincubation at room temperature for 30 min, the mixture was added to the 50 µl of CHO-K1 cell suspension.
(Reproduced with permission from ref. 6. Copyright 1992 Japan Society for Bioscience, Biotechnology, and Agrochemistry)

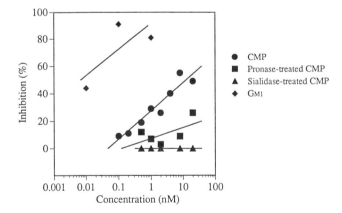

Figure 2. Competitive Binding Assay for CT
The mixture of sample solution and P-CT-B (1,000-fold diluted, List Biological Laboratories) was preincubated at room temperature for 30 min, and then added to a G_{M1}-coated microtiter plate (50 ng for each well). CT bound to G_{M1} was determined spectrophotometrically at 405 nm, after adding a solution of 2,2'-azino-bis-(3-ethylbenzothioazoline-6-sulfonic acid) as a substrate. A quantitative analysis of sialic acid using TBA method (25) indicated that sialidase-treatment completely eliminated sialic acids of CMP (6). The degree of hydrolysis with pronase-treated CMP was 71 % (6), which was measured by the TNBS method according to the NOVO (NOVO Industry) manual (26).
(Reproduced with permission from ref. 6. Copyright 1992 Japan Society for Bioscience, Biotechnology, and Agrochemistry)

Although terminal sialic acids are essential for the binding to CT, the sequence of the oligosaccharide chains also seems to establish the affinity for CT.

Table I. Inhibitory Effect of CT-derived Morphological Change of CHO-K1 Cells

Sample	Concentration (μg/ml)	Inhibition (%)
Control	-	0.0
CMP	20	81.0
	100	72.4
Sialidase-treated CMP*	1000	0.0
Pronase-treated CMP**	1000	24.1

*   Sialic acid was completely eliminated from CMP.
**   Degree of hydrolysis was 71%.

Contribution of Oligosaccharide Moiety and Polypeptide Portions to Inhibitory Activity. This study demonstrated clearly that CMP bound to CT. Removal of sialic acids completely suppressed the inhibitory activity, suggesting that the sialic acid seemed to be essential for the binding. Other glycoproteins containing sialic acid are able to interact with CT. For example glycophorin, the carbohydrate structure of which is identical to that of CMP, also inhibits the binding of the LT-B subunit to some extent (25%) (*15*). The binding mechanism between toxins and sugar sequences has been examined in detail using gangliosides with different oligosaccharide structures. It should be pointed out that sugar chains containing sialic acid are critical to the binding of several toxins (*16,17*). Ganglioside G_{M1} and G_{D1b} contain a partially common structural unit, Gal-β-(1-3)GalNAc-β-(1-4)[NeuNAc-α-(2-3)]Gal, that is highly specific for the complete binding of ganglioside to CT (*16*). CMP lacks this essential sequence, but contains a part of this sequence, NeuNAc-α-(2-3)Gal and Gal-β-(1-3)- GalNAc. Thus the oligosaccharide portions of CMP partially fits the receptor of CT.

The reduced inhibitory activity of CMP suggests that not only are the sialic acid residues but also the polypeptide portions or a limited chain length are responsible for the binding to CT. Shengrund *et al.* (*18*) described how polyvalent ligands bind to toxins more tightly than a monovalent ligand. It seems that CMP acts as a polyvalent ligand, because it contains two or more sugar chains containing sialic acid in a molecule. Therefore, it is not unreasonable to speculate that enzymatic digestion of CMP by pronase reduced its function as a polyvalent ligand and/or reduced the required length of the peptide chain, leading to the decrease of the inhibitory activity.

Preparation: pH-Dependent Molecular Weight Changes of GMP and Its Application to Large Scale Preparation.

Several methods for isolating CMP, which consists of glycosylated CMP (GMP) and non-glycosylated CMP, have been reported. CMP can be isolated with alcohol precipitation and ion-exchange chromatography from the supernatant of whey produced by heat coagulation of whey proteins (*19*). J.-M. Eustache (*20*) has isolated CMP from the supernatant of whey, by precipitation of its major proteins upon heating, and ultrafiltration through a UF module equipped with a membrane of MW cut off of 3,000 to retain CMP. The method seems reasonable because major impurities, such as whey proteins, are easily removed. The physical and nutritional properties of whey proteins coagulated upon heating are inferior to those of whey protein concentrates (WPC) produced by UF. Re-utilization of residual whey proteins is critical to reduce the production costs of CMP. A procedure for isolating CMP on pilot-plant scale was recently developed (*21*). The CMP was produced from rennet casein whey which was a by-product of manufacturing imitation cheese. In the course of the study, a pH dependent change in the turbidity of the CMP solution was found such that the CMP solution was turbid at neutral pH and became clear at acidic pH.

Morr and Seo (*22*) reported that the MW of CMP as determined by size exclusion chromatography was 33 kDa at neutral pH. This MW was considerably higher than the theoretical value of 9 kDa based on the amino acid sequence of CMP. These findings suggested that CMP formed oligomers in neutral solutions owing to its self association, and that the association was partially disrupted under acidic conditions.

Molecular Weights of CMP at Neutral and Acidic pH. The elution profiles of CMP obtained from whey on size exclusion chromatography are shown in Figure 3. Although the theoretical MW value of CMP was about 9 kDa, the retention time of major peak indicated that the apparent MWs of the main CMP fraction ranged from 20 to 50 kDa at pH 7.0; this was in agreement with Morr and Seo (*22*). On the contrary, at pH 3.5, five major peaks were detected with apparent MWs ranging from 10 to 30 kDa. The pH-dependent change in MW was completely reversible. Figure 4 shows the relationship between CMP-rejection (R_{CMP}) and pH of CMP solution during UF with either one of the two different UF-membranes GR61pp (MW 20 kDa cut off) or GR81pp (MW 8 kDa cut off). R_{CMP} was estimated with the following equation:

$$R_{CMP} = 1-(CMP \text{ content of permeate } / \text{ CMP content of retentate}).$$

More than 80% of the protein was rejected by GR81pp and GR61pp at pH ranging from 4.5 to 6.5, independent of MW cut off of membranes. At pH 3.5 R_{CMP} with GR61pp was 0.41, whereas much higher R_{CMP} was observed when the GR81pp membrane was used. Therefore, it is suggested that at pHs higher than 4.5 the CMP associates to form oligomers with a MW ranging from 20 to 50 kDa; on the other hand, at pH 3.5 the oligomers dissociated partially into smaller molecules. Since CMP is highly hydrophilic in nature, it is possible that CMP hydrates at pHs higher than 4.5 to form a molecule with a larger hydrodynamic effective volume. This hypothesis prompted the development of a novel procedure for isolating CMP. The procedure involved the use of UF and the control of the pH of whey.

CMP Isolation Procedure. Fifty L of 2% WPC solution or whey was used as raw material for isolating CMP. The procedure for isolating CMP by UF is shown in Figure 5. The pH of 2% WPC solution was adjusted to 3.5 with HCl and then passed through the UF membrane with a MW cut off of 50 kDa. The majority of whey proteins such as β-Lg, α-La, immunoglobulins and bovine serum albumin were

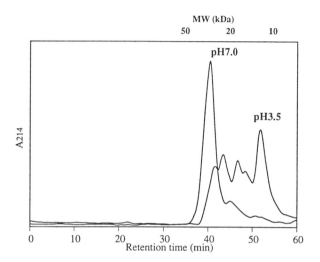

Figure 3. Size Exclusion Chromatography of CMP

Sample: 50 µl of sample solution (10 mg/ml)
Column: Superose 6 (12 x 30 min, Pharmacia)
Elution: 0.1 M Tris-HCl buffer (pH 7.0) or 0.1 M acetate buffer (pH 3.5)
Detection: UV (214 nm)

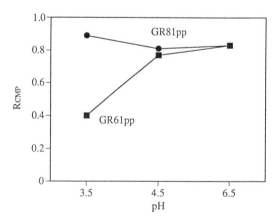

Figure 4. Relationship Between Rejection of CMP and pH

Sample: 0.2 % CMP solution (10 L)
UF-module: Lab-20 (DDS)
UF-membrane: GR61pp (MW 20 kDa cut off, 0.072 m2, DDS)
                      GR81pp (MW 8 kDa cut off, 0.072 m2, DDS)
Flow rate: 15 L/min
Temperature: 40 °C
Inlet-outlet pressure: 0.6-0.2 MPa

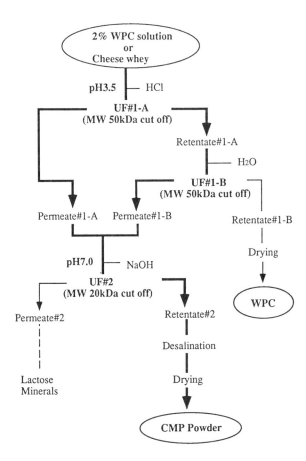

Figure 5. Isolation Procedure for CMP by Ultrafiltration

retained by the UF-membrane, whereas CMP, lactose and minerals were permeable (Permeate#1-A, 40 L). To recover the residual CMP in Retentate#1-A (10 L), 50 L of water was added to carry out continuous diafiltration. Permeate#1-A and #1-B were collected (90 L) and neutralized with NaOH. Because CMP associated in the neutral solution, the next UF step (UF#2) concentrated CMP into Retentate#2 (1 L); Permeate#2 contained only lactose and minerals. Retentate#2 was continuously desalinated by diafiltration and dried to obtain a CMP powder. Retentate#1-B was also desalinated and dried to re-use whey proteins as WPC.

Recovery, Purity and Amino Acid Composition of CMP. The elution profiles of the CMP powders obtained on size exclusion chromatography (23) are shown in Figure 6. Both CMP isolated from WPC and whey did not contain major

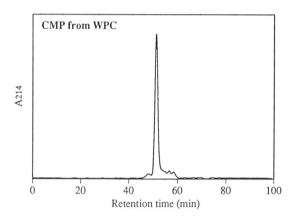

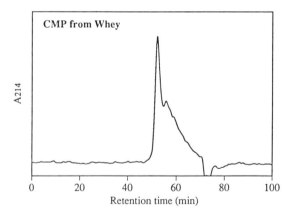

Figure 6. Size Exclusion Chromatography of CMP Obtained from WPC and Whey

Sample:	20 µl of sample solution (0.5 mg/ml)
Column:	Two coupled TSKgel G3000 PW$_{XL}$ (7.8 x 300 mm, Tosoh)
Elution:	55 % (v/v) acetonitrile containing 0.1 % (v/v) trifluoroacetic acid
Detection:	UV (214 nm)

whey proteins such as β-Lg, α-La and immunoglobulins. The purity of CMP obtained from WPC was high, whereas CMP from whey contained impurities which were probably proteose peptones. Table II summarizes the recovery and the purity of CMP. The recovery from whey was 100%, but contaminations of proteose peptones decreased the purity. Because the proteose peptones are mostly eliminated in WPC, higher purity is obtainable in CMP from WPC. The amino acid compositions of CMP listed in Table III were mostly similar to the values calculated from the amino acid sequence of CMP (*24*). CMP from whey has a very low level of Phe and a high content of branched chain amino acids. Further, CMP demonstrated biological functions such as inhibitory actions against oral microorganisms and toxins. CMP is a promising candidate for use in the ingredients of dietetic foods for phenylketonuria or cirrhosis patients, and in infant formula and weaning foods.

Table II. Recovery and Purity of CMP

Raw material	Recovery (%)	Purity (%)
WPC	63	81.0
Whey	100	44.3

Table III. Amino Acid Compositions of CMP

Amino acid*	Theoretical (%)**	Experimental (%)	
		From WPC	From whey
Asx	8.5	8.5	8.6
Thr	18.2	14.6	14.0
Ser	7.8	8.5	8.4
Glx	19.2	21.0	23.0
Pro	11.6	11.0	11.5
Gly	0.9	1.0	1.1
Ala	5.3	5.4	4.6
Val	8.9	7.7	7.3
Met	2.0	1.4	0.0
Ile	10.1	9.4	9.2
Leu	1.7	3.1	3.9
Tyr	0.0	0.3	0.0
Phe	0.0	0.6	0.9
Lys	5.7	5.7	5.6
His	0.0	0.5	0.5
Arg	0.0	1.0	1.0

*   Cys and Trp were not analyzed; they are not present in CMP (*24*).

**   The theoretical values were calculated, based on the primary structure of CMP (*24*).

Literature Cited

1. Mackinlay, A. G. and Wake, R. G. *Milk Protein II*; Mackenzie, H. A., Ed.; Academic Press: NY, **1971**, pp175-215.
2. Chernikov, M. P. and Stan, E. Ya *Physiological Activity of Products of Limited κ-Casein Proteolysis*; Book 2; XXI Intern. Dairy Congr.: Moscow,**1982**, pp161.
3. Stan, E. Ya, Groisman, S. D., Krasil'shchikov, K. B. and Chernikov, M. P. *Bull. Exp. Biol. Med.*, **1983**, *96*, 889.
4. Jollès, P., Levy-Toledano, S., Fiat, A.-M., Soria, C., Gilessen, P., Thomaidis, A., Dunn, F. W. and Casen, J. P., *Eur. J. Biochem.*, **1989**, *158*, 379.
5. Neeser, J.-R., Chambaz, A., Vedovo, S. D., Prigent, M.-J. and Guggenheim, B. *Infect. Immun.*, **1988**, *56*, 3201.
6. Kawasaki,Y., Isoda, H., Tanimoto, M., Dosako, S., Idota, T. and Ahiko, K. *Biosci. Biotech. Biochem.*, **1992**, *56*, 195.
7. Tanimoto, M., Kawasaki, Y., Shinmoto, H., Dosako, S., and Tomizawa, A. U. S. Pat., 5,075,424, **1991**.
8. Takeda, Y. *J. Clinic. Exp. Med.*, **1979**, *111*, 861.
9. Holmgren, J. *Nature*, **1981**, *292*, 413.
10. Heyningen, S. V. *Science*, **1974**, *183*, 656.
11. Holmgren, J., Elwing, H., Fredman, P. and Svennerholm, L. *Eur. J. Biochem.*, **1980**, *106*, 371.
12. Fournet, B., Fiat, A.-M., Alais, C., and Jollès, P. *Boichim. Biophys. Acta*, **1979**, *576*, 339.
13. Honda, T., Shimizu, M., Takeda, Y. and Miwatani, T. *Infect. Immun.*, **1976**, *14*, 1028.
14. Svennerholm, A.-M. and Holmgren, J. *Current Microbiol.*, **1978**, *1*, 19.
15. Sugii, S. and Tsuji, T. *FEMS Microbiol. Lett.*, **1990**, *66*, 45.
16. Fukuta, S., Magnani, J. L., Twiddy, E. M., Holmes, R. K. and Ginsburg, V. *Infect. Immun.*, **1988**, *56*, 1748.
17. Takamizawa, K. *Jpn. J. Dairy Food Sci.*, **1988**, *37*, A259.
18. Schengrund, C.-L. and Ringler, N. J. *J. Biol. Chem.*, **1989**, *264*, 13233.
19. Saito, T., Yamaji, A and Itoh, T. *J. Dairy Sci.*, **1991**, *74*, 2831.
20. Eustache, J.-M. U. S. Pat., 4,042,576, **1977**.
21. Tanimoto, M., Kawasaki, Y., Dosako, S., Ahiko, K. and Nakajima, I. *Biosci. Biotech. Biochem.*, **1992**, *56*, 140.
22. Morr, C. V. and Seo, A. *J. Food Sci.*, **1988**, *56*, 80.
23. Kawakami, H., Kawasaki, Y., Dosako, S., Tanimoto, M., and Nakajima, I. *Milchwissenschaft* (now printing).
24. Mercier, J. C., Bringnon, G. and Dumas, R. *Eur. J. Biochem.*, **1973**, *35*, 222.
25. Warren, L. *J. Biol. Chem.*, **1959**, *234*, 1971.
26. Novo Enzyme Information, November AF95/1-GB, **1971**.

RECEIVED December 1, 1992

Chapter 18

Potential Application of Some Synthetic Glycosides to Food Modification

K. Kinomura, S. Kitazawa, M. Okumura, and T. Sakakibara

Research Laboratories, Nippon Fine Chemical Company, Ltd., 5–1–1 Umei, Takasago-city, Hyogo 676, Japan

Alkyl glucosides and a novel glucoside containing a vinyl group as its aglycon have been synthesized by simple chemical reactions. The properties of these two kinds of products were investigated. They are found to be biodegradable and have low toxicity. The polymer properties of the vinyl type glucoside are also described.

Although the synthesis of glycosides having rather simple structures has been reported almost a century ago, few papers have mentioned potential applications of the glycosides as chemical materials. These compounds have previously been regarded as just a synthetic target during investigations of glycosylation. However, recent studies have suggested that glycosides offer an advantage as functional material. This paper will introduce some unique properties of a non-ionic surfactant and glucosyloxyethyl methacrylate (GEMA) (1), a novel vinyl monomer bearing a glucose residue. Their potential application to food modification will be discussed.

Properties of Alkyl Glucoside

Glycosides are carbohydrate derivatives which are usually hydrophilic, have low toxicity, and are biodegradable. For example, Octyl D-glucopyranoside (AG-8) is miscible in water at any concentration. Its toxicity to rats is lower than that of 1-octanol. Alkyl glucosides with more than a C6 alkyl chain length could serve as non-ionic surfactants because they have both hydrophobic alkyl groups and hydrophilic glucose moieties. These glycosides were first used as solubilizers of membrane proteins due to their property of minimally effecting the denaturation of protein (2-4). They are now applied more generally because of environmental and material safety reasons (5-6). The performance of alkyl glucosides in regard to their foaming ability, foam stability, and permeability is perhaps best represented by Decyl D-glucopyranoside (AG-10) (Table I). The stability of artificial additives is very important in food applications. According to thermogravimetric analyses, alkyl glucosides bearing from 6 to 18 even-carbon-numbered alkyl chains synthesized by Fischer alcoholysis (10), including direct coupling of 1-alcohol and glucose in the presence of acid catalyst, were decomposed and started to lose their weight at about 270°C under an air stream. There was no significant difference between heat stabilities of the compounds and their alkyl chain length. Their stability in aqueous solution was

Table I. Some Properties of Surfactants

	AG-8	AG-10	Sucrose Ester	LAS	SDS	Alkyl betain
foam height (mm)						
0 min	152	251	198	238	235	217
5 min	133	249	194	235	232	213
permeability (sec)	3 h	26	360	150	58	84
surface tention (dyn/cm)	43.6	28.7	33.1	30.7	35.5	32.2

Foam height: Loss-Miles (7) method 0.25% at 25°C, Permeability: Canvas disk method (8) 0.15% at 25°C, Surface tention: duNuoy method (9) at 20°C.

measured against sucrose dodecanoate, which is a chemically synthesized carbohydrate derivative approved by the Food and Drug Administration(FDA) as a food additive. AG-8 was not hydrolyzed with aqueous alkaline solution (pH 10.0) at 25°C in contrast to the sucrose ester being completely decomposed into fatty acid and sucrose after 8 h. At pH 2.0, acid hydrolysis was observed with both AG-8 and the sucrose ester. The rate of hydrolysis of AG-8 in aqueous acidic solution was much slower than that of the sucrose ester. After 8 h at 37°C, 5% of AG-8 was hydrolyzed compared to 31% for the sucrose ester.

Influence of Alkyl glucoside on Lipase Activity. A key feature of alkyl glucosides is that they possess no, or slight, harmful effects on biologically active proteins. The influence of surfactants on porcine lipase activity was investigated as a model to confirm this feature. Circular dichroism (CD) spectra were measured to determine whether a structural change occurred to the enzyme after incubation of the lipase in a buffered solution at pH 5.0 with and without surfactant at 37°C for an hour. In the case of AG-8 and sucrose dodecanoate (SE), no spectral changes were observed, while sodium dodecyl sulfate (SDS) affected the higher structure of the enzyme because a new negative Cotton effect (11) appeared at 209 nm (Figure 1). The enzyme activity was measured by hydrolysis of triheptyl glyceride as a substrate. Among the surfactants examined, AG-8 had a seemingly positive effect for enzyme activity (Figure 2). This was probably due to that the enzyme reaction occurring at the interface between the substrate and enzyme solution layer, and the surfactant's spread of the phase contact area by emulsion formation (12). The enzyme solution after several hours of incubation with AG-8 has no negative effect on the lipase activity.

A Novel Vinyl Monomer Bearing Carbohydrate Residue

Polymer bearing a carbohydrate residue on their side chain have been prepared in order to investigate their applicability as substances with high water-holding capacity and with pharmacological or biomedical materials (13-17). In our efforts to prepare such monomers, we synthesized several saccharide-bound monomers without any protective groups. The synthesis of this sort of monomers was carried out by coupling methyl α-D-glucopyranoside with a large excess of 2-hydroxyethyl methacrylate (2-HEMA) in the presence of 12-molybdophosphoric acid (PMo_{12}) and a polymerization inhibitor. A

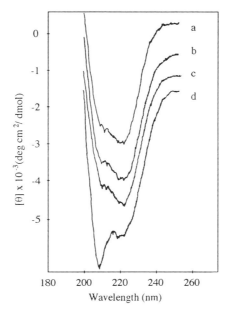

Figure 1. CD spectra of pocine lipase with and without surfactants. (a) without surfactant. (b) with octyl D-glucopyranoside (AG-8), 0.2%. (c) with sucrose dodecanoate (SE), 0.2%. (d) with sodium dodecyl sulfate (SDS), 0.2%.

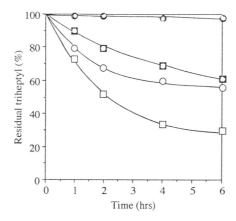

Figure 2. Lipase catalyzed hydrolysis of triheptyl in the absence and presence of surfactant at 35°C, pH 7.0: (○) control, (□) AG-8 at 0.2%, (●) SDS at 0.2%, and (■) Triton X-100 at 0.2%.

typical procedure for the preparation of GEMA is described below. Methyl α-D-glucopyranoside (44.7 g, 0.23 mol) was suspended in a mixture of 2-HEMA (300 g, 2.30 mol) and chlorobenzene (60 g) containing PMo_{12} (4.5 g, 2.50 x 10^{-3} mol) as a catalyst and 2,4-dinitrochlorobenzene (4.5 g) as an inhibitor. The suspension was then heated to 112°C and stirred and purged for 3 h with air. When thin layer chromatograms of the reacting material showed no change, the reaction mixture was cooled and then neutralized with sodium hydrogen carbonate. The crude products obtained by this method were purified by column chromatography over silica-gel with chloroform-methanol eluent (5:1 to 2:1 gradient elution). GEMA was obtained in 64% yield. The structure of these monomers were confirmed by ^{1}H and ^{13}C NMR and IR spectra. Additional confirmation was obtained by using the specific coloring reaction involving the phenol-sulfuric acid method, orcinol-sulfuric acid method, and bromine absorption (18-20).

Toxicity of Glucosyloxyethyl methacrylate (GEMA). Some toxicity studies of GEMA were performed, since the compound is expected to be used commercially. GEMA was decomposed completely and individually in 28 says by soil microorganisms that had been collected from 10 different sites in Japan. In the Ames test, GEMA showed no mutagenic activity towards five kinds of bacteria including *Esherichia coli* and *Salmonella typhimurium.* GEMA exhibited no antibacterial activity against the five bacteria. Oral acute toxicity test according to Organization for Economic Cooperation and Development (OECD) guideline revealed that GEMA did not affect the survival rate of Spraque-Dahlie (SD) rats at a dosage of 16,300 g/Kg. Compared with reported LD_{50} value of 2-HEMA (11,000 mg/Kg rat), GEMA was less toxic than its aglycon (21).

Polymerization of GEMA and the Properties of its Polymers

GEMA was readily polymerized at 60°C in aqueous solution containing ammonium persulfate as a radical initiator. The homopolymer thus prepared had an average molecular weight of 5 x 10^5 as determined by gel permeation chromatography. The polymer was soluble in water, N,N-dimethylformamide (DMF), and dimethyl sulfoxide (DMSO) but insoluble in other organic solvents, such as methanol, acetone, ether and chloroform. The glucose content in the polymer was reduced to about 1/5 in 6 days of digestion by soil bacteria at 28°C in buffered aqueous solution. GEMA was also copolymerized with other vinyl monomers, such as methyl methacrylate (MMA), acrylonitrile (AN), acrylamide (AAm), and styrene (St). To examine the monomer reactivity of GEMA with the other vinyl monomers in copolymer syntheses, GEMA and comonomers were polymerized in different proportions and the copolymers produced were separated by precipitating in acetone. Comparison of the reactivity of GEMA and MMA, indicated that they had almost the same reactivity, *i.e.*, the GEMA content in each poly(GEMA-co-MMA) were proportional to the GEMA contents in fed monomer mixtures (Figure 3). GEMA and MMA were randomly copolymerized. On the other hand, the curves obtained from the results of copolymerization of GEMA with AAm, AN, and St were shifted upward from the proportional line. This may due to the reactivity of GEMA to GEMA being greater than that of GEMA to the other three monomers. All 8 copolymers including poly(GEMA-co-MMA), poly(GEMA-co-AAm), poly(GEMA-co-AN), and poly(GEMA-co-St) comprising 40 mole% of GEMA and 60 mole% of GEMA respectively, were soluble in aprotic polar solvents such as DMF and DMSO; they were insoluble in water, methanol, and acetone. poly(GEMA-co-MMA) which was soluble in water was the exception. The other 6 copolymers swelled like a gel in the polar solvents.

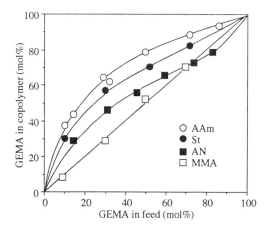

Figure 3. Copolymerization of glucosyloxyethyl methacrylate (GEMA) with other vinyl monomers. (○) with acrylamide (AAm), (●) with styrene (St), (■) with acrylonitrile (AN), and (□) with methyl methacrylate (MMA).

Surface Properties of GEMA Polymers. To identity whether glucose moieties exist on the surface of GEMA polymers, the surface free energy of poly(GEMA-co-MMA) at different molar proportion was calculated from the contact angle of a glycerol and iodomethane spot attached to their surface (Figure 4). Surface free energy, calculated according to the equations of Owens *et al.* (*22*), showed an increase from 45 to 52 erg/cm^2 at the GEMA increased. When the surface free energy was divided into dispersion and polar components, the polar force component increased continuously from 0 to 80 mole% of GEMA suggesting that the increase of surface free energy was due to the polar glucose moieties in GEMA. Additional support comes from electron spectroscopy for chemical analysis (ESCA) (*23*) spectra of poly(GEMA-co-St) surface. In the high binding energy region, a few new peaks at approximately 291 eV and 286 eV were observed. Intensities of each peak was related to GEMA content in the polymers. These two peaks were attributed to ethereal carbon and carbonyl carbon respectively, which were both present in GEMA molecule. Adhesion of a hydrophobic protein to the polymer surface was examined in an effort to determine GEMA application to biological systems. Human fibrinogen in 0.1M phosphate buffer were used to dip casted poly(GEMA-co-MMA) films in different monomer proportions. After 5 h at 37°C, the films were stained by bicinchoninic acid protein assay reagent (BCA protein assay kit, Wako Pure Chemical Co., Ltd.) and adhesive protein on their surface were determined (*24*) (Figure 5). The amount of adhesive fibrinogen correlated remarkably well with GEMA content in the polymers. The human fibrinogen did not adhere on the surface of poly(GEMA-co-MMA) if it contained more than 20 mole % of GEMA. The polymer can, therefore, be expected to be useful in biomedical material.

It may be concluded that glycosides obtained by the rather simple chemical reaction described above offer numerous biological applications. Table II summarizes some of the possible applications of these simple glycosides based on the properties mentioned in this report.

Table II. Potential Applications of Synthetic Glycosides

Aglycon	Applications
Alkyl groups	Foam entrainer of baking food Emulsifier Moisturizer
Viny group (monomer)	Protein modifier Deodorant
Vinyl group (polymer)	Hydro-gel, Water holding substance Biocompatible material

Although their safety *in vivo* needs further confirmation, at this present, they are basically low in toxicity and are biodegradable. These synthetic glycosides show tremendous potential for use in the foods.

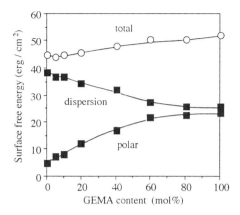

Figure 4. Effect of glucosyloxyethyl methacrylate (GEMA) content on the surface free energy of copolymer films including GEMA and methyl methacrylate (MMA).

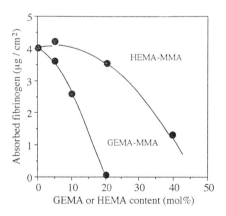

Figure 5. Amount of absorbed fibrinogen on copolymer surface including hydrophilic domain at 37°C. GEMA: glucosyloxyethyl methacrylate, HEMA: 2-hydroxyethyl methacrylate, MMA: methyl methacrylate, GEMA-MMA: poly(GEMA-co-MMA), and HEMA-MMA: poly(HEMA-co-MMA).

Literature Cited
(1) Kitazawa,S.; Okumura,M.;Kinomura,K.; Sakakibara,T. *Chem. Lettr.* **1990**, pp1733-1736.
(2) Spraque,S.G.; Camm,E.L.; Green,B.R.; Staehelin,L.A. *J. Cell. Biol.* **1985**,*100*, pp552-557.
(3) Bowes,J.M.; Stewart,A.C.; Bendall,D.S. *Biochim. Biophys. Acta.* **1983**,*725*, pp210-219.
(4) Knudsen,P.; Hubell,W.L. *Membr. Biochem.* **1978**,*1*,pp297-322.
(5) Roth,C.W.; Moser,K.B.; Bomball,W.A. U.S.Patent. **1980**, No.4223129.
(6) Liebowits,.; Mchugh,N.M. *U.S.Patent.* **1981**, No.4252656.
(7) Miles,G.D.; Ross,J. *J. Phys. Chem.* **1944**,*48*,pp280-290.
(8) American Society for Testing and Materials. *1980 Annual Book of ASTM Standards*; Part30; ASTM: Philadelphia, PA, 1980; pp184-186.
(9) *Surface and Colloid Science*; Matijevid ed.; Wiley-Interscience: New York, NY, 1969; vol. 1.
(10) Fischer, E. *Ber.* **1893**,*26*,p2400-.
(11) Bayley,P.M. *Amino-Acids, Peptides, Proteins.* **1972**,*4*,pp253-286.
(12) Gargouri,Y.; Julien,P.; Bois,A.G.; Verger,R.; Sarda,L. *J. Lipid. Res.* **1983**,*24*, pp1336-1342.
(13) Kawasaki,T.; Osaka,Y.; Yamaguchi,Y.; Ono,S. *Chem. Abstr.* **1984**,*101*, No.211648j.
(14) Ray-Chaundhui; Silip,K. *Chem. Abstr.* **1968**,*68*, No,40023v.
(15) Gruber,H. *Monatsh. Chem.* **1981**,*112*,pp273-285.
(16) Nishio,K; Nakaya,T.; Imoto,M. *Makromol. Chem.* **1974**,*175*,pp3319-3323.
(17) Carpino,L.A.; Ringsdorf,H.; Ritter,H. *Makromol. Chem.* **1796**,*177*,pp1631-1635.
(18) Dubois,M.; Gilles,J.; Hamilton,J.K.; Robers,P.A.; Smith,F. *Anal. Chem.* **1956**, *28*,pp350-356.
(19) Francois,C.; Marshall,R.D.;Neuberger,A. *Biochem. J.* **1962**,*83*,pp335-341.
(20) Critchfield,F.E. *Anal. Chem.* **1959**,*31*,pp1406-1408.
(21) Mir,G.N.; Laurence,W.H.; Autian,J. *J. Pharm. Sci.* **1973**,*62*,pp1258-1261.
(22) Owens,D.K.; Wendt,R.C. *J. Appl. Polym. Sci.* **1973**,*62*,pp335-341.
(23) Siegbahn,K. *Ann. Phys. (Paris).* **1968**,*3*,pp281-329.
(24) Smith,P.K.; Krohn,R.I.; Hermanson,G.T; Malia,A.K.; Gartner,F.H.; Provenzano,M.D.; Fujimoto,E.K.; Goeke,N.M.; Olson,B.J.; Klenk,D.C. *Anal. Biochem.* **1985**,*150*,pp76-85.

RECEIVED October 28, 1992

Chapter 19

Effect of Interdroplet Forces on Centrifugal Stability of Protein-Stabilized Concentrated Oil-in-Water Emulsions

Amardeep S. Rehill and Ganesan Narsimhan[1]

Department of Agricultural Engineering, Biochemical and Food Process Engineering, Purdue University, West Lafayette, IN 47907

A model for the centrifugal stability of protein stabilized concentrated oil-in-water emulsion is proposed. Interdroplet forces due to van der Waals, electrostatic and stearic interactions were evaluated. The proposed model is employed to predict the equilibrium profiles of continuous phase liquid holdup and film thickness for protein stabilized concentrated oil-in-water emulsion subjected to a centrifugal force field. The continuous phase liquid holdup as well as the film thickness and, consequently, the stability of emulsion were found to be higher for lower centrifugal accelerations, smaller emulsion drop sizes, higher protein concentrations, more favorable protein-solvent interactions and higher protein charge. The calculated profiles reflected jump transitions in film thickness as well as continuous phase liquid holdup as a result of centrifugal force exceeding the local maximum disjoining pressure.

Protein stabilized oil-in-water emulsions are found in various branches of food industry. These include milk, cream, ice-cream, mayonnaise, gravies and meat emulsions. These emulsions have extremely large interfacial area since oil is dispersed in the form of fine droplets. Consequently, formation of such emulsions usually results in an increase in the free energy. As a result, these emulsions are thermodynamically unstable. In other words, emulsions will eventually break due to coalescence of drops. As the individual droplets of the dispersed phase in an emulsion approach each other due to the relative motion brought about by Brownian and gravitational forces, they are attracted to each other by van der Waals forces. Without a stabilizer, the continuous phase between the two droplets drains, leading to their coalescence. Emulsifiers and stabilizers employed in food emulsions, however, slow down the overall rate of coalescence by providing an energy barrier to coalescence thereby increasing the shelf-life. This energy barrier is a result of repulsive forces between

[1]Corresponding author

0097–6156/93/0528–0229$06.00/0

emulsion droplets due to the adsorbed layer of stabilizers at the droplet surface. Caseinate, whey protein and egg protein are commonly employed to stabilize food emulsions. They tend to adsorb at the oil -in-water interface and modify the interparticle forces so as to provide kinetic stability to the emulsion. Many of the food emulsions are concentrated oil-in-water emulsions. Moreover, oil droplets flocculate due to creaming to form a concentrated cream layer in which the droplets are deformed and separated by thin films of continuous phase liquid. Interaction forces between droplets influence the drainage and stability of thin films which in turn determine the rates of coalescence between the droplets. It is, therefore, necessary to understand the nature of these interaction forces in order to predict long term stability or shelf life of food emulsions. Excellent reviews of protein stabilized emulsions have been written by Halling (1), Parker (2) and Narsimhan (3). In protein stabilized emulsions, the interaction forces between droplets depend on the surface concentration of adsorbed layer, pH, ionic strength, protein structure etc. The destabilizing influence of van der Waals attraction is overcome by repulsive interactions due to electrostatic and steric effects. Even though qualitative nature of these interactions is well known, very little quantitative information is available on these interactions in food emulsions. The centrifugation of emulsions has been used to obtain a measure of their stability (4) by observing the separation of coalesced phase as a function of time. Smith and Mitchell (5) proposed a method of calculating the maximum force between the droplets using centrifugation. Graham and Philips (6) investigated the effect of the structure of adsorbed protein layer on the stability of oil-in-water emulsion using centrifugation.

The objective of this paper is to illustrate the efficacy of inferring the interdroplet forces in a concentrated protein stabilized oil-in-water emulsion from the knowledge of the equilibrium profile of continuous phase liquid holdup (or, dispersed phase fraction) when the emulsion is subjected to a centrifugal force field. This is accomplished by demonstrating the sensitivity of continuous phase liquid holdup profile for concentrated oil-in-water emulsions of different interdroplet forces. A brief discussion of the structure of concentrated oil-in-water emulsion is presented in the next section. A model for centrifugal stability of concentrated emulsion is presented in the subsequent section. This is followed by the simulation of continuous phase liquid holdup profiles for concentrated oil-in-water emulsions for different centrifugal accelerations, protein concentrations, droplet sizes, pH, ionic strengths and the nature of protein-solvent interactions.

Structure of Concentrated Oil-in-Water Emulsion

Since the volume fraction of close packed spheres is 0.74, emulsion droplets in a concentrated emulsion are deformed whenever the dispersed phase volume fraction exceeds 0.74. The structure of such emulsions is very similar to polyhedral foams and has been well characterized (7). The emulsion droplets, on the average can be considered to be regular pentagonal dodecahedrons separated by thin films. A schematic of a dodecahedral droplet is shown in Figure 1. Every droplet is separated by twelve neighboring droplets as indicated by twelve faces of the polyhedron. Two neighboring droplets are separated by a thin film of continuous phase liquid. Each face of the polyhedron is the interface of thin film. Three adjacent thin films meet in a channel called plateau border. The continuous phase liquid is interconnected through a network of plateau borders. Schematic of the cross-section of the plateau border is also

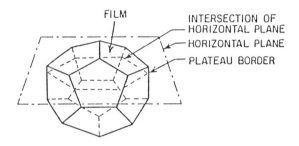

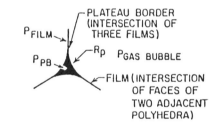

FIGURE 1.   Regular dodecahedral structure of a bubble and cross-sectional view of a plateau border.

shown in Figure 1. Since the film is planar, the pressure in the film is the same as that of an emulsion droplet. However, the pressure in the plateau border is less than that in the emulsion droplet by the capillary pressure due to the radius of curvature of the plateau border. This difference in the pressure is referred to as the plateau border suction. As a result of this suction, continuous phase liquid will flow from thin films to the neighboring plateau border. This will be followed by drainage of continuous phase through the network of plateau borders due to centrifugal force when the concentrated oil-in-water emulsion is centrifuged. In protein stabilized emulsions, the interface of thin film consists of interfacial adsorbed protein layer. When the thickness of draining film becomes of the order of a few hundred angstroms, the two phases of the film experience interaction forces due to intermolecular van der Waals, electrostatic and steric interactions. The draining film will eventually reach equilibrium when the repulsive interactions exactly counterbalance the plateau border suction. This repulsive interaction force per unit area of the film is usually referred as the 'disjoining pressure' since this force tends to 'disjoin' or pull the two faces of the film apart.

Model for Centrifugal Stability of protein Stabilized Concentrated Emulsions

Consider a concentrated emulsion consisting of equal size droplets. Even if the emulsion drops are not of the same size but are of narrow size distribution, there would be negligible segregation of drops according to size (due to creaming) since the emulsion is highly concentrated. Consequently, one can expect the drop size distribution to be independent of height. Since the continuous phase liquid holdup varies along the direction of centrifugal force, the emulsion can be assumed to be uniform across any cross section (perpendicular to the centrifugal force). Let the z coordinates be opposite to the direction of the centrifugal force (bottom to top) with $z = 0$ referring to the interface between the emulsion layer and the bottom continuous phase layer. At equilibrium, the continuous phase in the plateau border channels, being in hydrostatic equilibrium, would satisfy the equality (8),

$$\rho a_c + \frac{dp}{dz} = 0 \qquad [1]$$

where ρ is the density of the continuous phase, a_c is the centrifugal acceleration and is equal to $\omega^2 L$, where ω and L are the angular velocity of the centrifuge and the distance from the axis of rotation, respectively, and p is the pressure within the plateau border.

The pressure within the plateau border can be related to the pressure in the dispersed phase p_d and the plateau border suction p_c through

$$p = p_d + p_c \qquad [2]$$

where

$$\frac{dp_d}{dz} = -\rho_d a_c \qquad [3]$$

ρ_d being the density of the dispersed phase and

$$p_c = -\frac{\sigma}{r}$$ [4]

where r is the radius of curvature of the plateau border and σ is the interfacial tension.

$$\sigma = \sigma_o - \pi$$ [5]

σ_o and π being the interfacial tension between oil and water, and the surface pressure due to the adsorption of proteins and emulsifier, respectively. Combining equations 1-4, we obtain,

$$(\rho - \rho_d)a_c - \sigma\frac{d}{dz}(\frac{1}{r}) = 0$$ [6]

The radius of curvature of plateau border can be related to the area of the plateau border a_p and the film thickness x_F for a regular dodecahedral arrangement from geometric considerations by (9),

$$r = \frac{-1.732x_F + \left[(1.732x_F{}^2 - 0.644(0.433x_F - a_p)\right]^{1/2}}{0.322}$$ [7]

As pointed out earlier, the continuous phase liquid drains from thin films to plateau border due to plateau border suction. When the thickness of the films becomes of the order of a few hundred angstroms, the intermolecular forces due to van der Waals, double layer and steric interactions give rise to a disjoining pressure Π (defined as the excess pressure that tends to 'disjoin' or pull the two faces of the film apart) which counteracts the capillary pressure. Of course, this disjoining pressure is a strong function of film thickness. The variation of Π with x_F will depend upon the characteristics of interfacial protein adsorbed layer, pH, ionic strength, and concentration of adsorbed protein segments. The film will drain until the plateau border suction is exactly counterbalanced by the disjoining presure after which the film reaches equilibrium. At equilibrium,

$$\frac{\sigma}{r} = \Pi(x_F)$$ [8]

As pointed out earlier, the disjoining pressure is the sum of interdroplet forces due to van der Waals, electrostatic and steric interactions. Detailed discussion of the nature of these interactions and their effect on the disjoining pressure can be found elsewhere (3,8). Only the final expressions for the contributions of different interactions to the disjoining pressure are given below.

The disjoining pressure Π_{VW} due to van der Waals interactions is given by (8),

$$\Pi_{VW} = -\frac{A_{232}}{6\pi}\left[\frac{1}{(x_F - 2L_s)^3} - \frac{2}{(x_F - L_s)^3} + \frac{1}{x_F^3}\right]$$

$$-\frac{A_{132}}{6\pi}\left[\frac{2}{(x_F - L_s)^3} - \frac{2}{x_F^3}\right] - \frac{A_{131}}{6\pi x_F^3} \qquad [9]$$

when the film thickness $x_F \geq 2L_S$, L_S being the thickness of the adsorbed protein layer. In the above equation, the subscripts 1, 2 and 3 refer to the oil, protein layer and water respectively. The effective Hamaker constant A_{ijk} for the interaction between i and k through the intervening medium j is given by,

$$A_{ijk} = (A_{ii}^{1/2} - A_{jj}^{1/2})(A_{kk}^{1/2} - A_{jj}^{1/2}), \qquad [10]$$

A_{ii} being the Hamaker constant for the interaction of i through vacuum. When the film thickness $x_F \leq 2L_S$, the disjoining pressure Π_{VW} is given by

$$\Pi_{VW} = -\frac{A_{121}}{6\pi x_F^3} \qquad [11]$$

The disjoining pressure Π_{el} due to electrostatic interaction is given by (8),

$$\Pi_{el}(h) = 16nkT[\gamma^2 A_1^2 + 2\gamma^4 A_1(A_2 + A_1^3) + \gamma^6(2A_1 A_3$$

$$+ 8A_1^3 A_2 + 3A_1^6 + A_2^2)] \qquad [12]$$

where

$$A_1 = \frac{1}{\cosh(\kappa h/2)}, \qquad [13]$$

$$A_2 = -\frac{(\kappa h/2)\tanh(\kappa h/2)}{\cosh^3(\kappa h/2)} \qquad [14]$$

$$A_3 = \frac{A_1 - A_1^3}{4\cosh^2(\kappa h/2)} + \frac{3A_2}{4\cosh^2(\kappa h/2)}\left[1 - 4(\frac{\kappa h}{2})\tanh(\frac{\kappa h}{2})\right]$$

$$-\frac{A_1}{2\cosh^4(\kappa h/2)}\left[(\frac{\kappa h}{2})^2\right] \qquad [15]$$

$$\gamma = \tanh(y_s/4), \qquad [16]$$

$$y_s = ze\psi_s/kT \qquad [17]$$

In the above equations, h is the film thickness, n is the number concentration of $z:z$ symmetrical electrolyte and ψ_s is the surface potential. The surface potential ψ_s is the potential at the interface of stern and diffuse layers and is usually replaced by the zeta potential of the droplet determined from electrophoretic measurements. When the interface has an adsorbed layer of globular proteins, it may be reasonable to assume that the shear plane is located at the interface of protein layer. When $x_F > 2L_s$ the disjoining pressure Π_{el} can be evaluated by replacing ψ_s with ζ potential and taking h as $(x_F - 2L_s)$.

When the thickness of the draining film is less than twice the thickness of the adsorbed protein layer, (i.e. $2L_S$), the approaching faces of the film experience a steric interaction because of the overlap of the adsorbed protein layers.

When $L_S \leq x_F \leq 2L_S$, the disjoining pressure Π_{st} due to steric interaction is given by (3),

$$\Pi_{st} = \frac{2kT\,(0.5-\chi)\,\Gamma^2}{\overline{V}_1\,\rho_2{}^2\,L_S{}^2} \qquad [18]$$

where Γ is the surface concentration of adsorbed protein, k is the Boltzmann's constant, T is the temperature, χ is the Flory-Huggins parameter, $\overline{V}_1$ is the molar volume of the solvent and ρ_2 is the density of the adsorbed segments of protein.

When $x_F \leq L_S$, Π_{st} is given by (3),

$$\Pi_{st} = \frac{2kT\,(0.5-\chi)\,\Gamma^2}{\overline{V}_1{}^2\,\rho_2{}^2\,L_S{}^2} + \frac{2kt\nu}{x_F}\,, \qquad [19]$$

where ν is the number of adsorbed protein molecules per unit area.

The total disjoining pressure Π is the sum of the contributions due to van der Waals, double layer and steric interactions, i.e.

$$\Pi = \Pi_{VW} + \Pi_{el} + \Pi_{st} \qquad [20]$$

For a dodecahedral arrangement, the continuous phase liquid holdup ε is given by (9),

$$\varepsilon = Nn_F A_F x_F + Nn_p a_p l, \qquad [21]$$

where $N = (1-\varepsilon)/v$, v being the volume of the droplet is the number of droplets per unit volume, n_F (=6) and n_P (=10) are the number of films and plateau borders per droplet, A_F is the area of the film and l is the length of the plateau border.

From equation 6, one obtains,

$$\Pi(x_F) = \frac{\sigma}{r} = \frac{\sigma}{r_o} + (\rho - \rho_d)a_c z, \qquad [22]$$

where r_o is the radius of curvature of the plateau border at the interface of the emulsion and the continuous phase. If the emulsion drops are sufficiently small, then the capillary pressure inside a (spherical) droplet would far exceed the variation of hydrostatic pressure over a drop diameter. This would be valid if the emulsion drop radius R satisfies the following inequality,

$$R << \left[\frac{\sigma}{(\rho - \rho_d)a_c} \right]^{1/2} \qquad [23]$$

In such a case, the drops at the interface of emulsion and the continuous phase layer ($z = 0$) would essentially be spherical (10) so that r_o can be taken to be the radius of the droplet R. Therefore,

$$\Pi(x_F) = \frac{\sigma}{r} = \frac{\sigma}{R} + (\rho - \rho_d)a_c z, \qquad [24]$$

From the above equation, the variation of equilibrium disjoining pressure and the radius of curvature of plateau border with position for a concentrated emulsion can be obtained. If the polarizabilities of the oil, water and the adsorbed protein layer (the effective Hamaker constants), the net charge of protein molecule, ionic strength, protein-solvent interaction and the thickness of the adsorbed protein layer are known, the disjoining pressure $\Pi(x_F)$ can be related to the film thickness using equations 9 - 20. The variation of equilibrium film thickness with position in the emulsion can then be calculated. From the knowledge of r and x_F, the variation of cross sectional area of plateau border a_p and the continuous phase liquid holdup ε with position can then be calculated using equations 7 and 21 respectively. The results of such calculations for different parameters are presented in the next session.

Alternatively, experimental measurement of the variation of equilibrium continuous liquid holdup with position for a concentrated oil-in-water emulsion can be employed to infer the variation of disjoining pressure with film thickness. Since the continuous phase liquid holdup ε is known as a function of position, x_F, a_p and r can be calculated using equations 7, 21 and 24. Equation 24 will then yield the disjoining pressure Π at the film thickness x_F.

The overall disjoining pressure is the sum of the contributions from van der Waals, double layer and steric interactions. The contribution to the overall disjoining pressure from van der Waals attraction is negative whereas the contributions from the double layer and steric repulsions are positive. Consequently, the overall disjoining pressure can take both positive and negative values depending on the film thickness. The disjoining pressure would be positive for film thickness of the order of a few hundred angstroms as well as for very small film thickness (smaller than hundred angstrom) because of the predominant effects of double layer and steric repulsions respectively. For intermediate film thickness, however, van der Waals attraction would predominate so as to make the disjoining pressure negative. As can be seen from equation 24, the equilibrium film thickness decreases (corresponding to an increase in Π) away from the interface. This continuous decrease in the film thickness proceeds until Π reaches a local maximum at which height there would be a jump transition to very small thickness. Such a jump transition has been observed experimentally (5) for oil-in-water emulsions. For sufficiently large centrifugal speeds, the centrifugal force would exceed the maximum disjoining pressure at some height above which the thin films would collapse resulting in the coalescence of emulsion drops thus leading to the formation of a separate oil phase at the top. The maximum disjoining pressure Π_{max} is therefore given by,

$$\Pi_{max} \approx (\rho - \rho_d)a_c h,$$
[25]

where h is the height of the emulsion layer sandwiched between oil layer on top and continuous phase (water) layer at the bottom when the oil-in-water emulsion is subjected to a centrifugal acceleration a_c.

In a previous paper (8), Π_{max} was inferred for corn oil-in-water as well as toluene-in-water emulsions stabilized by bovine serum albumin (BSA). The effects of pH, ionic strength and BSA concentration on Π_{max} were investigated. Comparison of experimental maximum disjoining pressure with predicted Π_{max} indicated that steric interaction is the predominant mechanism of stabilization in such systems.

Simulation of Equilibrium Profiles of Film Thickness and Continuous Phase Liquid Holdup

As pointed out earlier, it is possible to infer the variation of interdroplet forces with film thickness in a concentrated oil-in-water emulsion from the experimental measurement of equilibrium profile of continuous phase liquid holdup (i.e. the variation of continuous phase liquid holdup with height) when subjected to a centrifugal force field. In order to investigate the sensitivity of the continuous phase liquid holdup profile (CPLHP) to the interdroplet forces, continuous phase liquid holdup profiles were evaluated for different centrifugal accelerations, protein concentrations, droplet-sizes, pH, ionic strengths and the nature of protein-solvent interactions. All calculations were performed for oil-in-water emulsion stabilized by BSA. The Hamaker constants for oil, protein layer and water were taken to be 1.05×10^{-20} J, 8.807×10^{-21} J and 2.43×10^{-20} J respectively. Π and r at different positions z were calculated using equation 24. The film thickness x_F was then evaluated using equations 9 - 20. a_p was calculated from equation 7. The continuous phase liquid holdup ε was then evaluated using equation 21. CPHLP for three different centrifugal accelerations is shown in Figure 2. The liquid holdup profile is steeper at higher centrifugal force fields. In other words, more drainage of continuous phase liquid occurs at higher centrifugal force fields. As can be seen from equation 24, the radius of curvature of plateau border r is smaller and the disjoining pressure Π is larger at higher centrifugal force fields. As a result, x_F and a_p are smaller at higher a_c thus leading to smaller values of ε. As pointed out earlier, the thin film exhibits jump transition whenever the force due to centrifugal acceleration exceeds the local maxima in the disjoining pressure. Such a jump transition in the liquid holdup can be observed in Figure 2. Similar jump transitions in film thickness can be seen in Figure 3. When the film thickness becomes equal to L_S , the thickness of the adsorbed protein layer, the steric repulsion would increase as the film thickness decreases because of the elastic

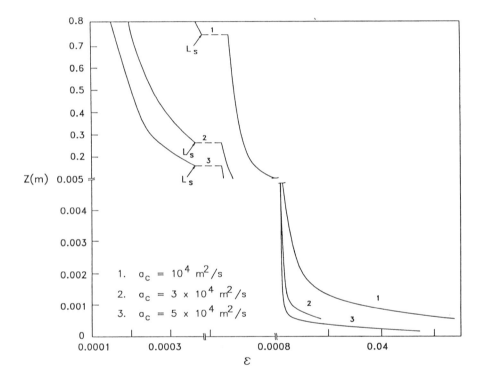

FIGURE 2.   Continuous phase liquid holdup profiles (CPLHP) for different
centrifugal accelerations for droplet size $R = 5x10^{-5}m$, surface
concentration $\Gamma = 5x10^{-6}kg/m^2$,   ionic   strength
$m = 0.1M$, thickness of adsorbed protein layer $L_s = 12x10^{-9}m$
and zeta potential $\zeta = 12mV$.

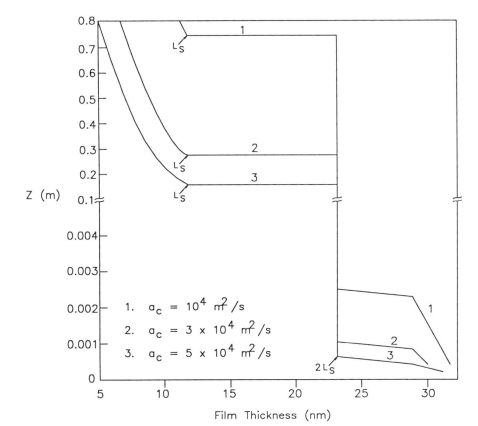

FIGURE 3. Variation of film thickness with emulsion height for different centrifugal accelerations for droplet size $R = 50x10^{-6}m$, surface concentration $\Gamma = 5x10^{-6}kg/m^2$, ionic strength $m = 0.1M$, thickness of adsorbed protein layer $L_s = 12x10^{-9}m$ and zeta potential $\zeta = 12mV$.

free energy due to the compression of the adsorbed protein layer [see equation 19]. From Figure 3, it can be seen that the film thickness reaches L_S at a shorter height for larger centrifugal accelerations. The effect of droplet size on CPLHP is shown in Figure 4. The continuous phase liquid holdup profile is found to be steeper with smaller values of ε for larger droplet sizes. For larger droplets, $\Pi(x_F)$ will be smaller and r will be larger [see equation 24]. As a result, x_F and a_p both will be larger. However, this increase in the film thickness and plateau border cross sectional area is offset by the decrease in the number of droplets per unit volume N so that ε decreases [see equation 21]. The effect of surface concentration of protein on CPLHP is shown in Figure 5. In these calculations it is assumed that the ζ potential is unaffected so that the surface concentration influences only the steric interaction. Consequently, the profile is identical until the steric interactions are present beyond which the liquid holdup is higher for larger values of Γ. This increase in liquid holdup is due to the fact that the equilibrium films are thicker at higher protein concentrations as a result of stronger steric repulsion. The effect of protein-solvent interaction (Flory Huggins χ parameter) on CPLHP is shown in Figure 6. For more favorable protein-solvent interactions (smaller χ), steric repulsion is higher so that the film thickness and consequently the liquid holdup are larger. The effects of ζ potential and ionic strengths on CPLHP are shown in Figures 7 and 8 respectively. Higher values of ζ potential (higher protein charge) and lower ionic strengths resulted in larger film thickness as well as liquid holdup because of stronger electrical double layer repulsions.

Conclusions

A model for the centrifugal stability of protein stabilized concentrated oil-in-water emulsion is proposed. This model could be employed to infer interdroplet forces in concentrated oil-in-water emulsions from the experimental measurement of profile of continuous phase liquid holdup (or, equivalently, dispersed phase fraction) in such systems. Interdroplet forces due to van der Waals, electrostatic and steric interactions were accounted for in the evaluation of disjoining pressure between the two faces of thin film. This model is employed to predict the equilibrium profiles of continuous phase liquid holdup and film thickness for different centrifugal accelerations, protein concentrations, droplet sizes, pH, ionic strength and the nature of protein-solvent interactions. The continuous phase liquid holdup as well as film thickness were found to be higher for lower centrifugal accelerations, smaller emulsion drop sizes, higher protein concentrations, more favorable protein-solvent interactions and higher surface potentials. Consequently, the concentrated oil-in-water emulsion was found to be more stable at lower centrifugal accelerations, smaller emulsion droplets, higher protein concentrations, more favorable protein-solvent interactions and higher protein charge. The calculated profiles reflected jump transitions in film thickness as well as continuous phase liquid holdup as a result of centrifugal force exceeding the local maximum disjoining pressure.

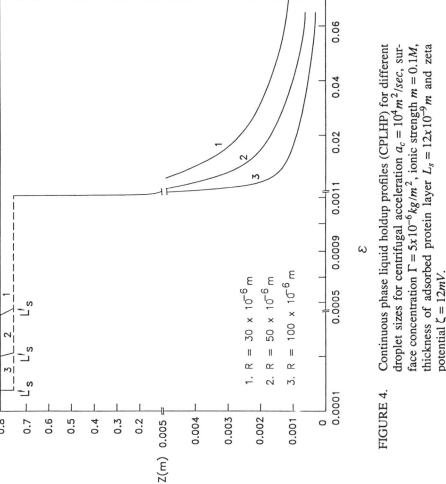

FIGURE 4.   Continuous phase liquid holdup profiles (CPLHP) for different droplet sizes for centrifugal acceleration $a_c = 10^4 m^2/sec$, surface concentration $\Gamma = 5x10^{-6} kg/m^2$, ionic strength $m = 0.1M$, thickness of adsorbed protein layer $L_s = 12x10^{-9} m$ and zeta potential $\zeta = 12mV$.

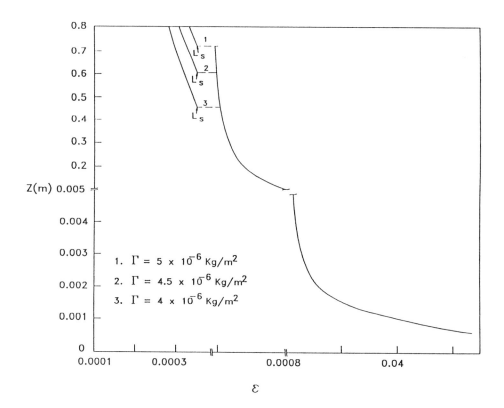

FIGURE 5.   Continuous phase liquid holdup profiles (CPLHP) for different
surface concentrations for droplet size $R = 50x10^{-6}m$, centrifugal   acceleration $a_c = 10^4 m^2/sec$,   ionic   strength
$m = 0.1M$, thickness of adsorbed protein layer $L_s = 12x10^{-9}m$
and zeta potential $\zeta = 12mV$.

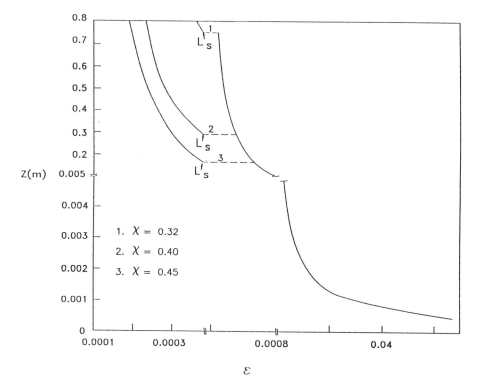

FIGURE 6.   Continuous phase liquid holdup profiles (CPLHP) for different χ parameters for droplet size $R = 50 \times 10^{-6} m$, centrifugal acceleration $a_c = 10^4 m^2/sec$, ionic strength $m = 0.1M$, thickness of adsorbed protein layer $L_s = 12 \times 10^{-9} m$ and zeta potential $\zeta = 12 mV$.

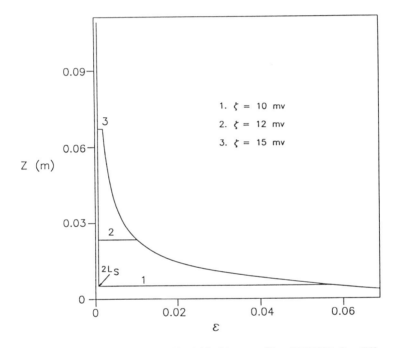

FIGURE 7.   Continuous phase liquid holdup profiles (CPLHP) for different
zeta potentials for droplet size $R = 50 \times 10^{-6} m$, centrifugal
acceleration $a_c = 10^4 m^2/sec$, ionic strength $m = 0.1M$, thickness of adsorbed protein layer $L_s = 12 \times 10^{-9} m$ and surface concentration $\Gamma = 5 \times 10^{-6} kg/m^2$.

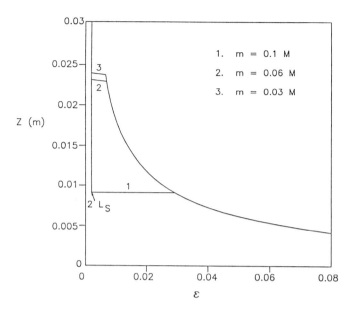

FIGURE 8.   Continuous phase liquid holdup profiles (CPLHP) for different ionic strengths for droplet size $R = 50x10^{-6}m$, centrifugal acceleration $a_c = 10^4 m^2/sec$, surface concentration $\Gamma = 5x10^{-6} kg/m^2$, thickness of adsorbed protein layer $L_s = 12x10^{-9}m$ and zeta potential $\zeta = 12mV$.

Literature Cited

1. Halling, P.J., Protein Stabilized Foams and Emulsions, *CRC Critical Reviews in Food Science And Nutrition*, 155 (1981).
2. Parker, N.S., Properties and functions of Stabilizing Agents in Food Emulsions, *CRC Critical Review in Food Science and Nutrition*, **25**,285 (1987)
3. Narsimhan, G."Emulsions" in *Physical Chemistry of Foods*, ed. Schwartzberg, H.G. and Hartell, R.W., Marcel-Dekker Inc. (1992).
4. Vold, R.D., Mittal, K.L. and Hahn, A.U.,"Ultracentrifugation Stability of Emulsions", *Surface and Colloid Sci.*, **10**, 45, Ed. E. Matijevic, Plenum Press, New York (1978).
5. Smith, A.L. and Mitchell, D.P.,"Centrifugation of Emulsions" in *Theory and Practice of Emulsion Technology*, Ed., Smith, A.L., Academic Press, New York (1976).
6. Graham, D.E. and Philips, M.C.,"The Conformation of Proteins at Interfaces and their Role in Stabilizing Emulsions", in *Theory and Practice of Emulsion Technology*, ed. Smith, A.L., Academic Press, New York (1976)..
7. Bikerman, J.J.,"*Foams*", Springer-Verleg, New York (1973).
8. Narsimhan, G., *Colloids and Surfaces*, **62**, 41 (1992).
9. Desai, D. and Kumar, R., *Chem. Eng. Sci.*, **37**, 1361 (1982).
10. Princen, H.M.,*Langmuir*, **2**, 519 (1986).

RECEIVED December 30, 1992

FOOD MICROBIOLOGY AND SAFETY ISSUES

Chapter 20

Determination of Shrimp Freshness Using Impedance Technology

Patti L. Wiese-Lehigh and Douglas L. Marshall[1]

Department of Food Science, Louisiana Agricultural Experiment Station, Louisiana State University Agricultural Center, Baton Rouge, LA 70803

This paper reviews seafood freshness determination methods and presents evidence showing that impedance can determine shrimp freshness. Seafood freshness detection methods include subjective sensory tests or objective chemical or microbiological tests. Most require raw products and are time consuming. Thus, a rapid method useful for both raw and cooked seafood is needed. The present study used impedance technology to rapidly measure shrimp freshness. Raw or thermally processed Gulf coast shrimp were stored at different temperatures and sampled periodically. Sampling consisted of removing shrimp from storage and homogenizing in deionized distilled water prior to transfer to an impedance measuring apparatus. Homogenate impedance values were determined in 30 min and showed increasing readings with increasing spoilage. Impedance was correlated with sensory tests and was affected by salt concentration.

The seafood industry is acutely aware of the need for maintaining product quality. There are changes in seafood that occur not only during but also before and after processing, especially during cold storage. Many of these changes affect nutrient content, texture, flavor, and color of seafoods. Each of these factors influence purchasing decisions by consumers (1).

Because of the global nature of the industry, uniform standards for seafood freshness are needed. The U.S. imports large quantities of shrimp from other nations. Some of these countries have advanced technology, while others are fairly primitive in operation and troublesome both with quality and health-related matters (2). Thus, quality standards may be difficult to implement on a worldwide basis.

---

[1]Corresponding author

Currently available methods for determining seafood freshness have been reviewed (*2,3,4,5*). In general, sensory and microbiological quality of seafoods are the standards most often utilized for determining freshness. There are, however, several limitations with these methods such as the need for trained technicians, problems with subjectivity, or long analysis times. Consequently, several new methods have been proposed using chemical indices of freshness that attempt to overcome these difficulties. The purpose of the present paper is to first review these methods with particular emphasis on shrimp freshness determinations and, secondly, to present experimental evidence showing that impedance can be used to denote shrimp freshness.

Freshness Detection Methods

Sensory. Sensory analysis is one of the oldest accepted methods for judging seafood quality and freshness. Although fairly accurate, it unfortunately requires trained, experienced panelists to attain accuracy. Perez-Villarreal (*6*) used evaluators experienced in sensory analysis of fish to judge freshness and quality. Results of the study showed sensory scores were better correlated with microbiological analyses, especially total viable counts, than with chemical analyses such as trimethylamine or adenine nucleotide decomposition (*6*).

Seafood companies that evaluate Gulf of Mexico shrimp normally use sensory evaluation, with no established standard, only the experience of the evaluator. The assessment includes odor, degree of melanosis (the darkening of tail segments called blackspot), color of the head fat (white or brown is freshest while yellow or orange indicates less fresh), and finally, head attachment (unfresh shrimp heads detach more readily) (Lusco, T., Lousiana Seafood Exchange, personal communications, 1991). Shrimp melanosis is caused by inherent enzymes that increasingly produce insoluble pigments during storage. Shrimp acceptability decreases as the browning of the surface increases (*7*). Chemical treatments for blackspot such as sulfite or resorcinol dips can stop melanosis or bleach prior darkening, thus masking spoilage (*7*).

Studies indicated that identification of spoiled seafood products by sensory evaluation is due to detection of putrefied odor associated with spoilage (*8*). Rapid detection of spoilage with sensory evaluation results from microbial action due temperature abuse or high initial levels of microorganisms (*9*).

Sensory methods have the ability to detect spoilage of either cooked or raw products. Because of this advantage, sensory methods are used as standards by agencies such as the U.S. Food and Drug Administration (FDA) for regulatory purposes. The shortcomings of such procedures are the subjective nature of the evaluations, which typically causes greater deviation than objective evaluations, and the need for trained evaluators.

Microbiological. Numerous factors affect how long seafood can be stored under refrigeration prior to spoilage. These factors can include the numbers and types of microorganisms present, the ability of the animal's "skin" to provide a protective barrier against microorganisms, the types of muscle found in fish and shellfish, and the activity of endogenous antimicrobial enzymes (10).

Many spoilage microorganisms are credited with the offensive putrefactive odor characteristic of seafood spoilage. Utilization of microbiological analysis to indicate the degree of spoilage is a frequently used method (8) because of the association of increasing microbial numbers with decreasing sensory quality. Organisms most prevalent in shrimp spoilage are Pseudomonas and Moraxella/Acinetobacter species (11,12). Because these organisms are Gram negative, mostly psychrophilic, and naturally present in cold waters, they will predominate on shrimp from these waters. For example, Alaskan Pandalid shrimp species spoil much faster than warm water shrimp possibly due to this "preinoculation" (13). Studies found that if the storage temperature was above refrigeration, spoilage bacteria shift from Pseudomonas to Moraxella, with higher temperatures favoring Proteus (13).

One advantage of microbial techniques is their ability to identify specific microorganisms that produce off odors and flavors. An individual organism can be studied to determine its singular ability to produce proteases and/or lipases that are considered primary causative agents of unacceptable flavor and odor (9). Another advantage is a strong correlation between microbial numbers and types with sensory evaluation. For example, when the numbers of spoilage bacteria such as Pseudomonas reach levels of approximately one million per gram or greater, spoilage can begin to be detected by sensory analysis (6). There are also disadvantages associated with microbial analyses. A major disadvantage is the need for trained technicians to prepare media and reagents, interpret data, and disposal. Another disadvantage is that thermal processing will destroy microorganisms (11,18), thus, negative results for cooked seafoods become meaningless if the raw product was spoiled prior to processing.

Chemical. Chemical indicators used for evaluating freshness include total volatile acids (TVA), total volatile bases (TVB), total volatile base nitrogen (TVBN), cadaverine, putrescine, and histamine. Of these, TVA and TVB are considered the best indicators of seafood decomposition (3,6,8). In most cases, spoilage was evaluated over periods greater than two weeks to allow for the accumulation of compounds to concentrations high enough to attain acceptable readings (3). Because of low sensitivity during early stages of spoilage, these methods are usually inadequate for the seafood industry due to short shelf lives of products.

Another chemical method for measuring freshness, that is more rapid, continuous, and less destructive than other methods is the detection of volatile trimethylamine (TMA), dimethylamine (DMA), monomethylamine (MMA), and ammonia (14,15). Trimethylamine oxide (TMAO) is a decomposition product of proteins as well as present in excretions of fish (16). Spoilage bacteria can reduce TMAO to TMA plus small amounts of DMA, MMA, and ammonia. Tissue TMA levels have been correlated with the pungent odor associated with spoiled seafood as well as total bacterial counts (14). Researchers incorporated a test strip (containing TMA dehydrogenase) for the semiquantitive determination of TMA in fish that utilized color comparisons between a reference of excellent quality and a sample. This was rapid (5 min) and correlated well with spectrophotometric procedures (*14*). Another method to detect the products of TMAO action is a gas sensor element of diindium trioxide (In_2O_3) treated with magnesium oxide (MgO). Unfortunately, a highly trained technician is required to assemble and operate the apparatus (*15*).   Seafoods can spoil via the autolytic action of inherent muscle tissue enzymes. An example is nucleotide degradation (Figure 1), where products of adenosine-5'-triphosphate (ATP) breakdown have been exploited in several freshness methods (*10,17,18,19,20*). The rates of the first 5 reactions are slow due to endogenous enzymes, while the rates of the last 2 reactions are rapid due to bacterial enzymes. The K value index relates the rate of the sum of inosine and hypoxanthine to that of ATP and its breakdown products (*19*) or the percentage of inosine and hypoxanthine among ATP-related compounds in fish muscle (*15*). Calculating K value is considered an accepted and trusted method for indicating fish freshness and is used regularly by the Japanese seafood industry (*19,23*). K value is calculated using the equation shown in Figure 2 (*20,22,23*). The importance of K value is that it includes intermediate compounds and end products (*20*). Several studies have shown a high correlation between K values and sensory evaluation of fish (*6,19,21*). Increasing K values indicate increasing spoilage. Further, the odors associated with spoiled fish are similar to the odors of high concentrations of nucleotide degradation products (*19,22*).

A method similar to K value is the K_1 value index that is used for freshness evaluation of fish that accumulate more inosine than hypoxanthine during spoilage (*24*). The K_1 value is illustrated in Figure 3 (*18,19,20,,24*). K_1 is used with salmon, halibut, yellowtail, and numerous other tropical fish, while K value is reliable for catfish, white flounder, ocean perch, English sole, and various other species (*24*). The K_1 value can be utilized for species specified for K value but not vice versa due to sensitivity of K_1 value to the build up of hypoxanthine and not the total sum of degradation products as needed with K value (*24*).

Procedures to measure K and K_1 values include high performance liquid chromatography (HPLC), simplified column chromatography, and enzymatic measurement with either an enzyme sensor system or an oxygen electrode (*19,25*). Each method requires highly trained technicians and substantial time input. Research found HPLC more accurate and trustworthy, but it is difficult to incorporate into a routine quality assurance program due to long analysis time (*21*). Other studies showed a drastic reduction in analysis time to perform a K_1 value measurement when using nucleoside oxidase. This enzyme increases the oxidation reaction of nucleotides in the absence of exogenous cofactors. It further catalyzes the oxidative coupling of phenolic compounds and 4-aminoantipyrine in proportion to the amount of nucleosides to form a color. The intensity of this color can be measured and correlated with the proportion of nucleosides oxidized. The method is rapid, yet requires a technician with expertise to gauge the reactions, as well as proper preparation of samples and reagents (*19*). Another new method to measure K_1 value more rapidly but as effectively as HPLC has been proposed (*21*). It

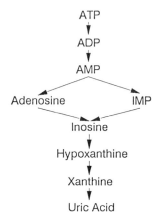

Figure 1. Nucleotide degradation pathway (ATP=adenosine-5'-triphosphate, ADP=adenosine-5'-diphosphate, AMP=adenosine-5'-monophosphate, IMP=inosine-5'-monophosphate).

$$K\,(\%) = \frac{\text{Inosine + Hypoxanthine}}{\text{ATP + ADP + AMP + IMP + Inosine + Hypoxanthine}} \times 100$$

Figure 2. K value equation.

$$K_1(\%) = \frac{\text{Inosine + Hypoxanthine}}{\text{IMP + Inosine + Hypoxanthine}} \times 100$$

Figure 3. K_1 value equation.

involves the use of a polarographic electrode to rapidly detect metabolic products produced as fish muscle degrades. The procedure was accurate and correlated well with HPLC and sensory evaluation. Since this method was simpler than the HPLC procedure, it may be more suitable routine quality control (*21*). Although K and K_1 values are useful for fish, they usually are not effective measures of shrimp spoilage because they do not measure the production of ammonia. A study used enzyme assays to determine the relative activities of adenosine deaminase and AMP deaminase in white shrimp tails. These enzymes produce over 50% of the total ammonia formed during spoilage. Activity of these enzymes varies with the type of shrimp being analyzed (*26*). For example, Alaskan shrimp (*27*) possess more active AMP deaminase and accumulate more IMP than Gulf coast shrimp (*28*) that have more active adenosine deaminase and accumulate more adenosine than Alaskan shrimp. This may explain why K and K_1 values are not particularly useful with shrimp.

Some studies found K values unable to detect cod and squid freshness. When fresh, these seafoods give K value readings indicative of spoiled products. Thus, alternatives to using K value for squid have centered on measuring the concentrations of the amino acids arginine and ornithine (*17*). In addition, the amount of the polyamine agmatine has been found to be a possible freshness indicator (*15,17*). Because K value does not give adequate measurement of spoilage during early stages, ^{31}Phosphorus-Nuclear Magnetic Resonance (^{31}P-NMR) has been used to detect initial spoilage within a few hours postmortem (*29*). This procedure estimates freshness by measuring metabolic changes of high energy phosphate compounds like phosphocreatine and its degradation products. Unfortunately, highly trained technicians are needed for reliable interpretation of results (*29*).

When appraising these techniques for freshness determination, there is clear evidence that although many of the methods are accurate, they frequently require uncooked grossly spoiled products and trained personnel for valid results. Another inconvenience is the inability to perform these tests outside of a laboratory setting. Many of the analyses take 1-2 d to obtain results. The seafood industry would benefit from a method that was rapid, sensitive during early stages of spoilage, simple to perform, and useful with both raw and cooked products.

Application of Impedance. Initial use of impedance centered on the conductive measurement of microbial metabolic products. As microorganisms grow, their metabolic products increase the conductivity of a medium. For example, the conductivity of putrefying defibrinated blood increased over time (*30*). Clinical microbiologists used impedance to detect urinary tract infections in half the time of standard methods (*31*).

Impedance is a function of conductance, capacitance, and applied frequency (*32*). Conductance is the reciprocal of resistance, so when resistance increases, conductance decreases. An increase in the number of charged compounds, especially smaller mobile ones that easily carry a charge, results in an increase in conductance or a decrease in the resistance of current flow through a solution. When metabolism alters slightly ionic compounds, such as lactose, to higher charged, smaller, and more mobile compounds, such as lactic acid, an electrical current can be carried more easily through a substance. Thus, as the concentration of higher charged compounds increases, an increase in conductance will occur. This implies that the substance being measured is a dielectric and is responsible for either the prevention or promotion of current flow. Thus, as conductance increases, impedance decreases. To make results more easily understood, impedance results are usually plotted inversely to mimic traditional microbial growth curves (*32*).

The remainder of this paper will provide evidence that impedance can be used to detect shrimp freshness. The theory for this study is simply that increasing spoilage, whether due to microorganisms or inherent enzymes, should increase the concentration of charged metabolic products that should be measured using readily available conductance or impedance equipment.

Materials and Methods

Preparation of Shrimp. Each study used fresh head on, white shrimp (Penaeus setiferus) obtained from a local wholesale dealer in Delcambre, Louisiana and transported immediately on ice to Baton Rouge. Total transport time was 1.5 h. In the laboratory, shrimp were randomly segregated into 50-g samples and placed into polyethylene bags that were then heat sealed. The bags were randomly separated and stored at various temperatures.

Effect of Sample Dilution. To determine the effect of sample dilution on impedance measurements, shrimp samples were stored at -20°C or 5°C for 21 d. Frozen samples were used to mimic fresh shrimp while refrigerated samples were used to represent spoilage over time. Duplicate 50-g samples were removed from storage every 7 d and used for impedance analysis. Each sample was removed from the bags, boiled for 5 min, cooled to room temperature, then transferred to a tared blender jar and diluted either 1:1 or 1:10 with sterile deionized, demineralized water. Samples were homogenized for 2 min on high. Controls consisted of water alone.

Impedance Monitoring. One-mL aliquots of the diluted samples were distributed into module wells of an impedance monitoring apparatus (Bactometer, bioMérieux Vitek Systems, Inc., Hazelwood, MO). Following well distribution, modules were inserted into the Bactometer incubator according to manufacturer's suggestions. Bactometer programming consisted of run times of 30 min at incubator temperatures of 25 and 40°C. Data generated from this procedure consisted of initial impedance values (I-value) with units in siemens.

Effect of Thermal Processing. Raw shrimp samples stored at 27°C for 48 h were removed at 12-h intervals and separated into two subsets. One set received a thermal treatment by boiling for 5 min followed by cooling to room temperature, while the other set received no heat treatment. Both sets were monitored using a 1:10 dilution and a 40°C monitoring temperature.

Effect of Storage Temperature. Two studies were conducted to determine the effect of storage temperature on impedance detection of spoilage. The first study was short term and involved storing shrimp samples at -20, 27, and 35°C for 9 h with sampling every 3 h. Prior to impedance analysis, each sample was boiled as previously described. This study was designed to mimic events that often occur during transportation of shrimp from harvest to shore or from wholesale to retail outlets. A long term second study used shrimp samples stored at -20 and 5°C for 15 d with sampling every 3 d. These samples were not boiled before impedance analysis. This study was designed to simulate retail or home refrigerated storage conditions. Impedance monitoring consisted of diluting samples 1:10 and a measuring temperature of 40°C. Both studies simultaneously used odor analysis on samples prior to impedance testing and before boiling by smelling bags immediately after opening. Aroma analysis consisted of two experienced sensory analysts who ranked the samples using a 10-point hedonic scale, with 1 being least offensive and 10 most offensive.

Results and Discussion

Effect of Sample Dilution and Measurement Temperature. Figure 4 illustrates the effect of dilution of samples stored at -20°C for 21 d. Results indicate that when samples were diluted 10-fold more, impedance values dropped approximately 600 siemens, regardless of the monitoring temperature. Samples analyzed at 40°C showed higher I-values than identical samples analyzed at 25°C. The magnitude of difference ranged from 90 siemens for samples diluted 1:10 compared with 170 siemens for samples diluted 1:1. During storage, the data demonstrated that I-values decreased slightly (negative slopes) by approximately 100 siemens. Similar trends were observed when shrimp samples were stored at 5°C for 21 d except that I-values increased during storage by approximately 500 siemens (Figure 5). Slope values of the curves did not differ appreciably. Measurement temperature differences were 190 siemens for samples diluted 1:10 compared with 310 for samples diluted 1:1.

The -20°C samples provided unspoiled controls that had small changes in readings over the 21-d period. Thus, storing fresh shrimp frozen to use as authentic fresh samples for impedance determinations appears valid. That samples diluted 1:10 gave similar trends as those diluted 1:1 and because 1:1 diluted samples were difficult to distribute to module wells, further experiments used only samples diluted 1:10. Since measurements at 40°C gave higher readings than at 25°C, due to temperature increase causing rises in impedance measurements, the former measuring temperature was used for the remaining experiments.

Shrimp samples held at 5°C showed increasing I-values during storage compared to frozen samples presumably due to increased breakdown of muscle tissue by inherent enzymes and/or increased microbial action.

Effect of Thermal Processing. The effect of boiling shrimp after removal from storage at 27°C on impedance measurements is shown in Figure 6. Results clearly indicate that boiling reduces I-value compared with raw samples, although both

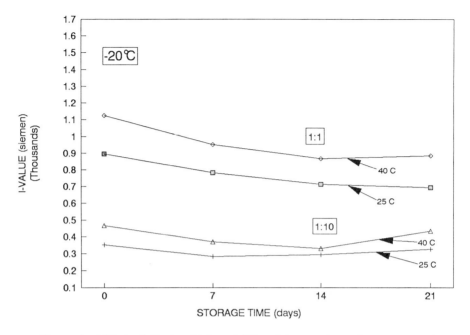

Figure 4. Effect of dilution (1:1 or 1:10) and temperature of measurement (25 or 40°C) on impedance of raw shrimp stored at -20°C.

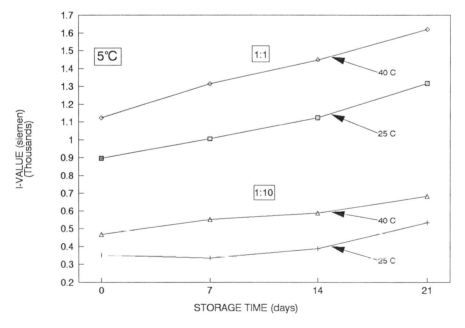

Figure 5. Effect of dilution (1:1 or 1:10) and temperature of measurement (25 or 40°C) on impedance of raw shrimp stored at 5°C.

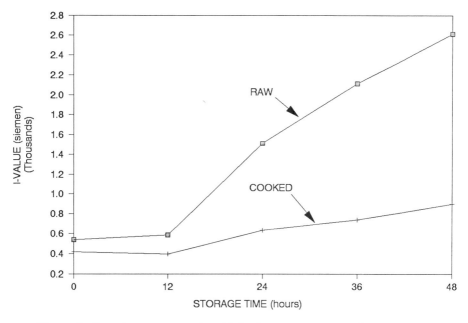

Figure 6. Impedance measured at 40°C of raw or cooked shrimp stored at 27°C.

curves show similar increases during storage. These data indicate that boiling will decrease impedance values probably due to denaturation of shrimp protein causing binding of small charged compounds and/or to leaching of small mobile decomposition products. Even with this impedance value decrease, it seems apparent that impedance measurement of spoilage can still be used with thermally processed shrimp unlike many other methods of spoilage detection. Methods such as K value (*15,17,19,20*), enzyme assays (*26*), or microbiological (*6,8*) can be rendered useless with thermal processing since heat destroys nucleotides, denatures enzymes, or kills microorganisms respectively.

Effect of Storage Temperature.   Figure 7 illustrates the effect of storage temperature on impedance during short term storage at -20, 27, and 35°C. As storage temperature increased, impedance values increased. Samples stored at 35°C had higher I-value readings than those stored at 27°C. When the samples were subjected to sensory analysis, similar trends were observed (Figure 8). Increasing temperature increased the degree of objectionable odors. It is interesting to note that unspoiled samples stored at -20°C remained unchanged during storage in relation to odor scores. Further, samples stored at the highest temperature gave higher impedance readings and higher odor scores presumably due to increased spoilage. Long term storage study of raw shrimp evaluated by impedance showed a similar tendency with higher readings of samples stored at 5°C compared with those stored at -20°C (Figure 9). Odor scores also reflected increased spoilage of the refrigerated sample compared to the frozen control (Figure 10).

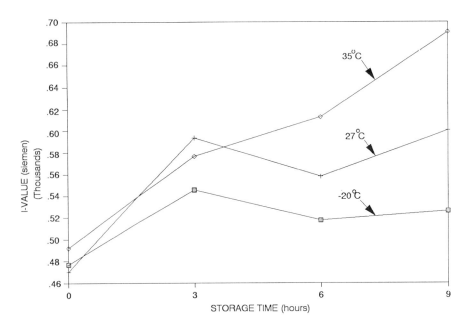

Figure 7. Effect of storage temperature (-20, 27, and 35°C) on raw shrimp that were cooked prior to measuring impedance at 40°C.

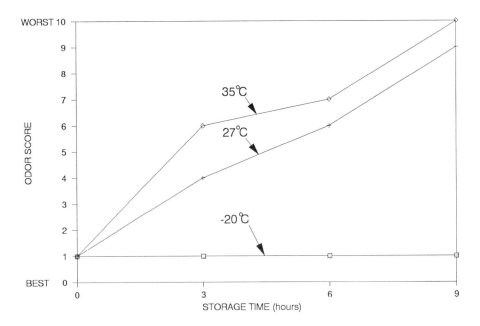

Figure 8. Effect of storage temperature (-20, 27, and 35°C) on odor analysis of raw shrimp.

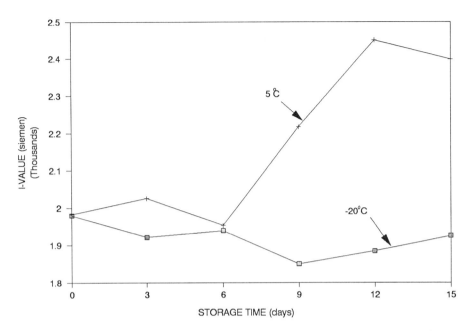

Figure 9. Impedance measured at 40°C of raw shrimp stored at -20 or 5°C.

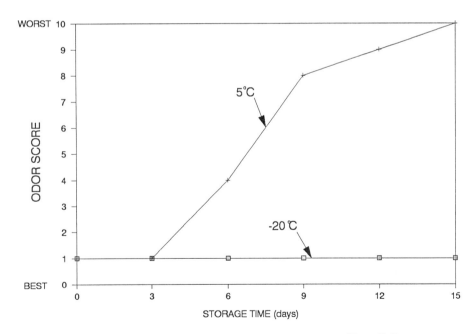

Figure 10. Odor analysis of raw shrimp stored at -20 or 5°C.

Conclusions

Several objective methods are available to determine the freshness of shrimp; however, many require a) raw products for analysis, b) complex chemistry and equipment for testing, or c) highly trained technicians. Additionally, some of these methods require extensive analysis time making results meaningless if product has already spoiled or has been released to consumers. Results for the impedance method discussed in the present paper demonstrate that spoilage of raw or thermally processed shrimp can be detected in 30 min with easy sample preparation. The limitations of this method revolve around the requirement for authentic fresh frozen standards as a basis for comparison. Further research is necessary to define the effect of seasonal and geographical variations, and species and size difference. The sum of these factors will likely affect the reproducibility of this method.

Acknowledgments

Approved for publication by the Director of the Louisiana Agricultural Experiment Station as Manuscript No. 92-21-6219.

Literature cited

(1)   Saguy, I.; and Karel, M. *J. Food Technol.* **1980**, *34*, pp. 78.
(2)   Buchanan, R.L. *J.Food Technol.* **1991**, *45*, pp. 157.
(3)   Hollingworth, Jr. T.A.; Kaysner, C.A.; Colburn, K.G.; Sullivan, J.J.; Abeyta, Jr., C.; Walker, K.D.; Torkelson, Jr., J.D.;Throm, H.R.; Wekell, M.M. *J. Food Sci.* **1991**, *56*, pp. 164.
(4)   Kinsella, J.E. *J. Food Technol.* **1988**, *42*, pp. 146.
(5)   Lovell, R.T. *J. Food Technol.* **1991**, *45*, pp. 87.
(6)   Perez-Villarreal, B. *J. Sci. Food Agri.* **1987**, *41*, pp. 335.
(7)   McEvily, A.J.; Iyengar, R.; and Otwell, S. *J. Food Technol.* **1991**, *45*, pp. 80.
(8)   Alur, M.D.; Venugopal, V.; Nerkar, D.P.; and Nair, P.M. *J. Food Sci.* **1991**, *56*, pp. 332.
(9)   Cox, N.A.; Lovell, R.T. *J. Food Sci.* **1973**, *38*, pp. 679.
(10)   *Microbiology of Marine Food Products*; Ward, D.R.; Hackney, C., Eds.; Van Nostrand Reinhold: New York, 1991; pp.65-87
(11)   Liston, J. In *Advances In Fish Science and Technology*; Connell J.J. and Staff of Torry Research Station, Eds.; Microbiology in fishery science; Fishing News Books Limited: Farmhan, Surrey, England 1980; pp. 138-157.
(12)   Nickelson, R.; Vanderzant, C. In *Proceedings of the First American Tropical and Subtropical Fisheries Technology Conference*; Cobb B.F. and Stochton A.B. Eds.; Bacteriology of Shrimp; Texas A&M Sea Grant College Program: College Station, TX, 1976; pp. 254-271.
(13)   Matches, J.R. *J. Food Sci.* **1982**, *47*, pp. 1044.
(14)   Wong, K.; Bartlett, F.; Gill, T.A. *J. Food Sci.* **1988**, *53*, pp. 1653.
(15)   Ohashi, E.; Takao, Y.; Fujita, T.; Shimizu, Y.; Egashira, M. *J. Food Sci.* **1991**, *56*, pp. 1275.

(*16*)   Lehninger, A.L. *Principles of Biochemistry*;   Worth Publishers, Inc.:   New York, 1982.

(*17*)   Ohashi, E.; Okamoto, M.; Ozawa, A.; and Fujita, T.   *J. Food Sci.*   **1991**, *56*, pp. 161-174.

(*18*)   *Biotechnology and Food Process Engineering*;   Schwartzberg, H.G.; Rao, M.A., Eds.; Marcel Dekker, Inc.:   New York, 1989.

(*19*)   Isono, Y.   *Agri. Bio. Chem.*   **1990**, *54*, pp. 2827.

(*20*)   Boyle, J.L.; Lindsay, R.C.; and Stuiber, D.A.   *J. Food Sci.*   **1991**, *56*, pp. 1267.

(*21*)   Luong, J.H.T.; Male, K.B.; and Huynh, M.D.   *J. Food Sci.*   **1991**, *56*, pp. 335.

(*22*)   Watanabe, E.; Nagumo, A.; Hoshi, M.; Konagaya, S.; Tanaka, M.   *J. Food Sci.*   **1987**, *52*, pp. 592.

(*23*)   Saito, T.; Arai, K.; and Matsuyoshi, M.   *Bull. Jap. Soc. Sci. Fish.*   **1959**, *24*, pp. 749.

(*24*)   Greene, D.H.; Babbitt, J.K.; and Reppond, K.D.   *J. Food Sci.*   **1990**, *55*, pp. 1236.

(*25*)   Karube, I.; Matsuoka, H.; Suzuki, S.; Watanabe, E.; Toyama, K.   *J. Agric Food Chem.*   **1984**, *32*, pp. 314.

(*26*)   Cheuk, W.L.; Finne, G.; and Nickelson II.R.   *J. Food Sci.*   **1979**, *44*, pp. 1625.

(*27*)   Stone, F.E.   *J. Milk Food Technol.*   **1971**, *35*, pp. 354.

(*28*)   Flick, G.J.; Lovell, R.T.   *J. Food Sci.*   **1972**, *37*, pp. 609.

(*29*)   Chiba, A.; Hamaguchi, M.; Kosaka, M.; Tokuno, T.; Asai, T.; Chichibu, S.   *J. Food Sci.*   **1991**, *56*, pp. 660.

(*30*)   Stewart, C.N.   *J. Exp. Med.*   **1899**, *4*, pp. 235.

(*31*)   Wheeler, T.G.; Goldschmidt, M.C.   *J. Clin. Microbiol.*   **1975**, *1*, pp. 25.

(*32*)   Eden, R.; Eden, G.   *Impedance Microbiology;* Letchworth: Research Studies Press Ltd.:   New York, 1984.

RECEIVED October 28, 1992

Chapter 21

Conjugated Dienoic Derivatives of Linoleic Acid

A New Class of Food-Derived Anticarcinogens

Sou F. Chin, Jayne M. Storkson, and Michael W. Pariza

Food Research Institute, University of Wisconsin—Madison, Madison, WI 53706

CLA is the acronym for a mixture of conjugated dienoic isomers of linoleic acid which occur naturally in food. Dairy products and other foods derived from ruminant animals are the most significant dietary sources of CLA. Synthetically prepared CLA has been shown to inhibit carcinogen-induced neoplasia in mouse epidermis and forestomach and rat mammary gland. The exact mechanism of anticarcinogenic action of CLA is still unclear. However, CLA exhibits several effects that could be related to its anticarcinogenic property. CLA acts as an antioxidant as evidenced in vitro and in vivo. CLA also inhibits the induction of ornithine decarboxylase by the epidermal tumor promotor, 12-0-tetradecanoylphorbol-13-acetate, apparently through the inhibition of protein kinase C. Following dietary administration of a mixture of CLA isomers, only the cis-9, trans-11 isomer is found in phospholipid. Thus, the cis-9, trans-11 CLA isomer, the major CLA isomer in the diet, may be the biologically active form.

The relationship between diet and cancer risk is extremely complex (1). Factors that appear to enhance carcinogenesis under one set of conditions may have no effect or even inhibit carcinogenesis under different conditions (2). The link between dietary fat and cancer is complicated by many factors, in particular total calorie intake and fatty acid composition (2). Among the fatty acids that comprise lipid, only linoleic acid is clearly linked to the enhancement of carcinogenesis in rat mammary gland (3), pancreas (4) and colon (5).

CLA, the acronym for a series of conjugated dienoic isomers of linoleic acid, occurs naturally in many foods, particularly dairy products and other foods derived from ruminant animals (6). Synthetically prepared CLA inhibits chemically-induced mouse epidermal and forestomach neoplasia (7, 8) and rat mammary neoplasia (9). Hence, the effect of CLA on carcinogenesis is opposite that of linoleic acid.

0097–6156/93/0528–0262$06.00/0

The purpose of this review is to consider these intriguing findings and propose mechanisms with respect to the formation and possible biological action of CLA.

CLA Formation in the Rumen. The fat depots of ruminant species contain mostly saturated fat. They are subject to little modification by dietary changes, including the feeding of relatively large amounts of unsaturated fats or oils. Dietary fats are modified in the rumen via hydrolysis and biohydrogenation reactions.

Garton *et al.* (*10*) demonstrated that rumen microbial suspensions could hydrolyze triglycerides. It was later established that virtually any ester link between fatty acid and glycerol was subject to hydrolytic cleavage by rumen organisms (*11*). As a consequence of the activity of the lipolytic enzymes, high levels of free fatty acids are produced in the rumen. The unsaturated fatty acids are substrates in biohydrogenation reactions.

The first step in the biohydrogenation of linoleic acid to stearic acid by the rumen microorganism, *Butyrivibrio fibrisolvens*, is the formation of the cis-9, trans-11 CLA isomer (Figure 1). This reaction is catalyzed by a membrane-bound enzyme, linoleate isomerase, which acts only on fatty acids possessing cis double bonds in positions Δ^9 and Δ^{12}, and a free carboxyl group (*12*).

It is not certain that the presence of CLA in tissue lipids is due entirely to the production of cis-9, trans-11 as an intermediate during the biohydrogenation of linoleic acid in the rumen. However, the amount of CLA in milk (*13*) and butter (*14*) is positively correlated to the level of dietary linoleic acid. Some long chain fatty acid intermediates reach the small intestine and are normally absorbed and deposited into adipose tissue (*15*). There is seasonal variation in CLA content of milk, with the highest values occurring usually in summer (*16*).

Significant amounts of CLA are found in muscle tissue from ruminant animals (Table 1). The CLA content for beef ranged from 2.9 to 4.3 mg CLA/g fat. Among ruminants lamb was the highest (5.6 mg CLA/g fat) and veal was the lowest (2.7 mg CLA/g fat). The cis-9, trans-11 isomer accounted for more than 79% of the total CLA isomers in meats.

Substantial amounts of CLA were found in cow's milk (5.5 mg CLA/g fat) (Table 2). More than 90% of the CLA isomer found in milk is the cis-9, trans-11 isomer. Total CLA content in other dairy products ranged from 3.6 to 7.0 mg CLA/g fat. Nonfat dairy dessert was the lowest (0.5 mg CLA/g fat). The CLA content in yogurt ranged from 1.7 to 4.8 mg CLA/g fat. Nonfat yogurt had the lowest CLA concentration (1.7 mg CLA/ g fat).

CLA content of unprocessed cheeses ranged from 2.9 to 7.1 mg CLA/g fat (Table 3). Cheeses such as Brick and Muenstre, aged 4 to 8 weeks, were among the highest. However, Parmesan and Romano cheeses, which were aged or ripened more than 10 months, had the lowest CLA content. The reason for low CLA content in aged cheese is unclear. However, during cheese ripening, hydrolysis of fatty acids by bacterial enzymes occurs producing free fatty acids and glycerol (*17*). Under such conditions, free fatty acids, including CLA, become very vulnerable to further oxidation. This might indirectly reduce the CLA concentration in aged cheeses.

CLA concentrations (average 5.0 mg CLA/g fat) in processed cheeses did not vary much among varieties and were comparable to unprocessed cheeses. The cis-9, trans-11 CLA isomer accounted for more than 83 % of the total CLA isomers in unprocessed and processed cheeses.

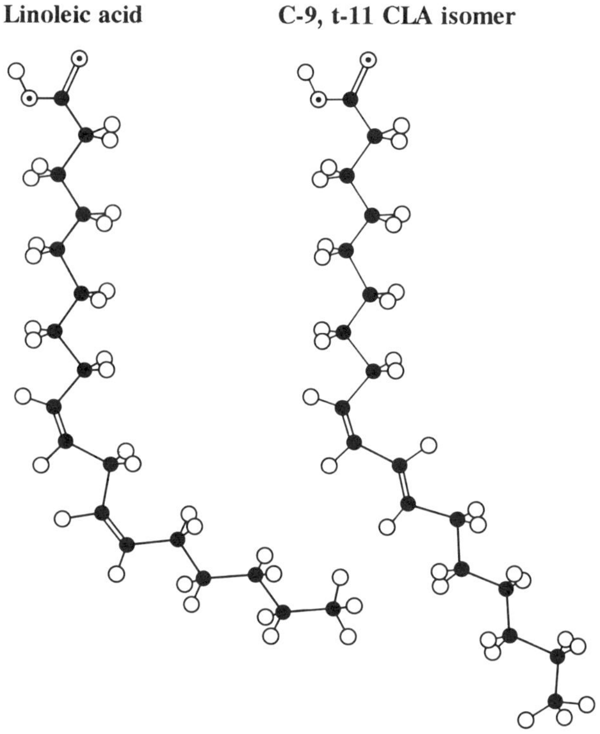

Linoleic acid   **C-9, t-11 CLA isomer**

Figure 1. The chemical structures of linoleic acid (cis-9, cis-12-octadecadienoic acid), and the cis-9, trans-11 CLA isomer (cis-9,trans-11-octadecadienoic acid).

(Reproduced with permission from the Food Research Institute Annual Report 1990. Copyright Food Research Institute 1991.)

Table 1. Conjugated Dienoic Isomers of Linoleic Acid in Animal Tissues

Foodstuff[a]	Number of samples	Total CLA[b] (mg/g fat)	c-9,t-11[c] (%)
Round beef	4	2.9 ± 0.09	79
Fresh ground round	4	3.8 ± 0.11	84
Fresh ground beef	4	4.3 ± 0.13	85
Veal	2	2.7 ± 0.24	84
Lamb	4	5.6 ± 0.29	92
Pork	2	0.6 ± 0.06	82
Chicken	2	0.9 ± 0.02	84
Fresh ground turkey	2	2.5 ± 0.04	76

SOURCE: Adapted from ref. 6.
[a]Samples were from commercially available, uncooked edible portions.
[b]Values are means ± standard error for the number of samples indicated.
[c]Values are means for the number of samples indicated. All standard error values are less than 3%. Data were expressed as % of total CLA isomers.

CLA Formation in Nonruminant Animals. CLA has been detected in the serum, bile and duodenal juice of humans (*18*). It has been confirmed that both the cis-9, trans-11 and trans-9, trans-11 CLA isomers are present in human depot fats (*19*).

CLA has also been identified in the tissue lipids of other nonruminant animals. The CLA concentration in nonruminant animals (chicken and pig) is considerably lower than ruminant animals (Table 1). The exception among nonruminants is turkey (2.51 mg CLA/g fat), which is about five fold higher than chicken or pork. More than 76% of the CLA isomer found in nonruminant tissues is cis-9, trans-11 isomer.

An important question is whether the CLA found in the tissues of non-ruminants is a consequence of dietary intake, or at least in part, due to the conversion of linoleic acid to cis-9, trans-11 CLA isomer by bacterial flora. In an effort to answer this question studies were undertaken in which rats were fed diet containing 5% corn oil alone or supplemented with 2.5% or 5% linoleic acid, and sacrificed at 2, 4, or 6 weeks. CLA concentrations in liver, lung, fat pad, muscle, kidney and blood serum increased as a function of linoleic acid feeding (*20*). A plateau appeared at 4 weeks, where CLA levels in the tissues of rats fed 5% linoleic acid were 5 to 10 times higher than those of controls. Similar results were observed in nonphospholipid and phospholipid fractions. This indicated that the bacterial flora of rats may convert linoleic acid to cis-9, trans-11 CLA. Hence, a similar study was conducted using gnotobiotic (germ free) rats. The animals were fed a diet fortified with 5% linoleic acid and sacrificed at 2 and 4 weeks. In this

Table 2. Conjugated Dienoic Isomers of Linoleic Acid in other Dairy Products

Foodstuff[a]	Number of samples	Total CLA[b] (mg/g fat)	c-9,t-11[c] (%)
Homogenized milk	3	5.5 ± 0.30	92
Condensed milk	3	7.0 ± 0.29	82
Cultured buttermilk	3	5.4 ± 0.16	89
Butter	4	4.7 ± 0.36	88
Butter fat	4	6.1 ± 0.21	89
Sour cream	3	4.6 ± 0.46	90
Ice cream	3	3.6 ± 0.10	86
Nonfat frozen dairy dessert	2	0.6 ± 0.02	n.d.[d]
Lowfat yogurt	4	4.4 ± 0.21	86
Custard style yogurt	4	4.8 ± 0.16	83
Plain yogurt	2	4.8 ± 0.26	84
Nonfat yogurt	2	1.7 ± 0.10	83
Frozen yogurt	2	2.8 ± 0.20	85
Yogurt pudding			
Milk chocolate:			
Sample 1	2	3.5 ± 0.06	76
Sample 2	2	2.5 ± 0.27	80
Double chocolate	2	3.1 ± 0.48	71
Vanilla	2	3.8 ± 0.10	84

SOURCE: Reprinted with permission from ref. 6. Copyright 1992.
[a] Samples were from commercially available.
[b] Values are means ± standard error for the number of samples indicated.
[c] Values are means for the number of samples indicated. All standard error values are less than 3%. Data were expressed as % of total CLA isomers.
[d] n.d. = not detectable.

Table 3. Conjugated Dienoic Isomers of Linoleic Acid in Unprocessed and Processed Cheeses

Foodstuff[a]	Number of samples	Total CLA[b] (mg/g fat)	c-9,t-11[c] (%)
Unprocessed cheese:			
Romano	2	2.9 ± 0.22	92
Parmesan	4	3.0 ± 0.21	90
Sharp cheddar	3	3.6 ± 0.18	93
Medium cheddar	4	4.1 ± 0.14	80
Cream	3	3.8 ± 0.08	88
Colby	3	6.1 ± 0.14	92
Mozzarella	4	4.9 ± 0.20	95
Cottage	3	4.5 ± 0.13	83
Ricotta	3	5.6 ± 0.44	84
Brick	2	7.1 ± 0.08	91
Natural Muenstre	2	6.6 ± 0.02	93
Reduced fat Swiss	2	6.7 ± 0.56	90
Blue	2	5.7 ± 0.18	90
Processed cheese:			
American processed	3	5.0 ± 0.13	93
Cheez Whiz	4	5.0 ± 0.07	92
Velveeta	2	5.2 ± 0.03	86
Old English Spread	2	4.5 ± 0.21	88

SOURCE: Reprinted with permission from ref. 6. Copyright 1992.
[a] Samples were from commercially available.
[b] Values are means ± standard error for the number of samples indicated.
[c] Values are means for the number of samples indicated. All standard error values are less than 3%. Data were expressed as % of total CLA isomers.

study, no increase in tissue CLA was found in rats fed linoleic acid fortified diets (Chin, S. F. and M. W. Pariza, University of Wisconsin-Madison, unpublished data), indicating that the bacterial flora of rats is capable of converting linoleic acid to cis-9, trans-11 CLA. Further study is needed to identify the bacteria responsible for this effect.

Absorption and Deposition of CLA in Animal Tissues. Miller *et al.* (*21*) described a method employing the methyl ester of conjugated dienes prepared from corn oil as tracers of fat metabolism. It was postulated that the conjugated dienoic isomers could be differentiated from other fatty acids in body fat by spectrophotometric absorbance at 233 nm.

Dietary administration of CLA resulted in an increase in conjugated diene concentration in the lipids of the intestinal mucosa and liver of rats (*22,23,24*). Similar results were also reported by Reiser (*25*) for rats fed CLA as either free fatty acids or in triglycerides. Maximum levels of CLA appeared in liver, blood and pooled organ (heart, lung and kidney) at 16 hrs after triglyceride ingestion, whereas after the ingestion of CLA as free fatty acid the maximum concentration occurred at 24 hrs. This indicates that CLA in triglycerides is absorbed more rapidly.

The accumulation of different CLA isomers in various tissues was also reported. When diets containing 1% of either cis, trans or trans, trans CLA isomers were fed to rats, conjugated dienes did not accumulate in testis or brain lipid. By contrast, CLA was incorporated into adipose tissue (*26*). Substantial deposition of conjugated dienes occurred in heart lipid when animals were fed cis, trans isomers, but no increase was observed when animals were fed trans, trans CLA isomers.

Ha *et al.* (*8*) administered chemically-synthesized CLA p. o. to mice and found that all isomers were deposited in triglycerides whereas only the cis-9, trans-11 isomer was incorporated into phospholipids. The CLA used contained 8 isomers with cis-9, trans-11 and trans-10, cis-12 representing 45 and 47%, respectively.

Why cis-9, trans-11 alone is found in phospholipids is unclear. However, the cis-9, trans-11 isomer exhibits a configuration that is most similar to linoleic acid (Figure 1). We have proposed that the cis-9, trans-11 isomer may be the biologically active anticarcinogenic CLA isomer.

Anticarcinogenic Activity of CLA. Anticarcinogenic activity of synthetically prepared CLA was first tested in the two-stage mouse epidermal carcinogenesis model. In this study, 7 days, 3 days, and 5 min prior to 7, 12-dimethylbenz[a]anthracene (DMBA) application, CLA was applied directly on the shaved backs of individual mice at doses of 20, 20, and 10 mg, respectively (*7*). Control mice were treated similarly with linoleic acid or acetone. All mice were given TPA to effect tumor promotion. It was found that CLA treated mice developed only about half as many papillomas and exhibited a lower tumor incidence than control mice.

In another study CLA was found to be effective in inhibiting benzo[a]pyrene (BP)-induced forestomach neoplasia in mice (*8*). In this study, female ICR mice were given 0.1 ml CLA or linoleic acid plus 0.1 olive oil, or 0.1 ml olive oil alone plus 0.85% saline as control, by stomach tube four and two days prior to administration of BP. Treatment with CLA reduced forestomach neoplasia by 46-67% relative to animals given linoleic acid or olive oil.

CLA was also studied as an anticarcinogen for DMBA-induced mammary neoplasia (*9*). Rats were fed AIN-76A basal diet alone or supplemented with 0.5, 1 or 1.5% synthetically prepared CLA. Diets were fed 2 weeks before DMBA administration and continued until the end of the experiment. CLA at 0.5 and 1% reduced adenocarcinomas by 32 and 56%, respectively. Feeding CLA at 1.5% resulted in 60% reduction in adenocarcinomas. This study indicated that CLA is more effective than any other fatty acid in modulating mammary tumor development (*27*).

CLA as An Antioxidant. The complete mechanism of anticarcinogenic activity of CLA is not known. Some of the CLA effect is believed due to its antioxidant properties. For example, use of a water/ethanol system that is incubated at 40°C under air for 14 days, showed CLA reduced the oxidation of linoleic acid by 86% (*8*). Under the same conditions α-tocopherol reduced oxidation by only 63% and butylated hydroxytoluene (BHT) reduced oxidation by 92%. Dose-response studies were conducted, and it was found that the optimal ratio for CLA to protect linoleic acid from oxidation is 1:1000 (CLA:linoleic acid).

Antioxidant activity was also tested in a liver microsome system. In this study, mice were treated by oral intubation (2 times/wk) with 0.2 ml olive oil alone or containing CLA (0.1 ml), linoleic acid (0.1 ml), or dl-α-tocopherol (10 mg). Four weeks after the first treatment, liver microsomes were prepared and subsequently subjected to oxidative stress using a non-enzymatic iron-dependent lipid peroxidation system. Microsomal lipid peroxidation was measured as thiobarbituric acid-reactive substance (TBARS) production using malondialdehyde as the standard. It was found that pretreatment of mice with CLA or dl-α-tocopherol significantly decreased TBARS formation in mouse liver microsomes ($p < 0.05$) (Sword, J. T. and M. W. Pariza, University of Wisconsin, unpublished data).

A recent study was carried out with rats fed diets containing vitamin E, butylated hydroxyanisole (BHA) or CLA for 1 or 6 months. Feeding with CLA reduced the production of TBARS in the mammary gland (*9*). The maximum antioxidant effect was observed using 0.25% CLA in the diet. A suppressive effect of CLA on TBARS formation was not detected in the liver. In contrast to CLA, both vitamin E and BHA proved to be effective antioxidants in the liver as well as in the mammary gland systems. Additionally, CLA does not induce glutathione-S-transferase activity in liver or mammary gland, suggesting that CLA is "recognized" biochemically as a nutrient rather than a toxicant (*9*).

It is not known why the rat mammary gland is more responsive to CLA-mediated inhibition of lipid peroxidation than the liver, especially since CLA treated mouse liver microsomes exhibit inhibition of lipid peroxidation. Based on these observations it might be assumed that CLA affords different degrees of protection from oxidation in different tissues and species. In order to answer these questions, we are studying liver microsomes from rats and mice fed identical CLA-treated diets.

Effect of CLA on Protein Kinase C (PKC) Activity. 12-0-tetradecanoylphorbol-13-acetate (TPA) administered by gavage induced the activity of ornithine decarboxylase (ODC) in a similar manner to that observed in mouse skin (*28*). The peak activity was about 5-times control and occurred 6 hrs after TPA intubation. Treating mice with CLA (100 mg p. o., twice per week) for 1, 2 or 4 weeks progressively reduced the TPA induction of ODC activity in forestomach (*28*). There is good evidence that the induction of ODC is controlled by PKC, so this finding indicates that in the forestomach the inhibition of ODC activity may be through the inhibition of PKC.

PKC is a member of a large family of proteins that are activated by diacylglycerol resulting from the receptor-mediated hydrolysis of inositol phospholipid. PKCs play an important role in the transfer of information from a variety of extracellular signals across the cell membrane with the final outcome being the regulation of many intracellular processes. Evidence accumulating indicates that PKC is the principal target receptor for TPA.

We observed PKC-like activity in mouse forestomach extract (*29*). The activity was activated by TPA plus phosphatidylserine in the absence of calcium. Protein associated with this activity was partially-purified by DEAE-cellulose chromatography. In mice pretreated with CLA 24 hours prior to sacrifice, the PKC-like activity was refractory to activation by TPA and phosphatidylserine. This observation indicates that cis-9, trans-11 incorporated into phospholipid might directly affect the interaction of TPA with PKC. Further, since PKC controls superoxide generation (*30*), CLA might in this way serve as an indirect antioxidant.

Conclusion

CLA is a naturally occurring constituent of the diet that is consumed on a regular basis. The principal dietary sources of CLA are animal and dairy products. In general, tissues from ruminants contain considerably more CLA than tissues from nonruminants. CLA inhibits neoplasia in several models. In fact, CLA is the only fatty acid that has been shown clearly to reduce the development of neoplasia in experimental animals. Free radical-mediated cell damage is thought to be important in the tumor prevention phase of carcinogenesis. The cis-9, trans-11 CLA isomer is incorporated into the phospholipids of cell membranes. Thus, the antioxidant effect of CLA may be of significance in protecting cell membranes from oxidation-induced free radical damage.

Antioxidant action alone, however, may not fully account for the anticarcinogenic mechanism of CLA. PKCs are key enzymes in the signal transduction pathway and principal receptor(s) of TPA. The inhibition of PKC activation by CLA in forestomach suggests that CLA may be involved in regulating cell division, a necessary step if initiated cells are to progress toward malignancy. Further studies on the anticarcinogenic mechanism of CLA are needed to determine the possible role of CLA in reducing cancer risk in humans.

Literature Cited

1) Doll, R.; Peto, R. *JNCI*, **1981**, *66*, pp. 1191-1308.
2) Pariza, M.W. *Ann. Rev. Nutrit.* **1988**, *8*, pp. 167-183.
3) Ip, C.; Carter, C.A.; Ip, M.M. *Cancer Res.* **1985**, *45*, pp. 1997-2001.
4) Roebuck, B.D.; Longnecker, D.S.; Baumgartner, K.J.; Thron, C.D. *Cancer Res.* **1985**, *45*, pp. 5252-5256.
5) Sakaguchi, M.; Minoura, T.; Hiramatsu, Y.; Takada, H.; Yamamura, M. *Cancer Res.* **1986**. *46*, pp. 61-65.
6) Chin, S. F.; Liu, W.; Storkson, J.M.; Ha, Y.L.; Pariza, M.W. *JFCA* .**1992**, (In press).

7) Ha, Y.L.; Grimm, N.K.; Pariza, M.W. *Carcinogenesis*, **1987**, *8*, pp. 1881-1887.
8) Ha, Y.L.; Storkson, J.M.; Pariza, M.W. *Cancer Res.* **1990**, *50*, pp. 1097-1101.
9) Ip, C.; Chin, S.F.; Scimeca, J.A.; Pariza, M.W. *Cancer Res.* **1991**, *51*, pp. 6118-6124.
10) Garton, G.A.; Hobson, P.N.; Lough, A.K. *Nature* **1958**, *182*, pp. 1511-1512.
11) Garton, G.A. In *World Review of Nutrition and Dietetics;* The digestion and absorption of lipids in ruminant animals. Bourne, G.H. Ed, Karger, Basel, **1967**, *7*, pp 225-250.
12) Kepler, C.R.; Hirons, K.P.; McNeill, K.P.; Tove, S.B. *J. Biol. Chem.* **1966**, *241*, pp. 1350-1354.
13) Strocchi, A.; Capella, P.; Carnacini, A.; Pallotta, U. *Ind. Agri.* **1967**, *5*, pp. 177-185.
14) Bartlet, J.; Chapman, D.G. *J. Agric. Food Chem.* **1961**, *9*, pp. 50-53.
15) Bickerstaffe, R.; Noakes, D.E.; Annison, E.F. *Biochem. J.* **1972**, *130*, pp. 607-617.
16) Riel, R. R. *J. Dairy Sci.* **1963**, *46*, pp. 102-106.
17) Kosikowski, F. Cheese and fermented milk foods; Fundamentals of cheese making and curing; Brooktondale, New York. **1982**, pp. 91-108.
18) Cawood, P.; Wickens. D.G.; Iversion, S.A.; Braganza, J.M.; Dormandy, T.L. *FEBS Lett.* **1983**, *162*, pp. 239-243.
19) Ackman, R.G.; Eaton, C.A.; Sipos, J.C.; Crewe, N.F. *Can. Inst. Food Sci. Technol. J.* **1981**. *14*, pp. 103-107.
20) Chin, S.F.; Liu, W.; Albright, K.; Pariza, M.W. *FASEB J.* **1992**, *6*, pp. A1396.
21) Miller, E.S.; Barnes, R.H.; Kass, J.P. and Burr, G.O. *Proc. Soc. Exper. Biol. Med.* **1939**, *41*, pp. 485-488.
22) Barnes, R.H.; Miller, E.S.; Burr, J.O. *J. Biol. Chem.* **1941**, *140*, pp. 233-240.
23) Barnes, R.H.; Miller, E.S.; Burr, J.O. *J. Biol. Chem.* **1941**, *140*, pp. 247-253.
24) Barnes, R.H.; Miller, E.S.; Burr, G.O. *J. Biol. Chem.* **1941**, *140*, pp. 773-778.
25) Reiser, R. *Proc. Soc. Exper. Biol. Med.* **1950**, *74*, pp. 666-669.
26) Aaes-Jorgensen, E.J. *Nutr.* **1958**, *66*, pp. 465-483.
27) Reddy, B. S.; Burill, C.;Rigotty, J. *Cancer Res.* **1991**, *51*, pp. 487-491.
28) Benjamin, H.; Storkson, J.M.; Albright, K.; Pariza, M.W. *FASEB J.* **1990**, *4*, pp. A508.
29) Benjamin, H.; Storkson, J. M.; Liu, W.; Pariza, M.W. *FASEB J.* **1992**, *6*, pp A1396.
30) Merrill, A.H. *Nutrition Rev.* **1989**, *47*, pp. 161-169.

RECEIVED February 6, 1993

Chapter 22

Preharvest Aflatoxin Contamination
Molecular Strategies for Its Control

D. Bhatnagar, P. J. Cotty, and T. E. Cleveland

Southern Regional Research Center, Agricultural Research Service, U.S. Department of Agriculture, 1100 Robert E. Lee Boulevard, New Orleans, LA 70124

Aflatoxins are carcinogens produced by *Aspergillus flavus* and *A. parasiticus* when these fungi infect crops before and after harvest, thereby contaminating food and feed and threatening both human and animal health. Traditional control methods (such as the use of certain cultural practices, pesticides and resistant varieties), which effectively reduce populations of many plant pests in the field, have not been effective in controlling aflatoxin-producing fungi. Our research, therefore, consists of acquiring knowledge of: 1) the molecular regulation of aflatoxin formation within the fungus, 2) environmental factors and biocompetitive microbes influencing growth of *A. flavus* and aflatoxin synthesis in crops, and 3) enhancement of host plant resistance to aflatoxin accumulation through understanding the biochemistry of host plant resistance responses. This understanding is expected to lead to development of biocontrol strategies and/or, in longer term research, development of elite crop lines "immune" to aflatoxin producing fungi.

Aflatoxins are extremely potent naturally-occurring carcinogens. These toxins occur in feed for livestock and in food for human consumption. The two fungi that produce aflatoxin, *Aspergillus flavus* and *A. parasiticus*, can grow and produce aflatoxin on a number of substrates, but aflatoxin contamination is a serious concern on diverse substrates such as corn, peanuts, cottonseed, and tree nuts. *A. flavus* appears to be the primary aflatoxin-producing fungus on these commodities; *A. parasiticus* also occurs frequently on peanuts. Both of these fungi produce a family of related aflatoxins; the aflatoxins most commonly produced by *A. flavus* are B1 and B2 (Figure 1) and *A. parasiticus* produces two additional aflatoxins, G1 and G2. Aflatoxin B1 is the most carcinogenic of the aflatoxins and is thus receiving the most attention in mammalian toxicology. The aflatoxin family of compounds can form "adducts" with animal and human DNA (*1, 2*) and can cause

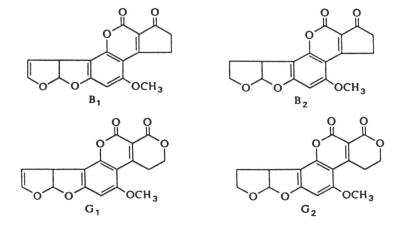

Figure 1.   Chemical structures of aflatoxins B1, B2, G1, G2.

primary liver cancer in certain animal systems (*3, 4*). The carcinogenicity of aflatoxins in human systems is less clear (*5*), but certain associations between aflatoxin intake by human populations and primary liver cancer have been reported (*3, 6*). Although aflatoxin contamination occurs world-wide (*7*), the extent of contamination in many countries is unknown because of reluctance to report its occurrence. With the implications to human and animal health worldwide, intense efforts are underway to remove these compounds from animal and human food supplies. The U.S. Food and Drug Administration (FDA) prohibits interstate commerce of dairy feed grain containing more than 20 ppb aflatoxin and the sale of milk containing more than 0.5 ppb aflatoxin.

Several conventional agronomic practices influence preharvest aflatoxin contamination of crops; these include: use of pesticides, altered cultural practices (such as irrigation), and use of resistant varieties (*8, 9, 10, 11*). However, such procedures have only a limited potential for reducing aflatoxin levels in the field. Detoxification (*2*) and absorptive removal (*13*) of aflatoxins from already-contaminated foods and feeds are two promising methods currently under extensive investigation; these should prove to be important control measures for immediate application, since less progress has been made in the development of practical methods to prevent preharvest aflatoxin contamination of foods and feeds. However, prevention of aflatoxin contamination before harvest is probably the best long-term approach; this strategy would obviate the need to detoxify large quantities of aflatoxin-contaminated seed material and avoid the uncertainties of gaining approval from regulatory agencies for the use of detoxified seed for animal or human food.

Domestic growers and food processors are being placed under increased pressure from consumer groups, merchants, and regulatory agencies to eliminate mycotoxins from food and feed. Therefore, there is an increasing need to develop new technology to reduce and eventually eliminate preharvest aflatoxin contamination. Elimination of preharvest contamination might be achieved with the development of novel biotechnological approaches (*14, 15*). To develop such approaches, additional knowledge is required in two broad areas: 1) fundamental molecular and biological mechanisms that regulate the biosynthesis of aflatoxin by the fungus and the ecological and biological factors that influence toxin production in the field; and 2) biochemistry of host-plant resistance to aflatoxin and/or aflatoxigenic molds. Knowledge in these areas will aid in development of novel methods to manipulate the chain of events in aflatoxin contamination. This chapter reviews the information developed towards achieving that goal.

Aflatoxin Biosynthesis

Chemistry and Biochemistry

Previous studies have determined that aflatoxins are biosynthesized by the polyketide metabolic pathway and are considered secondary metabolites because these compounds have no known function in the fungi that produce them (for review see *16*). The generally accepted scheme for aflatoxin biosynthesis is: polyketide precursor --> norsolorinic acid, NOR --> averantin, AVN --> averufanin, AVNN --> averufin, AVF --> hydroxyversicolorone, HVN --> versiconal hemiacetal acetate, VHA --> versicolorin A, VERA --> sterigmatocystin, ST --> O-methylsterigmatocystin, OMST --> aflatoxin B1, AFB1 (Figure 2). Recently, versicolorin B has been demonstrated to be a precursor of versicolorin A (*17, 18*). A branch point in the pathway has been established, following VHA production (Figure 3), leading to different aflatoxin structural forms B1 and B2 (*19-22*).

Initial attempts to purify enzymes from *A. parasiticus* mycelial extracts that catalyze aflatoxin synthesis were unsuccessful because these enzymes are present in relatively low concentrations and are extremely short-lived (*23*). Subsequently, several techniques were examined for disruption of large quantities of mycelia to obtain active and stable cell-free preparations (*23, 24*); pertinent enzymes were recovered from cell-free extracts after grinding mycelia under liquid nitrogen (*24*). The optimum age of mycelial cultures for recovery of aflatoxin pathway enzymes was determined to be between 72 and 84 h (*24a, 25*).

Several specific enzyme activities have been associated with precursor conversions in the aflatoxin pathway (*16, 18, 19, 22, 23, 26, 27*); some of these activities have been partially purified (*19, 28*). Bhatnagar et al (*29*) and Keller et al (*30*) have purified two distinct methyltransferases (168 KDa and 40 KDa) to homogeneity; both catalyze ST --> OMST conversion. Yabe et al. (*31*) identified two distinct methyltransferase activities in cell-free extracts of *A. parasiticus* strain NRRL 2999 which migrated with the 180 KDa and 210 KDa fractions on a gel filtration column; the former activity (180 KDa) could correspond to the protein (168 KDa) purified by Bhatnagar et al. (*29*) because it catalyzed ST to OMST conversion, whereas the 210 KDa fraction methylated only demethylsterigmatocystin (DMST) to yield ST. In addition, a 38 KDa reductase that catalyzes the reduction of NOR to AVN has been purified (*32*); an isozyme (48 KDa) of the reductase has also been purified (D. Bhatnagar, unpublished observations). It has been postulated (*16*) that alternate pathways may exist at several steps in the aflatoxin pathway, hence, different enzymes with similar catalytic functions may be isolated from pertinent fungal cells. A cyclase has also been purified from *Aspergillus parasiticus* which is involved in aflatoxin biosynthesis (*33*).

It has been demonstrated that independent reactions and different chemical precursors involved in AFB1 and AFB2 syntheses are catalyzed by common enzyme systems (*19, 21, 27, 30, 34*); AFB1 precursors are the preferred substrates for the relevant enzymes (*19*). A desaturase activity has been demonstrated in cell-free fungal extracts at the branch (Figure 3) in the AFB1/B2 biosynthetic pathways (*18*).

Genetics of Aflatoxin Biosynthesis

Classical genetic investigations. Early genetic investigations of *A. flavus* and *A. parasiticus* were hampered by the lack of any known means of sexual reproduction in these imperfect fungal species. Furthermore, a complex vegetative

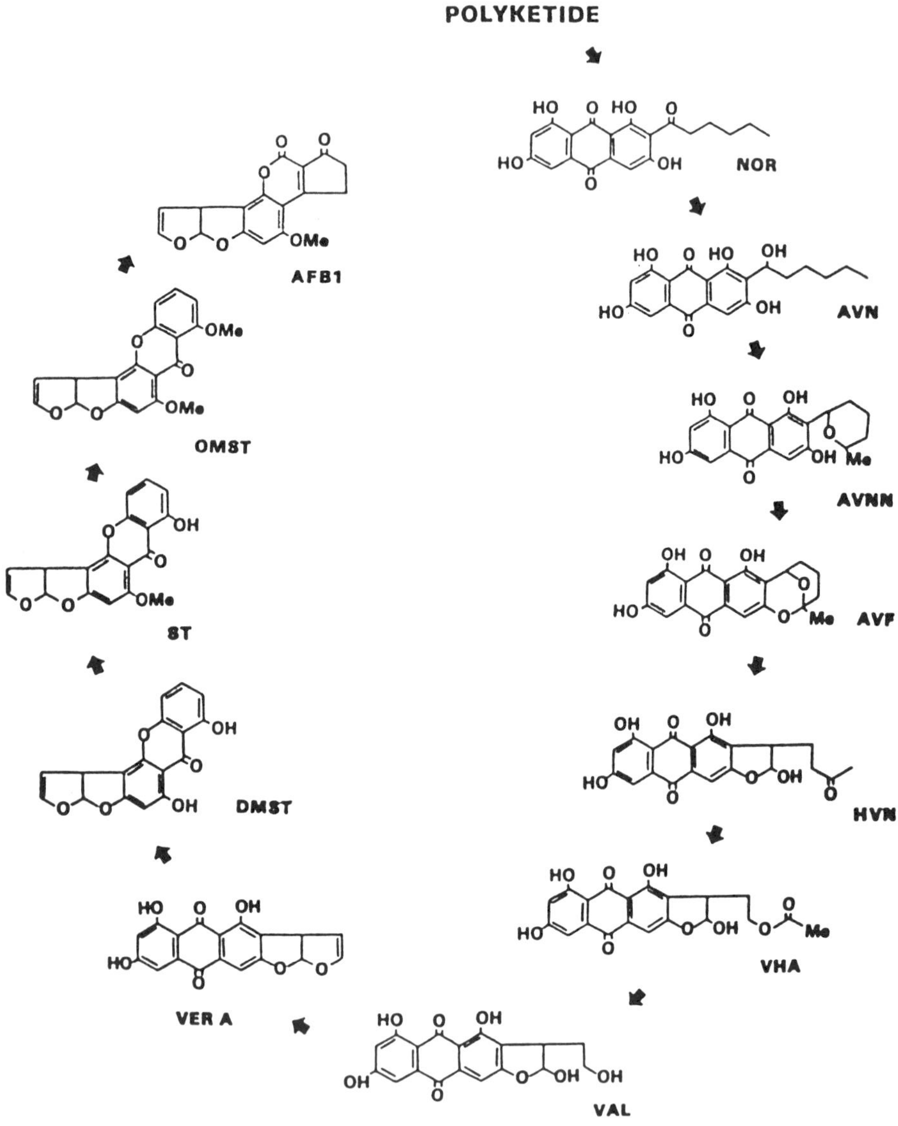

Figure 2. Scheme for aflatoxin B1 biosynthetic pathway.
NOR, norsolorinic acid; AVN, averantin; AVNN, averufanin; AVF, averufin; HVN, hydroxyversicolorone; VHA, versiconal hemiacetal acetate; VAL, versiconal alcohol; VER A, versicolorin A; DMST, demethylsterigmatocystin; ST, sterigmatocystin; OMST, O-methylsterigmatocystin; AFB1, aflatoxin B1.

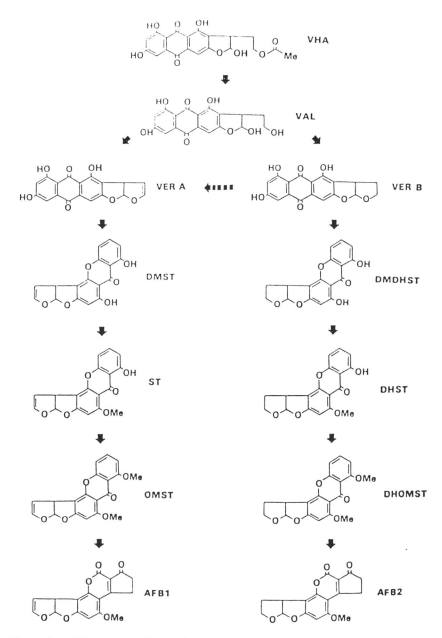

Figure 3.   Divergent pathways for aflatoxin B1 and B2 biosynthesis. VHA, versiconal hemiacetal acetate; VAL, versiconal alcohol; VER A, versicolorin A; DMST, demethylsterigmatocystin; ST, sterigmatocystin; OMST, O-methylsterigmatocystin; AFB1, aflatoxin B1; DMDHST, demethyl-dihydro-ST; DHST, dihydro-ST; DHOMST, dihydro-OMST; AFB2, aflatoxin B2.

compatibility system prevents hyphal anastamosis and nuclear exchange between all but the most closely related strains (*34a*.) However, mutants of *A. parasiticus* and *A. flavus* strains have been analyzed using the parasexual cycle (*35* and *36* for review). The genetics of *A. flavus* is better understood than that of *A. parasiticus*, and over 30 genes have been mapped to 8 linkage groups (*35, 37*). A recent technology helpful in resolving karyotypes and defining genetic maps of imperfect fungi has been pulsed-field gel electrophoresis. Karyotype analysis for several *A. flavus* and *A. parasiticus* strains show that there are 6-8 chromosomes ranging in size from approximately 3 to ≥ 7 mb (*38*). Strains that were used for linkage group studies of *A. flavus* have eight chromosomes; assignment of the eight linkage groups to these chromosomes is currently being determined. In addition, chromosome length polymorphisms and the presence of small chromosomes were observed in several *A. flavus* strains (*38*).

Molecular genetics of aflatoxin biosynthesis. With strains made available by Papa, research work of Payne and Woloshuk (*37*) has demonstrated that selectable markers could be transferred from one strain of *A. flavus* to another (perhaps to a more desirable strain for studying aflatoxin biosynthesis) through parasexual recombination. Genes in the aflatoxin biosynthetic pathway might be readily identified by their ability to complement strains of *A. flavus* with specific blocks in the pathway transformation (*37*). Genetic transformation systems have also been developed for *A. parasiticus* (*39, 40*). Individual conserved structural genes have been cloned from both species (*41, 42*).

Using complementary DNA hybridization with *A. flavus*, Payne and coworkers (*43*) have isolated a gene (afl-2) which appears to be directly involved in aflatoxin biosynthesis; the gene was identified by its ability to restore aflatoxin-producing capacity to a non-aflatoxin producing strain. A homogenous gene has also been isolated from *A. parasiticus* (*44*). Linz and coworkers have also identified two genes associated with aflatoxin biosynthesis in *A. parasiticus*: (a) the nor-1 gene, associated with the conversion of norsolorinic acid to averantin (*45*), and (b) the ver-1 gene, associated with the conversion of versicolorin A to sterigmatocystin (*46*). Molecular cloning of genes related to aflatoxin biosynthesis has also been attempted by using differential screening with different cDNA probes and colony hybridization procedures (*47*).

An alternative to identifying aflatoxin genes through transformation and complementation is to identify and characterize catalysts involved in aflatoxin biosynthesis and subsequently to isolate the genes coding for these enzymes. The purified proteins are used to obtain antibodies as immunoscreening probes and amino acid sequences for oligonucleotide probes; contemporary methods are available for screening and cloning genes of interest through pertinent antisera and oligonucleotide gene probes. cDNA libraries have been constructed (*34, 48*) that consist of cloned cDNA's synthesized with *A. parasiticus* mRNA's as templates and various nucleic acid polymerases; mRNA's were isolated from late-growth-phase mycelia (*24a*) during production of aflatoxin pathway enzymes. A 1.5 kb genomic DNA (pF9-1) from *A. flavus* NRRL 3557 identified with an oligonucleotide based on the N-terminus amino acid coding sequence of the 40 kDa methyltransferase (*30*); the enzyme is active in aflatoxin biosynthesis. Genomic DNA hybridizing to pF9-1 is present in isolates of *A. flavus* and *A. parasiticus* and some other members of *Aspergillus* section *Flavi*, but is not present in more distantly related aspergilli (*49*). Recently, using antibodies against the 40 kDa methyltransferase protein as probes, the same gene for the ST to OMST conversion activity has been cloned and its sequence determined (*50*).

Preharvest Control of Aflatoxin Contamination

Conventional technology for preharvest control of aflatoxin. Conventional methods that are currently being utilized successfully to control many plant pathogens in the field have not been effective in field control of aflatoxin-producing fungi that infect peanut, corn and cottonseed, the three major U.S. crops susceptible to preharvest aflatoxin contamination. Plant breeding programs have provided germplasm that shows less susceptibility to aflatoxin contamination (*8, 15, 51-53*), but none of the genetic material was "immune" to either infection or aflatoxin contamination by *A. flavus* or *A. parasiticus*. However, the results of these extensive plant breeding trials at least have provided hope that certain traits exist in plant varieties that could be combined by traditional plant breeding or perhaps by genetic engineering techniques to provide "elite" varieties with improved levels of resistance to aflatoxin contamination.

Cultural practices such as irrigation are effective in reducing aflatoxin contamination of peanut and corn (*54*), but this practice is not always available or cost effective to growers. Other conventional disease control practices, such as the use of fungicides, are largely ineffective in controlling *A. flavus* infection of crops when utilized at concentrations that are cost effective as well as environmentally safe.

Insecticidal control of the pink bollworm is yet another management practice used by growers in Arizona; pink bollworm exit holes in cotton bolls might provide portals of entry for *A. flavus* (*55*). Unfortunately, it is not economically feasible to achieve 100% control of the pink bollworm in cotton through the high-frequency use of insecticides, and even relatively low levels of infestation by this insect pest is well correlated to high levels of *A. flavus* infection and subsequent aflatoxin contamination.

Thus, conventional control practices are available that reduce aflatoxin levels in the field, but at a substantial and often unacceptable cost to the grower. However, the partial effectiveness of these control practices has suggested to researchers that "weak links" exist in the chain of events leading to aflatoxin contamination that could perhaps be exploited even more effectively to interrupt the contamination process.

Nonconventional methods. Novel biotechnological methods are needed to control aflatoxin contamination since conventional methods are only partially effective. Furthermore, conventional methods are not expected to achieve the extremely low or negligible levels of aflatoxin required to meet regulatory guidelines for the sale of commercial food and feed.

Control of Aflatoxin Contamination through Biotechnology

Three generic approaches for all aflatoxin-susceptible crops could be used to exclude toxigenic fungi from their environmental niches and to regulate aflatoxin biosynthesis:

1. Replace aflatoxigenic strains with nonaflatoxigenic strains in the field (a biocompetitive approach),

2. Incorporate into plant varieties, perhaps through genetic engineering, specific antifungal genes expressed in the specific plant tissues, e. g. seed tissues contaminated by aflatoxigenic strains (a host-plant resistance approach), and

3. Inhibit the biosynthetic or secretory processes responsible for aflatoxin contamination. This last approach could be the ultimate generic control method since the aflatoxin biosynthetic pathway is common to all aflatoxin contamination problems.

The novel biocompetitive strategies outlined here will complement conventional agronomic techniques which result in only partially preventing aflatoxin contamination.

Use of Biocompetitive Agents. The use of microbes to control aflatoxin contamination has been suggested repeatedly (56-59). This approach seeks to utilize organisms either to degrade aflatoxins after contamination has occurred (59) or to prevent infection and/or aflatoxin production by A. *flavus* through either antibiosis or competitive exclusion (56-58). Degradation of aflatoxins by microbes is, in most cases, not economically viable because adequate microbial action would be associated with unacceptable reductions in commodity quality. Microbes directed at preventing infection and/or contamination are, however, potentially useful. Microbes may be more environmentally acceptable, have a longer period of efficacy and be more readily distributed than agrochemicals. Furthermore, protective biocontrol agents may retain full efficacy when host defenses are least efficient as with damaged and stressed plant parts, i.e., in areas where the conditions are optimum for aflatoxin contamination to occur. Additionally, microbes applied early in the season may remain associated with the crop from early development through harvest and processing, control methods would, therefore, be applicable to both pre-harvest and post-harvest aflatoxin contamination.

Native strains of A. *flavus* have been discovered (55, 60) from agricultural fields that produce little or no aflatoxin. These strains maintain aggressiveness while lacking significant aflatoxin producing ability, implying that the aflatoxin trait is independent of aggressiveness during the invasion of plant tissues. Results of testing native non-toxigenic strains of A. *flavus* in biocontrol applications have recently been reported in detail from our laboratory (*55, 61-63*). Theoretically, native non-toxigenic strains of *A. flavus* could have the ability to compete with other aflatoxigenic *A. flavus* strains under identical agronomic and ecological conditions. The biocompetitive property of non-aflatoxigenic strains was first demonstrated by simultaneously inoculating developing cotton bolls with various aflatoxigenic (Strains A and B) and non-aflatoxigenic (Strain C) fungal strains via simulated pink bollworm exit holes (Table I) (*61, 64, 65*). Simultaneous inoculations resulted in significant (over 10 fold) reductions in toxin content of the seed at maturity. When nontoxigenic strains were inoculated 24 hr before toxigenic strains, contamination was either prevented or reduced over 100 fold (data not shown) (*62, 63*). These results indicate that non-toxigenic strains of *A. flavus* which occur naturally in agricultural fields may be potentially useful in controlling aflatoxin contamination.

Table I. Prevention of aflatoxin contamination of cottonseed by toxigenic strains of *Aspergillus flavus* with a strain of the fungus which does not produce aflatoxins (Cotty, 1989. Proc. 38th Oilseed Processing Clinic. pg. 30)

Strains inoculated[a]	Aflatoxin B1 in cottonseed (PPB)
Toxigenic strain A alone	72,000
Toxigenic strain B alone	17,000
Non -toxigenic strain C	0
Strain A plus Strain C	6,000
Strain B plus Strain C	0

[a] Immature bolls were inoculated via simulated pink bollworm exit holes in the greenhouse. Equal quantities of spores of each strain were used. Seed were harvested and analyzed after boll opening.

The above success of using native nontoxigenic strains of *A. flavus* to reduce aflatoxin contamination by toxigenic strains could encourage the development of "superior" biocontrol strains through genetic engineering. Engineered strains could potentially be constructed to obviate any concern of a non-toxigenic biocontrol strain acquiring toxigenicity through anastomosis with toxigenic strains in the field. This might be achieved through elimination (e.g., by homologous recombination) of two or more unlinked genes for specific enzymes essential to the aflatoxin biosynthetic pathway or through insertion of genes (e.g., antisense genes, *66*) which could either directly, or through their products, interfere with toxin synthesis. The elimination or modification of aflatoxin pathway genes for production of superior biocompetitive strains will necessitate cloning and characterization of these genes. Cloning of aflatoxin pathway genes is now feasible and, in some cases, has been accomplished recently (see "Aflatoxin Biosynthesis," previous section).

There are probably several fungal virulence factors that determine the ability of an *A. flavus* strain to infect and spread through host tissues. Engineered strains could also be constructed to augment aggressiveness traits and optimize infection site occupation and competitiveness while minimizing tissue disruption. Before aggressiveness traits can be selected for and manipulated in the fungal genome, they must be first identified. *A. flavus* cell wall degrading enzymes, such as pectinases, have been identified in this laboratory and have been shown to be associated with *A. flavus* aggressiveness and infection of cotton bolls (*67-67b*). Preliminary investigations have identified at least one fungal pectinase that is strongly correlated with aggressiveness during invasion of cotton bolls (*67b*). The discovery of enzymes involved in aggressiveness could be important in efforts to genetically regulate aggressiveness traits in the fungus to produce "superior" biocompetitive agents.

Enhancement of Resistance in Plants against Aflatoxigenic Fungi. Advantages and disadvantages must be considered in the conventional plant breeding and new plant transformation technologies for enhancement of host-plant resistance; plant breeding has the advantage of being a known technology, whereas plant transformation techniques have not become completely routine for incorporation of desirable genetic traits/genes into commercially important crops. In addition, plant breeding is an empirical approach that does not depend upon the identification of the biochemical mechanism or function of the trait being sought, and depends only on the ability to screen efficiently for the desirable trait (for example, disease resistance against a particular fungal pest). In contrast, plant transformation technology depends on identification and cloning of the desirable gene for incorporation into the plant genome. However, one of the disadvantages of the plant breeding approach, besides being time consuming, is that it is often impossible to transfer only the desirable genes of interest into the plant; many times genes of interest are closely linked to a multiple of other genetic traits/genes, perhaps some desirable, some neutral and some undesirable. The strength of the plant transformation approach is that it can be accomplished relatively quickly (if the technology is available) with only the selected genes of interest, provided they have been identified and cloned. Also, unlike the plant breeding approach, plant transformation technology allows genes to be transferred across the species barrier.

Before resistance traits can be enhanced either by plant breeding or through genetic engineering, specific chemicals linked to resistance must be identified. A further requirement of the genetic engineering approach is that a specific gene(s) for the trait must be identified, cloned and stably inserted into the plant genome. Corn kernels from varieties with varying levels of resistance/susceptibility to *A. flavus* contained chitinases and glucanases (*68*); these are hydrolytic enzymes often implicated in the lysis of fungal cell walls, and plant resistance to fungal pests (*69*). Although potential antifungal enzymes have been identified in corn kernels, a correlation between kernel resistance/susceptibility to *A. flavus* and/or aflatoxin contamination (*70*) and levels of hydrolytic antifungal enzymes in the kernels is lacking. Both large (*71*) and small molecular weight compounds (*72*) were detected in developing cottonseed and in cotton leaves, respectively, that inhibit aflatoxin biosynthesis in *A. flavus* liquid fermentations. In other research, investigators have shown that cotton and peanut contain constitutive levels of antifungal compounds and that under certain conditions these tissues can respond to invading aflatoxigenic molds by producing phytoalexins (73-79). The discovery of constitutive and *de novo*-produced antifungal compounds (phytoalexins) in crops subject to aflatoxin contamination suggests that endogenous resistance mechanisms exist that could be enhanced through conventional plant breeding or new genetic engineering methods. Exposure of cotton-leaf tissue to the fungus as well as to certain volatile compounds derived from cotton were shown to elicit sesquiterpenoids, such as the cadalenes and their oxidized products, lacinilenes (78, 80) which are considered phytoalexins in tissues remote from the source of these volatiles (Tables II, III). Certain volatiles have multiple effects on the physiology of *A. flavus* ranging from aflatoxin inhibition to aflatoxin stimulation and fungal growth inhibition (72, 79).

Table II. Effects of 2- and 7-Days Incubation of *A. flavus* in Contact with Cotton Leaf Volatiles (Adapted from 79)

Cotton Cultivar		Mycelial Dry Weight as a Percent of Control
2-Days		
8160[a]	Non-wounded	90.7 ± 9.1
8160	Wounded	32.4 ± 8.6
SJ-2[b]	Non-wounded	115.3 ± 6.7
SJ-2	Wounded	32.4 ± 3.2
7-Days		
8160	Non-wounded	100.6 ± 3.3
8160	Wounded	96.6 ± 4.0
SJ-2	Non-wounded	179.3 ± 5.6
SJ-2	Wounded	178.3 ± 1.9

[a] Glandless variety
[b] Glanded variety

In peanut, small molecular weight phytoalexins, the stilbenes (76, 77) are induced by the presence of invading aflatoxigenic fungi and appear to be correlated with ability to resist fungal attack.

Table III. Radial Growth of *A. flavus* as a Percent of Control After Two Days in Contact with Some Selected Volatiles (Adapted from *79*)

Volatile Component	Level of tested component (μl)				Volatile Concentration μmol/μl
	1	3	5	10	
Aldehydes					
hexanal	84 ± 5[a]	76 ± 3	76 ± 2	0 ± 0	8.3
trans-2-hexenal	0 ± 0	0 ± 0	0 ± 0	0 ± 0	8.6
2,4-hexadienal	53 ± 3	0 ± 0	0 ± 0	0 ± 0	9.0
2-hexenal, diethylacetal	98 ± 2	0 ± 0	0 ± 0	0 ± 0	4.9
heptanal	67 ± 5	58 ± 3	49 ± 6	0 ± 0	7.4
trans-2-heptenal	82 ± 3	0 ± 0	0 ± 0	0 ± 0	7.6
octanal	114 ± 7	88 ± 5	50 ± 3	46 ± 3	6.5
trans-2-octenal	77 ± 3	0 ± 0	0 ± 0	0 ± 0	6.7
nonylaldehyde	75 ± 4	60 ± 3	46 ± 3	0 ± 0	5.8
trans-2-nonenal	82 ± 2	0 ± 0	0 ± 0	0 ± 0	6.0
N-decylaldehyde	96 ± 2	92 ± 8	91 ± 3	91 ± 3	5.3
dodecyl aldehyde	136 ± 5	112 ± 4	104 ± 3	104 ± 2	4.5

[a]   Radial growth (cm) of the fungus in a Petri plate on solid medium; Mean ± SD for 3 replicates/tested level.

Genes for some of the potential antifungal hydrolases (chitinases and glucanases) and for the biosynthetic enzymes catalyzing synthesis of certain phytoalexins (e.g., stilbenes) have been cloned (Table IV) and could serve as tools in genetic engineering for resistance against toxigenic fungi. For example, the gene for resveratrol synthase, a key enzyme catalyzing biosynthesis of resveratrol (a stilbene phytoalexin), has been cloned from peanut (*81*). Cloning of the resveratrol synthase gene from peanut has important implications in genetic engineering of plants other than peanut, since, if this gene is incorporated into other plants, stilbenes could be synthesized from precursors (4-coumaroyl-CoA and malonyl-CoA) commonly available in various plant species. Genes for other key enzymes (e.g., phenylalanine lyase) (*82*) involved in formation of certain phenylpropanoid precursors of stilbenes could also be useful in genetic engineering to enhance resistance in peanut, perhaps with genetic engineering techniques to enhance gene expression, such as insertion of multiple gene copies and use of more powerful plant gene promoters or enhancers.

Table IV.   Potential fungal growth inhibitors from cotton, corn (or other related grains) and peanut (Adapted from *14*)

Hydrolases	Lytic Peptides	Phytoalexins
Glucanases[a] (CO[b],C)	Zeamatins (C)	Stilbenes[c] (P)
Chitinases[a] (P,C)	Thionins[a] (OG)	Lacinilenes (CO)

[a]   Direct gene products and some have been cloned from plants.
[b]   CO =cotton; C =corn; P =peanut; OG = other grains.
[c]   Not direct gene product; however, gene for enzyme (resveratrol synthase) catalyzing synthesis of stilbene phytoalexin in peanut has been cloned.

Use of Natural Product Inhibitors to Control Aflatoxin Contamination. There are several plant-derived, natural product inhibitors of aflatoxin synthesis and this subject has been reviewed extensively (*83*).   Inhibitors with unknown mechanisms of action have been discovered in our laboratory (*79, 84*) that could be subject to biotechnological utilization through direct application to crops in the field or through the molecular design of ecologically safe pesticides based on the chemical structures of these inhibitors.   Certain of these natural product inhibitors, that occur naturally in crops commonly contaminated with aflatoxin producing fungi, could serve as markers for enhancement of aflatoxin resistance traits in plants through classical plant breeding or contemporary molecular engineering techniques.

Examples of natural products that may have potential in augmenting host plant resistance against *A. flavus* infection are certain plant derived volatile compounds (*78, 85*) as described earlier.

Other naturally derived aflatoxin inhibitors obtained from the "neem" tree have been investigated in our laboratory (*84*). *Azadirachta indica* Juss. commonly known as "margosa" or "neem" is an ornamental tree of Asia and Africa that produces natural products having reputed value for their medicinal, antiviral, antibacterial, insecticidal, antifungal and antinematode properties (*86, 87*).   Several active principles from different parts of the neem tree have been reported (*88*).   Our investigation (*84*) examined the effects of these neem leaf components in neem leaves on aflatoxin biosynthesis by either *Aspergillus parasiticus* or *A. flavus*.

Neem extracts when added to fungal growth media prior to inoculation (Table V), did not affect fungal growth (i.e., mycelial dry weight), but essentially blocked (> 98%) aflatoxin biosynthesis at concentrations greater than 10% (vol/vol). The inhibitory effect was somewhat diminished (60-70% inhibition) in heated leaf extracts.   Volatile components of the extracts were analyzed using capillary gas chromatography/mass spectrometry, and the bioactivity of the neem leaf volatiles was assessed by measuring the fungal growth and aflatoxin production by the fungus grown on agar medium in a Petri plate and exposed to an atmosphere containing volatiles from neem leaf extracts.   Volatiles from blended leaf extracts, however, did not affect either aflatoxin synthesis (21.3 µg total aflatoxin in control and 20.4 µg in treated) or fungal growth. Therefore, the inhibitory component was a soluble ingredient which may be somewhat heat liable.

Table V. Effect of Concentration of Neem Leaf Extract in the Incubation Medium on Aflatoxin B1 Biosynthesis (Adapted from *84*)

Concentration of extract (vol/vol)	Aflatoxin B1 (% of control)[a]			
	Blended extract	Blended extract (autoclaved)	Heat extracted	Heat extracted (autoclaved)
0	100 [b]	100	100	100
1	15.1	48.9	52.3	60.2
5	6.5	36.2	35.2	43.4
10	2.6	24.6	18.7	35.2
20	2.0	17.8	9.2	29.8
50	1.8	16.3	6.4	31.2

[a]   The pooled mean standard error in the results was ± 12.4% (n=3).
[b]   100% refers to 20.6 µg aflatoxin B1 produced/g mycelial dry weight.

Inhibition of aflatoxin biosynthesis by neem extracts in fungal cells appear to occur in the very early stages of the biosynthetic pathway (i.e., prior to norsolorinic acid synthesis) because after the initiation of secondary metabolism, the inhibitory effect of the neem leaf constituents was lost (*84*).

If the inhibitory factor in neem leaf extracts could be effective in field studies, these extracts could be used in controlling the preharvest aflatoxin contamination of food and feed commodities. Therefore, greenhouse experiments were conducted to test the effectiveness of these extracts in developing cotton bolls (*89*). In separate treatments, a spore suspension of *A. flavus* (control), the aqueous neem leaf extract plus a spore suspension of *A. flavus,* or the extract followed by an *A. flavus* cotton spore suspension after 48 hrs were injected onto the surfaces of locks of developing cotton bolls (30-day post anthesis). Thirteen days after the treatments, the seeds from the locules were harvested and both fungal growth and aflatoxin production were determined. Fungal growth was unaffected by the treatments but the seeds from locules receiving both neem leaf extracts and *A. flavus* simultaneously exhibited 16% inhibition of aflatoxin production, while the seeds in locules receiving *A. flavus* spores 48 hrs after neem extract was added exhibited >98% inhibition in aflatoxin production. From these results it appears that the aflatoxin inhibiting factor in the neem leaf extracts may need to translocate from the fibrous locule surface to the seed prior to the fungal inoculation for maximal effect. Experiments are underway using individual, separated components from the neem leaf extract to determine the component(s) responsible for the bioactivity described. Effective ways of delivering this bioactive natural product to the cotton seed in developing cotton bolls are also being designed. The practical application of this discovery will be utilized in field trials in attempts to eliminate the preharvest aflatoxin contamination of cottonseed as well as other crops.

Summary

Several approaches are being explored and developed using new methods in biotechnology to eliminate pre-harvest aflatoxin contamination of food and feed. These approaches resulted from recent information acquired on: 1) non-aflatoxigenic A. *flavus* strains that prevent aflatoxin contamination of cottonseed when co-inoculated with aflatoxigenic strains, 2) molecular mechanisms governing aflatoxin biosynthesis, and 3) plant-derived metabolites that inhibit aflatoxin biosynthesis.

Agricultural fields often harbor strains of A. *flavus* that produce little or no aflatoxin during invasion of cotton bolls; these strains greatly reduced aflatoxin contamination when the bolls were co-inoculated with aflatoxigenic strains. Therefore, non-toxigenic strains are potential biocontrol agents.

Significant progress has been made in identifying enzymes that are specific to the aflatoxin pathway and those that are involved in strain aggressiveness. The cloning of aflatoxin pathway genes and aggressiveness genes is now feasible. Efforts are underway in this laboratory to clone aflatoxin pathway genes to be used as molecular tools in the production of stable aflatoxin non-producers by genetic engineering for future use in biocontrol applications. Similarly, aggressiveness genes could be cloned and used as tools in fungal genetic engineering for optimization of strain competitiveness.

Volatile compounds originating from the aflatoxin susceptible crop, cotton, and other plant-derived compounds that inhibit aflatoxin production have been identified. The various compounds are being tested for their potential to enhance host plant resistance by inhibition of fungal growth/aflatoxin production.

Experience in our laboratory suggests a combined approach utilizing both host defense augmentation and biological control will be necessary to complement existing conventional methods in the eventual elimination of aflatoxin from the food and feed supply.

Literature Cited

1. Hsu, I. C.; Metcalf, R. A.; Sun, T.; Welsh, J. A.; Wang, N. J. and Hams, C. C. *Nature* **1991**, *350*, pp. 427-428.
2. Bressac, B.; Kew, M.; Wands, J. and Ozturk, M. *Nature* **1991**, *350*, pp. 429--431.

3. Groopman, J. D. and Sabbioni, G. "Detection of aflatoxin and its metabolites in human biological fluids" In *Mycotoxins, Cancer and Health*; Bray, G. A. and Ryran, D. H., Eds.; Pennington Center Nutrition Series; Louisiana State University Press: Baton Rouge, LA, **1991**, *Vol. 1*, pp. 18-31.

4. Wogan, G. N. "Aflatoxins as risk factors for primary hepatocellular carcinoma in humans" In *Mycotoxins, Cancer and Health;* Bray, G. A. and Ryan, D. H., Eds.; Pennington Center Nutrition Series; Louisiana State University Press: Baton Rouge, LA, **1991**, *Vol. 1*, pp. 3-17.

5. Campbell,T. C.; Chen, J.; Liu, C.; Li, J. and Parpia, B. *Cancer Res.* **1990**, *50*, pp. 6882-6893.

6. Yeh, F.- S.; Ju, M. C.; Mo, C.- C.; Lou, S.; Tong, M. J. and Henderson, B. E. *Cancer Res.* **1989**, *49*, pp. 2506- 2709.

7. Jelinek, C. F.; Pohland, A. E. and Wood, G. E. *J. Assoc. Of. Anal. Chem.* **1989**, 72, pp. 223-230.

8. Widstrom, N. W. "Breeding strategies to control aflatoxin contamination of maize through host plant resistance" In *Aflatoxin in maize: A proceedings of the workshop*; Zuber, M. S., et al., Eds.; CIMMYT: Mexico City, Mexico, **1987**, pp. 212-220.

9. Scott, G. E. and Zummo, N. *Crop Science* **1988**, 28, pp. 504-507.

10. Darrah, L. L. and Barry, B. D. "Reduction of preharvest aflatoxin in corn" In *Pennington Center Nutrition Series, Mycotoxins, Cancer and Health*; Bray, G. A. and Ryan, D. H., Eds.; LSU Press: Baton Rouge, LA, **1991**, *Vol. 1*, pp. 283-310.

11. Lillehoj, E.B. "Aflatoxin: genetic mobilization agent" In *Mycotoxins in ecological systems*.; Bhatnagar, D., Lillehoj, E. B. and Arora, D. K., Eds.; Marcel Dekker, Inc.: New York, NY, **1991**, pp. 1-22.

12. Park, D. L.; Lee, L. S.; Price, R. L. and Pohland, A. E. *J. Assoc. Of. Anal. Chem.* **1988**, 71, pp. 685-703.

13. Phillips, T.D.; Sarr, B. A.; Clement, B. A.; Kubena, L. F. and Harvey, R. B. "Prevention of aflatoxicosis in farm animals via selective chemisorption of aflatoxin" In *Pennington Center Nutrition Series, Mycotoxins, Cancer and Health*; Bray, G. A. and Ryan, D. H., Eds.; LSU Press: Baton Rouge, LA, **1991**, *Vol. 1*, pp. 223-237.

14. Cleveland, T. E. and Bhatnagar, D. "Molecular strategies for reducing aflatoxin levels in crops before harvest" In *Molecular Approaches to Improving Food Quality and Safety*; Bhatnagar, D. and Cleveland, T. E., Eds.; Van Nostrand Reinhold: New York, NY, **1992**, pp. 205-228.

15. Payne, G. A. *Critical Rev. Plant Sci.* **1992**, 10, pp. 423-440.

16. Bhatnagar, D.; Ehrlich, K. C. and Cleveland, T. E. "Oxidation-reduction reactions in biosynthesis of secondary metabolites" In *Mycotoxins in ecological systems*; Bhatnagar, D.; Lillehoj, E. B. and Arora, D. K., Eds.; Marcel Dekker, Inc.: New York, NY, **1991**, pp. 255-286.

17. McGuire, S. M.; Brobst, S. W.; Graybill, T. L.; Pal, K. and Townsend C. A. *J. Am. Chem. Soc.* **1989**, 111, pp. 8308-8309.

18. Yabe, K.; Ando, Y. and Hamasaki, T. *Agric. Biol. Chem.* **1991**, 55, pp. 1907-1911.
19. Bhatnagar, D.; Cleveland, T. E. and Kingston, D. G. I. *Biochemistry* **1991**, 30, pp. 4343-4350.
20. Cleveland, T. E.; Bhatnagar, D.; Foell, C. J. and McCormick, S. P. *Appl. Environ. Microbiol.* **1987**, 53, pp. 2804-2807.
21. Yabe, K.; Ando, Y. and Hamasaki, T. *Appl. Environ. Microbiol.* **1988**, 54, pp. 2101-2106.
22. Yabe, K.; Ando, Y. and Hamasaki, T. *J. Gen. Microbiol.* **1991**, 137, pp. 2469-2415.
23. Dutton, M. F. *Microbiol. Rev.* **1988**, 52, pp. 274-295.
24. Bhatnagar, D.; Cleveland,T. E. and Lillehoj, E. B. *Mycopathologia* **1989**, 107, pp. 75-83.
24a. Cleveland, T. E.; Lax, A. R.; Lee, L. S. and Bhatnagar, D. *Appl. Environ. Microbiol.* **1987**, *53*, pp. 1711-1713.
25. Chuturgoon, A. A. and Dutton, M. F. *Mycopathologia* **1991**, 113, pp. 41-44.
26. Hsieh, D. P. H.; Wan, C. C. and Billington, J. A. *Mycopathologia* **1989**, 107, pp. 121-126.
27. Anderson, J. A.; Chung, C. H. and Cho, S.-H. *Mycopathologia* **1990**, 111, pp. 39-45.
28. Chuturgoon, A. A.; Dutton, M. F. and Berry, R. K. *Biochem. Biophys. Res. Comm.* **1990**, 166, pp. 38-42.
29. Bhatnagar, D.; Ullah, A. H. J. and Cleveland, T. E. *Preparative Biochemistry* **1988**, 18, pp. 321-369.
30. Keller, N. P.; Dischinger, H. C.; Bhatnagar, D.; Cleveland, T. E. and Ullah, A. H. J. *Appl. Environ. Microbiol.* **1992**, 59, (in press).
31. Yabe, K.; Ando, Y.; Hashimoto, J. and Hamasaki, T. *Appl. Environ. Microbiol.* **1989**, 55, pp. 2172-2177.
32. Bhatnagar, D. and Cleveland, T. E. *FASEB J.* **1990**, 4, pg. A2164.
33. Lin, B.-K. and Anderson, J. A. *Arch. Biochem. Biophys.* **1992**, 293, pp. 67-70.
34. Bhatnagar, D. and Cleveland, T. E. "Aflatoxin biosynthesis: developments in chemistry, biochemistry and genetics" In *Aflatoxin in corn: New perspectives*; Shotwell, O. L. and Hurburgh, Jr., C. R., Eds; Iowa State University Press: Ames, IA, **1991**, pp. 391-405.
34a. Bayman, P. and Cotty, P. J. *Can. J. Bot.* **1991**, *69*, pp. 1707-1711.
35. Bennett, J. W. and Papa, K. E. "The aflatoxigenic *Aspergillus* spp." In *Advances in plant pathology: Genetics of Plant Pathogenic Fungi*; Sidhu, G. S., Ed.; Academic Press: New York, NY, **1988**, *Vol. 6*, pp. 263-280.
36. Keller, N. P.; Cleveland, T. E. and Bhatnagar, D. "A molecular approach towards understanding aflatoxin production" In *Mycotoxins in ecological systems*; Bhatnagar, D., Lillehoj, E. B. and Arora, D. K., Eds.; Marcel Dekker, Inc.: New York, NY, **1991**, pp. 287-310.
37. Payne, G. A. and Woloshuk, C. P. *Mycopathologia*, **1989**, 107, pp. 139-144.

38. Keller, N. P.; Cleveland, T. E. and Bhatnagar, D. *Current Genetics* **1992**, 21, pp. 371-315.
39. Skory, C. D.; Horng, J. S.; Pestka, J. J. and Linz, J. E. *Appl. Environ. Microbiol.* **1990**, 56, pp. 3315-3320.
40. Horng, J. S.; Chang, P.-K.; Pestka, J. J. and Linz, J. E. *Mol. Gen. Genet.* **1990**, 224, pp. 294-296.
41. Horng, J. S.; Linz, J. E .and Pestka, J. J. *Appl. Environ. Microbiol.* **1989**, 55, pp. 2561-2568.
42. Seip, E. R.; Woloshuk, C. P.; Payne, G. A. and Curtis, S. E. *Appl. Environ. Microbiol.* **1990**, 56, pp. 3686-3692.
43. Payne, G. A.; Nystrom, G. J.; Bhatnagar, D.; Cleveland, T. E. and Woloshuk, C. P. *Appl. Environ. Microbiol.* **1993**, 59, pp. 156-162.
44. Chang, P.-K.; Cary, J.; Bhatnagar, D.; Cleveland, T. E.; Bennett, J. W.; Linz, J.; Woloshuk, C. and Payne, G. A. *93rd ASM Annual General Meeting Abstracts* **1993**, (in press).
45. Chang, P.-K.; Skory, C. D. and Linz, J. E. *Curr. Genet.* **1992**, 21, pp. 231-233.
46. Skory, C. D.; Chang, P.-K.; Cary, J. and Linz, J. E. *Appl. Environ. Microbiol.* **1992**, 58, pp. 3527-3537.
47. Feng, G. H.; Chu, F. S. and Leonard, T. J. *Appl. Environ. Microbiol.* **1992**,58, pp. 455-460.
48. Cary, J. W.; Cleveland, T. E. and Bhatnagar, D. *FASEB J.* **1992**, 6, pg. A228.
49. Keller, N. P.; Cary, J. W.; Cleveland, T. E.; Bhatnagar, D. and Payne, G. A.*FASEB J.* **1992**, 6, pg. A228.
50. Yu, J.; Cary, J. W.; Bhatnagar, D.; Cleveland, T. E. and Chu, F. S. *93rd ASM Annual General Meeting Abstracts* **1993**, (in press).
51. Gorman, D. P. and Kang, M. S. *Plant Breeding* **1991**, 107, pp. 1-10.
52. Scott, G. E.; Zummo, N.; Lillehoj, E. B.; Widstrom, N. W.; Kang, M. S.; West, D. R.; Payne, G. A.; Cleveland, T. E.; Calvert, 0. H. and Fortnum, B. A. *Agronomy J.* **1991**, 83, pp. 595-598.
53. Mehan, V. K.; McDonald, D. and Ramakrishna *Peanut Sci.*, **1986**, 13, pp. 7-10.
54. Payne, G. A.; Cassel, D. K. and Adkins, C. R. *Phytopathology* **1986**, 76, pp. 679-684.
55. Cotty, P. J. *Phytopathology*, **1989**, 79, pp. 808-814.
56. Ashworth, L. J.; McMeans, J. L. and Brown, C. M. J. *Stored Prod. Res.* **1969**, 5, pp. 93-202.
57. Kimura, N. and Hirano, S. *Agric. Biol. Chem.* **1988**, 52, pp. 1173-1178.
58. Wicklow, D. T. and Shotwell, O. L. *Can. J. Microbiol.* **1983**, 29, pp. 1-5.
59. Bhatnagar, D.; Lillehoj, E. B. and Bennett, J. W. "Biological detoxification of mycotoxins" In *Mycotoxins and Animal Foods.*; Smith, J. E. and Henderson, R. S., Eds.; CRC Press, Inc.: Boca Raton, FL, **1991**, pp. 815-826.
60. Joffe, A. Z. *Nature* **1969**, 221, pg. 492.
61. Cotty, P. J. *Plant Disease* **1989**, 73, pp. 489-492.

62.   Cotty, P. J. *Plant Disease* **1990**, 74, pp. 233-235.
63.   Cotty, P. J.; Bayman, P. and Bhatnagar, D. *Phytopathology* **1990**, 90, pg. 995.
64.   Cotty, P. J. and Lee, L. S. *Trop. Sci.* **1989**, 29, pp. 273-277.
65.   Cotty, P. J. *Plant Disease* **1991**, 75, pp. 312-314.
66.   Rothstein, S. J.; DiMaio, J.; Strand, M. and Rice, D. *Proc. Natl. Acad. Sci.* **1987**, 84, pp. 8439-   .
67.   Cleveland, T. E. and McCormick, S. P. *Phytopathology* **1987**, 77, pp. 1498-1503.
67a. Cleveland, T. E. and Cotty, P. J. *Phytopathology* **1991**, *81*, pp. 155-158.
67b. Brown, R. L.; Cleveland, T. E.; Cotty, P. J. and Mellon, J. E. *Phytopathology* **1992**, *82*, pp. 462-467.
68.   Neucere, J. N.; Cleveland, T. E. and Dischinger, C. J. *Agric. Food Chem.* **1991**, 39, pp. 1326-1328.
69.   Boiler, T. "Induction of hydrolases as a defense reaction against pathogens" In *Cellular and Molecular Biology of Plant Stress*; Key, J. L. and Kosaye, T., Eds.; A. R. Lisa: New York, NY, pp. 247-262.
70.   Lisker, N. and Lillehoj, E. B. "Prevention of mycotoxin contamination (principally aflatoxins and *Fusarium* toxins) at the preharvest stage" In *Mycotoxins and Animal Foods*; Smith, J. E. and Henderson, R. S., Eds.; CRC Press: Baton Rouge, LA, **1991**, pp. 689-719.
71.   McCormick, S. P.; Bhatnagar, D.; Goynes, W. R. and Lee, L. S. *Can. J. Bot.* **1988**, 66, pp. 998-1002.
72.   Zeringue, H. J. and McCormick, S. P. *Toxicon* **1990**, 28, pp. 445-448.
73.   Bell, A. A. and Stipanovic, R. D. *Mycopathologia* **1978**, 65, pp. 91-106.
74.   Bell, A. A. *Ann. Rev. Plant Physiol.* **1981**, 32, pp. 21-81.
75.   Bell, A. A. "Morphology, Chemistry and genetics of *Gossypium* adaptations to pests" In *Phytochemical Adaptions to Stress, Recent Advances in Phytochemistry* Timmermann, B, N.; Steelink, C. and Loewus, F. A., Eds.; Plenum Press: New York, NY, *Vol. 18*, **1983**, pp. 197-229.
76.   Cole, R. J.; Sanders, T. H.; Dorner, J. W. and Blankenship, P. D. "Environmental conditions required to induce preharvest aflatoxin contamination of groundnuts: summary of six years research" In *Aflatoxin contamination of groundnuts*; ICRISAT publication: Patancheru, India, **1989**, pp. 279-287.
77.   Wotton, H. R. and Strange, R. N. *Appl. Environ. Microbiol.* **1987**, 53, pp. 270-273.
78.   Zeringue, H. J., Jr. *Phytochemistry* **1987**, 26, pp. 1357-1360.
79.   Zeringue, H. J., Jr. and McCormick, S. P. *J. Am. Oil Chem. Soc.* **1989**, 66, pp. 581-585.
80.   Zeringue, H. J., Jr. *Appl. Environ. Microbiol.* **1991**, 57, pp. 2433-2434.
81.   Lanz, T.; Schroder, G. and Schroder, J. *Planta* **1990**, 181, pp. 169-175.
82.   Habereder, H. G.; Schroder, G. and Ebel, J. *Planta* **1989**, 177, pp. 58-65.
83.   Zaika, L. L. and Buchanan, R. L. *J. Food Protec.* **1987**, 50, pp. 691-708.

84. Bhatnagar, D. and McCormick, S. P. *J. Am. Oil Chemists' Soc.* **1988**, 65, pp. 1166-1168.

85. Wilson, D. M.; Gueldner, R. C.; McKinney, J. K. and Lieusay, R. H. *J. Am. Chem. Soc.* **1981**, 58, pg. 959.

86. Singh, U. P.; Singh, U. B. and Singh, R. B. *Mycologia* **1980**, 72, pg. 1077.

87. Jacobson, M. "The Neem Tree: National Resistance Par excellence" In *National Resistance of Plants to Pests*; American Chemical Society Publications: pg. 225.

88. Satyavati, G. V.; Raine, M. K. and Sharma, M. "Medicinal Plants of India: *Azadirachta (Meliaceae)*; Indian Council of Medical Research Publication: New Delhi, India, *Vol. I*, pp. 487.

89. Zeringue, H. J., Jr. and Bhatnagar, D. *J. Am. Oil. Chemists' Soc.* **1990**, 67, pp. 215-216.

RECEIVED February 8, 1993

Chapter 23

Effects of Ionizing Radiation Treatments on the Microbiological, Nutritional, and Structural Quality of Meats

Donald W. Thayer, Jay B. Fox, Jr., and Leon Lakritz

Food Safety Research Unit, Eastern Regional Research Center, Agricultural Research Service, U.S. Department of Agriculture, 600 East Mermaid Lane, Philadelphia, PA 19118

Treating fresh or frozen meats with ionizing radiation is an effective method to reduce or eliminate several species of food-borne human pathogens such as *Salmonella, Campylobacter, Listeria, Trichinella,* and *Yersinia.* It is possible to produce high quality, shelf-stable, commercially sterile meats. Irradiation dose, processing temperature, and packaging conditions strongly influence the results of irradiation treatments on both microbiological and nutritional quality of meat. These factors are especially important when irradiating fresh non-frozen meats. Radiation doses up to 3.0 kGy have little effect on the vitamin content, enzyme activity, and structure of refrigerated non-frozen chicken meat, but have very substantial effects on food-borne pathogens. Some vitamins, such as thiamin, are very sensitive to ionizing radiation. Thiamin in pork is not significantly affected by the FDA-approved maximum radiation dose to control *Trichinella*, but at larger doses it is significantly affected.

Treating fresh or frozen meats with ionizing radiation in the form of gamma rays from cobalt-60 or cesium-137, accelerated electrons of 10 MeV or lower energy, or X-rays of less than 5 MeV can reduce the populations or eliminate many food-borne pathogens and extend the shelf life of the product. This manuscript describes the effects of ionizing radiation on the microbiological, nutritional, and structural quality of meat (edible tissue of vertebrate animals) and discusses how and why processing variables may dramatically alter the results of the treatments.

Appropriateness of Technology

The appropriateness of any meat processing technology is determined by its ability to control food-borne pathogens and spoilage microorganisms without adversely affecting the wholesomeness, nutritive value, and organoleptic properties. The technology must also be economically competitive. The effects of ionizing radiation on food-borne pathogens and on the meat itself depend on the absorbed radiation dose, irradiation temperature, irradiation atmosphere, packaging, dose rate, storage time and temperature before cooking and consumption, and cooking method.

Food irradiation cannot substitute for proper food sanitation, packaging, refrigerated storage, and cooking. The primary reason for treating fresh or frozen meat with ionizing radiation is to eliminate food-borne pathogens. Extending shelf life of fresh, non-frozen meats may result, but is secondary in importance. Shelf-stable meats that can be stored at room temperature without refrigeration can be produced. Sterile, refrigerated meats suitable for consumption by immuno-compromised hospital patients may be prepared. Each of these products requires specific processing conditions; ionizing radiation treatments are often self-limiting because of changes in the organoleptic properties of the treated meat at higher dose levels.

The following are recommended as tests of the wholesomeness of irradiated foods (1): failure to induce gene mutations in bacteria or in cultured mammalian cells; failure to alter DNA repair in mammalian cells; failure to induce recessive lethal mutations in *Drosophila*; and failure to exhibit evidence of treatment related toxicological effects in 90-day feeding studies with a non-rodent species and a rodent species that include *in utero* exposure of the fetuses to the test material. These tests are inherently different from toxicological tests of food additives because an irradiated meat cannot be included in the diet of test animals in amounts that are greatly in excess of those usually found in the diet without causing toxicity from excessive protein consumption.

All meats are considered to be a good source of high quality protein, and red meats a source of the B-complex vitamins. Tests for the effects of ionizing radiation treatments on the nutritive value of meat may include analyses of amino acids, fatty acids, and vitamins and include the effects of storage time and temperature and cooking on these nutrients in the irradiated product. Gross tests of the food value of the meat, such as for the protein efficiency ratio of the irradiated meat, may reveal subtle changes in the irradiated meat. The treatment may also change the sensory properties of the meat. Some of these, such as texture, color, firmness, softness, juiciness, chewiness, taste, and odor are closely related to possible changes in nutrient value. Enzymatic and ultrastructural studies may be performed to help define the nature of the textural changes in irradiated meats. These sensory changes cannot be too severe or the treated meat will have no economic value.

Each of the effects of ionizing radiation on red meat, poultry meat, or food-borne pathogens, whether beneficial or adverse, is predictable from the known chemistry of meat and fundamental principles of radiation chemistry.

Predictability

The initial reaction of ionizing radiation is most likely to occur with water because approximately 70% of both meat and food-borne pathogens are water. The initial reactions occur with an electron, whether the source of the radiation is a photon or an accelerated electron, because photon energies from X-rays of less than 5 MeV or gamma-rays from isotopic sources such as ^{137}Cs or ^{60}Co primarily interact with water through the Compton effect (*2*). When the energy of a photon exceeds the binding energy of an electron, it is ejected and the Compton photon is scattered. A small fraction of the photon energy is converted into the kinetic energy of the freed electron. In water that freed electron becomes solvated very quickly. Klassen (*3*) has described the ionization of water at 25°C by the absorption of 100 eV as follows:

$$4.14\ H_2O = 2.7\ e_{aq}^- + 2.7\ H^+ + 2.87\ OH^-$$
$$+\ 0.43\ H_2 + 0.61\ H_2O_2 + 0.026\ HO_2$$

The primary products are the hydrated electron (e_{aq}^-), the proton, and the hydroxyl radical. Lesser but significant amounts of hydrogen and hydrogen peroxide are produced. Enough hydrogen is produced to warrant consideration when canning meats that will be sterilized with ionizing radiation. The maximum radiation dose approved in the United States to eliminate food-borne pathogens from fresh or frozen poultry is 3.0 kGy (*4*). Since 3 kGy equals the absorption of 3 kJ/kg, the amount of energy absorbed by the meat is 1.87 X 10^{16} eV/g. A G value of 2.7 at 25°C indicates that 0.84 nmole of product will be formed in meat that has received 3 kGy of ionizing radiation, and the temperature of the meat will increase 0.72°C. The actual yield of radiolytic products in meat will be much less because of many competing reactions for the active species. Both the temperature and the physical state of the product during irradiation affect the results. These effects cause the G value for formation of the hydrated electron to decrease from 2.7 at 25°C to 0.3 at -5°C (*5*). Oxygen rapidly reacts with both the hydrated electron and the proton, eliminating them from further reactions.

Other than water, protein is the major constituent of meat averaging nearly 21% in beef or chicken meat, with fat varying from 4.6 to 11.0% in beef and from 2.7 to 12.6% in chicken. The principal radiolytic reactions of aqueous solutions of aliphatic amino acids are reductive deamination and decarboxylation. Alanine yields NH_3, pyruvic acid, acetaldehyde, propionic acid, CO_2, H_2, and ethylamine (*6*). Sulfur-containing amino acids are especially sensitive to ionizing radiation. Cysteine can be oxidized to cystine by the hydroxyl radical or it can react with the hydrated electron and produce hydrogen sulfide, alanine, and hydrogen (2). Methionine reacts with the hydroxyl radical producing homocysteine, aminobutyric acid, methionine sulfoxide, and dimethyldisulfide (6). Merritt (7) reported that methyl mercaptan, ethyl mercaptan, dimethyl disulfide, benzene, toluene, carbonyl sulfide, and hydrogen sulfide are typical volatile products of irradiated meats. These products could be responsible for off odors and flavors. Ionizing irradiation reacts with proteins in aqueous solution, breaking hydrogen bonds, which results in the unfolding of the molecule; fragmentation, aggregation, and loss of active groups may occur. Peptide radicals are formed by the reaction of e_{aq} with the peptide bond (8). Myoglobin is the most important porphyrin protein in meat. The oxygenation of deoxymyoglobin in meat results in a change from purplish red to red, which is associated with normal meat bloom (8). When vacuum-packaged beef is irradiated, its color changes to brown and represents a change to the trivalent iron of metmyoglobin (9) and oxidation by the hydroxyl radical resulting in the loss of oxygen. When exposed to oxygen, metmyoglobin is gradually converted back to the oxymyoglobin. When cooked beef is irradiated in the absence of air, the brown metmyoglobin is converted into a reduced form of the pigment that is red possibly by reaction with the solvated electron (10).

One of the primary reactions of ionizing radiation with saturated fatty acids is decarboxylation and alkane formation (2). Dimers tend to be produced by reaction of ionizing radiation with unsaturated fatty acids (2). When meats are irradiated C_1-C_{17} n-alkanes, C_2-C_{17} n-alkenes, and C_4-C_8 iso-alkanes are the predominant products from the lipid fraction (10). Irradiation of the lipoprotein fraction of meat results in the formation of the following volatile compounds: C_1-C_{17} n-alkanes, C_2-C_{17} n-alkenes, dimethyl sulfide, benzene, and toluene (10).

Foodborne Pathogens

The temperature and atmosphere of irradiation often affect the control of food-borne pathogens in or on meat. Very low radiation doses are required to inactivate the protozoan *Toxoplasma gondii*, the nematode *Trichinella spiralis*, and the cestoda *Cysticercus bovis* and *Cysticercus cellulosae* in meat (11, 12, 13) (Table I). Toxoplasmosis is one of the most common infections of man; though usually subclinical in the adult, pneumonitis sometimes occurs. It is, however, highly opportunistic and may cause much more severe problems with immuno-compromised patients. *Toxoplasma gondii* can be transmitted to man through the ingestion of undercooked pork. If pork is irradiated as permitted since 1985 in the United States (14, 15) for the control of *T. spiralis*, then *T. gondii* will also be inactivated. *T. spiralis* also reaches man through the ingestion of undercooked pork, resulting in trichinosis, which can range in severity from asymptomatic to death. The cestoda *C. bovis* and *C. cellulosae* are the larval stages of the beef and pork tape worms, respectively, and can be inactivated by low doses of ionizing radiation (Table I).

The radiation doses required to inactivate 90% (D_{10}-value) of the colony forming units of 6 common food-borne pathogens associated with meat are presented in Table II. The values for the inactivation range from 0.16 kGy for the vegetative cells of *Campylobacter jejuni* in beef to 3.56 kGy for the inactivation of *Clostridium botulinum* endospores in vacuum-packed enzyme-inactivated chicken meat at a temperature of -30°C. If irradiated chicken were to receive exactly 3.0 kGy then, in theory, the population of the very mild pathogen *A. hydrophila* would decrease by 15.8 logs and the most resistant strain of *Salmonella enteritidis* (*20*) by 3.9 logs. *Salmonella sp.* are a very important potential contaminant of all meats but especially of poultry carcasses (*22*) and have been the subject of many studies for potential uses of food irradiation. The D_{10}-value reported in Table II for *C. botulinum* is a special case in that the value refers to an irradiation temperature of -30°C delivered *in vacuo* to enzyme-inactivated chicken to produce a shelf stable product. A 12D dose of 42.7 kGy is required to assure the elimination *C. botulinum* endospores. The high D_{10} value for *C. botulinum* is the result of the greater resistance of the bacterial endospore to ionizing radiation and the effect of subfreezing temperatures during irradiation. It was noted earlier that the G value for e_{aq} decreases dramatically in ice (*5*). If the lethal action of ionizing radiation were primarily due to direct interaction with the pathogen, we would not expect to find a temperature dependence of the irradiation process nor would we find an effect of irradiation atmosphere on the inactivation process. Thayer and Boyd (*23*) reported highly significant effects for both temperature and atmosphere during irradiation of *Salmonella typhimurium* in mechanically deboned chicken meat over the temperature range of -20 to +20°C. In the presence of air, a dose of gamma radiation of 3 kGy destroyed 4.8 and 6.4 logs of colony-forming units at -20 and +20°C, respectively. Hydroxyl radicals are identified as the major damaging species for the inactivation of *Escherichia coli* ribosomes and tRNA in aerated solutions (*24*). Though data are lacking for other bacterial species, it seems probable that the hydroxyl radical has a major role in their inactivation. This conclusion is supported by the increased resistance to ionizing radiation of many bacterial species at subfreezing temperatures where the mobility and reactivity of free radicals is known to be severely restricted (*25*). Advantage can be taken by the food processor of the decreased mobility of free radicals at subfreezing temperatures to help prevent changes in organoleptic properties and vitamin losses in meats.

Radiation-Sterilized Chicken Meat

Thayer et al. (*26*) reported the results of nutritional, genetic, and toxicological studies of enzyme-inactivated, radiation-sterilized chicken meat. The study included four enzyme-inactivated chicken meat products: 1) a frozen control, 2) a thermally processed product (115.6°C), 3) a gamma-sterilized product, and 4)

Table I. Control of Food-borne Protozoans, Nematodes, and Cestoda by Irradiation

Pathogen	Radiation Dose to Inactivate kGy	Ref.
Toxoplasma gondii	0.25	11
Trichinella spiralis	0.3	12
Cysticercus bovis	0.4 to 0.6	13
Cysticercus cellulosae	0.4 to 0.6	13

Table II. Control of Food-Borne Bacterial Pathogens by Treatment of Meat with Ionizing Radiation

Pathogen	Irradiation Temperature °C	Substrate	D_{10}-Value (kGy)	Ref.
Aeromonas hydrophila	2	Beef	0.14-0.19	16
Campylobacter jejuni	0 - 5	Beef	0.16	17
Clostridium botulinum	-30	Chicken	3.56	18
Listeria monocytogenes	2 - 4	Chicken	0.77	19
Salmonella sp.	2	Chicken	0.38-0.77	20
Staphylococcus aureus	0	Chicken	0.36	21

an electron-sterilized product. The radiation-sterilized products were given a minimum radiation dose of 46 kGy and a maximum of 68 kGy at -25 $\pm$ 15°C with gamma rays from ^{60}Co or 10 MeV electrons. All of the meat was enzyme-inactivated by heating to an internal temperature of 73-80°C. The frozen control, the thermally sterilized product, and the gamma-sterilized meats were vacuum canned. The chicken intended for sterilization by electrons was vacuum packed in laminated foil pouches. The entire study required 230,000 broilers (135,405 kg of enzyme-inactivated meat) and four production runs. Because each product was analyzed chemically for nutrients before use in the animal feeding studies, an unusual opportunity existed to compare the effects of ionizing radiation to freezing and thermal processing. No evidence of genetic toxicity or teratogenic effects was observed when any of the four products was included as 35 or 70% of the diet of mice, hamsters, rats, and rabbits. No treatment-related abnormalities or changes were observed in dogs, rats, or mice fed any of the four test products as 35% of their diet during multigenerational studies (26). No treatment effect was found for any amino acid or fatty acid in the four test meats (27). No treatment effect was found for the contents of pyridoxine, niacin, pantothenic acid, biotin, choline, vitamin A, vitamin D, and vitamin K. The percentage of thiamin in the thermally processed and gamma-sterilized meats was approximately 32% lower than that in the frozen control. The percentages of riboflavin and folic acid were significantly higher in the electron-sterilized product than in the frozen control. The percentages of vitamin B_{12} were significantly higher in the gamma-sterilized and in the thermally processed meats than in the frozen control meat. Thus, the only identifiable significant adverse change was a decrease in thiamin in the thermally and gamma-processed meats.

Vitamins

The results obtained with a sterilization dose administered to chicken meat at subfreezing temperature can be compared to those obtained with radiation doses of less than 10 kGy administered to non-frozen chicken. Fox et al. (28) investigated the effects of ionizing radiation treatments (0, to 6.65 kGy) at temperatures from -20 to +20°C on the content of niacin, riboflavin, and thiamin in chicken breasts and of cobalamin, niacin, pyridoxine, riboflavin, and thiamin in pork chops. Chicken breasts irradiated to 3.0 or 6.65 kGy at 0°C and then cooked had thiamin losses of 8.6 and 36.9%, respectively, when compared to unirradiated cooked chicken breasts. The 8.6% loss at the maximum dose currently approved for the poultry irradiation is not significant from a nutritional standpoint, because chicken contributes only 0.53% of the total thiamin in the American diet (29). The loss of thiamin in chicken irradiated in air at 0°C to a dose of 6.65 kGy (28) was slightly larger than that of the chicken meat treated *in vacuo* to a radiation dose of between 46 to 68 kGy at -25$\pm$15C (27). The increased loss can be attributed to a large extent to the difference in irradiation temperature. At doses of 3 kGy or less there was no loss of either niacin or riboflavin from irradiated chicken (28). De Groot et al. (30) also found no significant changes in nutritive value or in the vitamin content of chicken meat irradiated to 0, 3, and 6 kGy. Large variations in values were noted between samples, and the temperature of irradiation was not stated. Fox et al. (28) reported that when pork chops were irradiated to 1.0 kGy at 0°C and then cooked, there was a 17.5% loss of thiamin. These authors concluded that if all pork produced in the United States were to be irradiated, which is extremely unlikely, the American diet would lose about 1.5% of its thiamin, since pork contributes about 8.78% to total thiamin consumption (29). In contrast to the minor loss of thiamin at a dose of 1 kGy, pork chops irradiated to 6.65 kGy at 0°C lost 65.5% of their thiamin content when cooked. The reason for the large difference in the rates of thiamin loss in irradiated pork and chicken is currently under study.

Sensory Panel Tests

Klinger et al. (31) reported that extensive taste panel tests of chicken breast meat or leg meat irradiated to 3.7 kGy and cooked by boiling in water showed no loss in sensory quality immediately after treatment. The sensory quality of the irradiated chicken deteriorated during refrigerated storage over a period of 3 to 4 weeks. Irradiated chicken breast meat was acceptable for about three weeks; however, quality of unirradiated chicken was retained for only about four days during chilled storage.

Meat Enzyme Activity

Lakritz et al. (32) reported that radiation doses of less than 10 kGy (at 0 to 4°C) produced minimal changes in the micro structure of bovine longissimus dorsi muscle. At doses of 30 kGy or higher, myofibril fragmentation and decreased tensile strength were noted. Lakritz and Maerker (33) reported reductions of 8% and 42% in the activities of lysosomal enzymes and acid phosphatase of irradiated (10 kGy) bovine longissimus dorsi muscle tissue.

Conclusion

Knowledge of radiation chemistry and meat chemistry allows us to predict the effects of irradiation treatments on both meat and its accompanying microflora. Predictably, higher temperatures of irradiation correspond with greater destruction of structure, nutrients, and microflora. Inhibition of the movement of free radicals by freezing the meat prior to irradiation greatly reduces secondary reactions and provides better nutrient retention. Bacterial pathogens, however, can still be killed by direct action of the ionizing radiation, producing high quality, sterile meats.

Acknowledgments

Portions of this paper were presented at the 203rd ACS National Meeting, San Francisco, CA, April 5-10, 1992. The authors thank D. O. Cliver, V. H. Holsinger, R. J. Maxwell, and S. C. Thayer for their reviews of the manuscript.

Literature Cited

1. Takeguchi, C. A. FDA By-Lines 1981, *4*, 206-210.
2. von Sonntag, C. *The Chemical Basis of Radiation Biology*; Taylor and Francis: London, GB, **1987**, pp 515.
3. Klassen, N. V. In *Radiation Chemistry*; Farhataziz, and Rodgers, M. A. J., Ed.; VCH Publishers: New York, NY, **1987**, pp 29-64.
4. Anonymous, Fed. Reg. 1990, *55*, 18538-18544.
5. Taub, I. A. In *Preservation of Food by Ionizing Radiation* Vol II; Josephson, E. S., and Peterson, M. S., Ed., CRC Press, Inc.: Boca Raton, FL, **1982**, pp 125-166.
6. Urbain, W. M. In *Radiation Chemistry of Major Food Components*; Elias, P. S., Cohen, A. J., Eds.; Elsevier Scientific Publishing Company: Amsterdam, The Netherlands, 1977; pp 63-130.
7. Merritt, Jr. C. Food Irradiation Information, 1980, *10*, 20-33.
8. Simic, M. G. In *Preservation of Food by Ionizing Radiation*; Josephson, E. S, Peterson, M. S., Eds.; CRC Press, Inc.: Boca Raton, Florida, 1982, Vol. 2; pp 1-73.
9. Bernofsky, C., Fox, J. B. Jr., Schweigert, B. S. *Arch. Biochem. Biophys*., **1959**, *80*, 9-21.
10. Taub, I. A., Robbins, F. M., Simic, M. G., Walker, J. E., Wierbicki, E. *Food Technol.*, **1979**, *38*,184-192.
11. Dubey, J. P., Brake, R. J., Murrell, K. D., Fayer, R. *Am. J. Vet. Res.* **1986**, *47*, 518-522.
12. Brake, R. J., Murrell, K. D., Ray, E. E., Thomas, J. D., Muggenburg, B. A., Sivinski, J. S. *J. Food Safety* **1985**, *7*, 127-143.
13. King, B. L., Josephson, E. S. In *Preservation of Food by Ionizing Radiation*, Josephson, E. S. and Peterson, M. S., Ed.; CRC Press: Boca Raton, Florida, 1983, 1983, Vol. 2; pp. 245-267.
14. Anonymous. Federal Reg. 1985, *50*, 29658-29659.
15. Anonymous. Federal Reg. 1986, *51*, 1769-1770.
16. Palumbo S. A. et al. J. Food Prot. 1986, *49*, 189-191.
17. Lambert, J. D., Maxcy, R. B. J. Food Sci. 1984, *49*, 665-667, 674.
18. Anellis, A., Shattuck, E., Morin, M., Sprisara, B., Qvale, S., Rowley, D. B., Ross, E. W. Jr., *Appl. Environ. Micro.* **1977**, *34*, 823-831.
19. Huhtanen, C. N. et al. J. Food Prot. 1989, *52*, 610-613.
20. Thayer, D. W. et al. J. Indust. Microbiol. 1990, *5*, 383-390.
21. Thayer, D. W., Boyd, G. J. Food Sci. 1992, *57*, 848-851.

22. Dubbert, W. H. Poultry Sci. 1988, 67, 944-949.
23. Thayer, D. W., Boyd, G. Poultry Sci. 1991, 70, 381-388.
24. Singh, H., Vadasz, J. A., Int. J. Radiat. Biol. 1983, 43, 587-597.
25. Taub, I. A., Kaprielian, R. A., Halliday, J. W., Walker, J. E., Angelini, P., Merritt, C. Jr. Radiat. Phys. Chem. 1979, 14, 639-653.
26. Thayer, D. W., Christopher, J. P., Campbell, L. A., Ronning, D. C., Dahlgren, R. R., Thomson, G. M., Wierbicki, E. J. Food Prot. 1987, 50, 278-288.
27. Thayer, D. W. J. Food Qual. 1990, 13, 147-169.
28. Fox J. B. Jr., Thayer, D. W., Jenkins, R. K., Phillips, J. G., Ackerman, S. A., Beecher, G. R., Holden, J. M., Morrow, F. D., Quirbach, D. M. Int. J. Radiat. Biol. 1989, 55, 689-703.
29. Block, G., Dresser, C. M., Hartman, A. M., Carroll, M. D. Am. J. Epidemiology. 1985, 122, 13-26.
30. De Groot, A. P., Dekker, van der M., Slump, P., Vos, H. J., Willems, J. J. L. 1972, Report No. R3787. Central Institute for Nutrition and Food Research, The Netherlands.
31 Klinger, I., Fuchs, V., Basker, D., Juven, B. J., Lapidot, M., Eisenberg, E. Isr. J. Vet. Med. 1986, 42, 181-192.
32. Lakritz, L., Carroll, R. J., Jenkins, R. K., Maerker, G. Meat Sci. 1987, 20, 107-117.
33. Lakritz, L., Maerker, G. Meat Sci. 1988, 23, 77-86.

RECEIVED February 8, 1993

Chapter 24

Antimicrobial Peptides and Their Relation to Food Quality

Peter M. Muriana

Department of Food Science, Purdue University, West Lafayette, IN 47907

Antimicrobial proteins are produced by bacteria (bacteriocins), frogs (magainins), insects (cecropins), and mammals (defensins). A common characteristic among these agents is their proteinaceous composition and antimicrobial activity. In spite of distinct differences in genetic organization (eukaryotic vs. prokaryotic) of these inhibitory agents, there are definite similarities in biological activity which relates to their molecular structure. The lactic acid bacteria (LAB), long known for their acidifying properties and applications in food fermentations, are now becoming widely recognized for the production of bacteriocins. The widespread application of LAB in cultured/fermented foods, their recognition as safe "food-grade" organisms, and their presence as part of the human intestinal flora has facilitated the application of their bacteriocins as antimicrobial agents in food. LAB bacteriocins would likely be more acceptable than antimicrobial peptides from other sources, however, they are mainly effective against Gram-positive organisms. Eukaryotic antimicrobial peptides are unique in that they are also inhibitory towards Gram-negative bacteria, fungi, and protozoa. Information on antimicrobial peptides from various sources may provide a greater understanding of their mode of action and facilitate their application as "biopreservatives" in food.

Problems associated with processed foods often result from the introduction of new foods/processes that have been identified from market trends. The current trend for fresh, minimally-processed foods requires the assurance that these foods are also safe from spoilage or pathogenic organisms. Newly evolving strains of pathogenic bacteria may also test the limits of traditional processing methods. In the interest of "food safety", researchers have embarked on the exploitation of various inhibitory agents to improve the multi-tiered protective aspects of what is generally referred to as the "barrier concept." The barrier concept implies that one antimicrobial barrier (acidification, preservatives, salt, vacuum packaging, etc) may be overcome, but this would be less likely if a product included several barriers. Antimicrobial peptides are among those barriers that have acquired a strong interest as potential antimicrobials in food.

0097–6156/93/0528–0303$06.00/0

Characteristics of Antimicrobial Peptides

The most well-known bacterial antimicrobial proteins are the colicin-family of bacteriocins which have been characterized among *Escherichia coli* and related organisms (1). These bacteriocins are typically of large size (30-95 kDa) and have specific receptors for their uptake on sensitive target cells. The colicins characteristically have 3 functional domains on the mature bacteriocin, one which is involved with cellular translocation, another with receptor recognition, and still another with catalytic activity. The various colicins also have a similar genetic organization comprising a 3-gene operon: one gene encodes the bacteriocin, another encodes the bacteriocin immunity protein, and another for a bacteriocin release protein. Unlike the colicins, most bacteriocins from lactic acid bacteria and eukaryotic sources are small ($<$ 8 kDa) antimicrobial peptides.

Bacteriocins of Lactic Acid Bacteria. Bacteriocins produced by lactic acid bacteria can be classified into several categories based on simplicity, size, and/or composition (Table I). Both large and small bacteriocins of LAB have been associated with macromolecular

Table I. Classes of Bacteriocins Found Among the Lactic Acid Bacteria

Bacteriocin Class	Examples
1. Antimicrobial Protein ($>$ 10 kDa)	helveticin J
2. Antimicrobial Peptide ($<$ 10 kDa)	
a. Lantibiotic	nisin
	lacticin 481
b. Simple Peptide	lactacin F
	leucocin A-UAL187
	lactococcin A

complexes ($>$ 100-300 kDa) in culture supernatants whereas the purified (but active) monomer is much smaller in size. It remains to be shown whether such complexes result from physical interactions with components in the culture supernatant or are secreted in membrane vesicles as has been observed for other bacteriocins (2) or enzymes (3). Many of the antimicrobial peptides of lactic acid bacteria are small, hydrophobic peptides (1.4-7 kDa) and have more in common with cecropins and magainins than other antimicrobial peptides of bacterial origin. There are now many bacteriocins which have been either fully or quasi-characterized among the LAB (Table II). Other food starter cultures, such as the *Propionibacteria* also produce bacteriocins (4, 5). The timeliness of many of these studies has been in response to the fervor created by the recognition of the need for improving "food safety."

Table II. Antimicrobial Peptides of Lactic Acid Bacteria

Bacterial Genus	Bacteriocin	Mass (kDa)	Reference
Carnobacterium	carnobacteriocins		
	- A1	4.9[a]	6
	- A2	5.1[a]	6
	- A3	5.1[a]	6
Lactobacillus	lactacin F	6.3[a]	7
	lactacin B	6.2	7a
	reutericin 6	ND[b]	8
	gassericin A	ND	9
	acidophilucin A	ND	10
	lactocin S	ND	11
	plantaricin A	ND	12
	brevicin 37	ND	13
Lactococcus	lacticin 481	1.7	14
	lactococcin A	5.8[a]	15
	lactococcin B	5.3[a]	16
	lactococcin M	4.3[a]	16
	lactococcin G	4.4[a]	16a
	bacteriocin S50	ND	17
	lactococcin	2.3	18
	dricin	ND	19
	diplococcin	5.3	20
Leuconostoc	leuconocin S	ND	21
	leucocin A-UAL187	3.9[a]	22
	mesenterocin 5	4.5	23
Pediococcus	pediocin PA-1	4.6[a]	24
	pediocin AcH	2.7	25

[a]Determined by protein or nucleic acid sequence analysis.
[b]ND, not determined.

Lantibiotics. Certain antimicrobial peptides have been identified which contain "unusual" amino acids such as lanthionine, β-methyllanthionine, dehydroalanine, and β-methyldehydroalanine. Due to the predominance of lanthionine they have been collectively referred to as "lantibiotics" (26). Among the lactic acid bacteria, two bacteriocins have been identified as lantibiotics, nisin and lacticin 481. Nisin, the first bacteriocin from lactic acid bacteria to be identified, characterized, and successfully used in food was structurally determined in 1971 (27). Ingram (28) proposed a mechanism for nisin synthesis, indicating that dehydro-amino acid residues could result from dehydration of serine or threonine, while lanthionine-related residues may result from subsequent reaction of the dehydro-amino acids with cysteine. Cloning and sequencing of the nisin structural gene has confirmed the proposed predecessor molecule (pre-pro-nisin) from which active nisin is obtained (29, 30). However, the inability to obtain expression of active nisin from cloned nisin gene sequences has been attributed to the absence of genes encoding the post-translational processing enzymes (29, 30, 31, 32, 33). Recent studies have further demonstrated the involvement of a novel 70-kb transposon, Tn5276, in the transposition of a sucrose-nisin gene cluster among strains of *Lactococcus lactis* (32, 33).

The inability to obtain complete protein sequence analysis of purified bacteriocins has been reason to suspect the presence of N-blocked peptide sequences (34) or lantibiotic residues (14). Recently, Piard et al. (14) have shown from partial sequencing and composition analysis that lacticin 481, a broad spectrum bacteriocin produced by *L. lactis* 481, also contains lanthionine residues. The early widespread interest in nisin and nisin-producing strains had given the impression that lantibiotics may be characteristic of bacteriocins of lactic acid bacteria. However, recent studies with other LAB bacteriocins suggest that simple peptide bacteriocins may prevail among the LAB.

Simple Peptide Bacteriocins. Interest in LAB bacteriocins has prompted the application of molecular analyses to identify the genetic and/or protein sequences of these compounds. Such efforts have demonstrated the absence of lanthionine-related residues in leucocin A-UAL 187 (22) and lactacin F (35), for which the first 13 and 25 amino acid residues, respectively, were determined by sequence analysis; the absence of lanthionine-related residues was confirmed by composition analysis. Similarly, the entire 54- and 39-amino acid sequences of lactococcin A (15) and $G\alpha_1$ (16a), respectively, were also obtained by direct sequencing.

In contrast to the difficulty in obtaining expression with cloned nisin genes, expression of active bacteriocin has been readily demonstrated from cloned gene sequences of other lactic bacteriocins (Table III). For instance, expression of lactacin F was obtained from a cloned 2.2-kb *Eco*RI fragment (7). Similarly lactococcins A, B, and M were expressed from cloned 1.2-, 1.3-, and 1.8-kb DNA fragments (15, 16, 36). Evidence from either protein sequence/composition analysis and/or expression of active bacteriocin from small fragments of cloned DNA are indicative that such peptides are likely to be simplistic in both genetic organization and peptide structure (relative to nisin). These types of bacteriocins would therefore offer a greater likelihood of success in applying biotechnology to genetically engineer a "modified" bacteriocin, whether for basic research or practical applications.

Eukaryotic Antimicrobial Peptides. Antimicrobial peptides are not solely derived from microorganisms - they are also part of the immune defense system of amphibia, insects, and mammals (Table IV). Unlike the antimicrobial peptides of lactic acid bacteria that inhibit mainly Gram-positive organisms, the eukaryotic antimicrobial peptides are also inhibitory towards Gram-negative organisms, fungi, and protozoa. Such a broad inhibitory spectrum has prompted suggestions as to potential applications as medical

Table III. Cloned Bacteriocin Genes from Lactic Acid Bacteria

Bacteriocin	Gene	Cloned Fragment Size (kb)	Bacteriocin Expressed?	Reference
Carnobaterium				
carnobacteriocin A1	UN[a]	2.0	ND[b]	6
Lactobacillus				
helveticin J	*hlvJ*	4.0	Yes	34
lactacin F	*laf*	2.2	Yes	7
Lactococcus				
bac$_{WM4}$	UN	18.4	Yes	37
nisin	*spaN*	5.0	ND	29
"	*nisA*	4.3	ND	30
"	UN	5.5	No	31
lactococcin A	*lcnA*	1.2	Yes	15
lactococcin B	*lcnB*	1.3	Yes	16
lactococcin M	*lcnM,lcnN*	1.8	Yes	16
Leuconostoc				
leucocin A-UAL 187	*lcnA*	2.9	No	22
Pediococcus				
pediocin PA-1	*pedA*	5.6	Yes	38, 38a

[a]UN, un-named.
[b]ND, not-determined.

therapeutics, both topically and internally applied, and as antimicrobials against pathogens in foods.

The best characterized antimicrobial peptides among amphibia are the magainins (39). Zasloff (40) first became curious as to the reason why frogs of the genus *Xenopus* did not become susceptible to bacterial infection when placed in holding tanks after undergoing surgical operations. Investigations revealed the production of several antimicrobial peptides (magainins, PGLa, XPF, CPF) by the granular gland in the skin of *Xenopus*; the peptides were inhibitory against Gram-positive and Gram-negative bacteria and fungi (41, 42). These antimicrobial peptides, along with a novel 24-amino acid peptide, PGQ, have also been found in extracts of *Xenopus* stomach tissue where they are synthesized by granular glands analagous to those found in the skin (42).

Demonstration of the ability to induce the production of antimicrobial peptides among insects is perhaps even more significant since they are not capable of eliciting a

Table IV. Eukaryotic Antimicrobial Peptides

Source	Peptide	Size (#AAs)	Mass (Da)	Reference
Amphibians	Magainin	23	2410	39
	XPF	25		41
	PGLa	21		41
	CPF	27		39
	PGQ	24	2457	42
Insects	Cecropins	35-39	~4000	43
	Melittin	26	2847	44
	Apidaecins	18	2100	45
	Royalisin	51	5523	46
Mammals	Defensins	29-34	~4000	47

cell-mediated or humoral immune response in the same manner as amphibia or higher animals (i.e., they do not produce antibodies). Using a model insect system, Boman and coworkers (48, 49, 50) found that when bacteria were injected into pupae of the Cecropia moth, they could induce the synthesis of more than a dozen antibacterial proteins. A family of related proteins first isolated from the Cecropia moth were therefore designated cecropins; various cecropins have now been isolated from at least six genera of insects (43). Antimicrobial peptides have also been found among honeybees. Melittin (44) is found in honeybee venom while apidaecins (45) are antimicrobial peptides whose synthesis can be induced by injection of sublethal doses of *E. coli* into adult honeybees. Furthermore, Fujiwara et al. (46) found a potent antibacterial peptide in royal jelly (royalisin) which is secreted from the pharyngeal glands of the honeybee.

Antimicrobial peptides are also an integral part of mammalian cell-mediated immunity. They are produced by phagocytic cells, particularly neutrophils, and are designated defensins (47). Although defensin molecules from humans, guinea pigs, rabbits, and rats may vary between 29-34 amino acid residues and differ slightly in sequence, they invariably show conservation in six cysteine residues which are required for a defined pattern of disulfide bonds (47).

Inhibitory Activity of Antimicrobial Peptides from LAB and Eukaryotic Sources. The key feature which draws attention to these peptides in respect to food applications is their ability to inhibit undesirable organisms, either spoilage or pathogenic organisms. Although the peptides from eukaryotic sources appear far-removed from immediate application in foods, it is important to study their mechanism(s) of action in relation to those of lactic acid bacteria to better understand commonalities in structure-function relationships. Such commonalities may serve as a base from which to initiate molecular approaches for examining mechanisms of action and/or generate engineered bacteriocins with improved antimicrobial characteristics.

As noted earlier, nisin was the first bacteriocin among the lactic acid bacteria to show broad-spectrum inhibitory activity and to be successfully applied in foods. Nisin is inhibitory to Gram-positive bacteria including *Bacillus*, *Clostridium*, *Listeria*, *Staphylococcus* spp., as well as LAB which may often constitute spoilage bacteria in foods. Many bacteriocin-producing species of LAB have been identified (Table II) demonstrating a wide range of inhibitory activity. For instance, most bacteriocin-producing strains of *Lactobacillus acidophilus* (51), and the *Lc. cremoris* bacteriocins, lactococcin A (15) and diplococcin (20), inhibit only other *Lactobacillus* or *Lactococcus* spp., respectively. Other LAB bacteriocins, such as the plantaricins (12, 52), lactocin S (53), and brevicin 37 (13) demonstrate broad-spectrum inhibitory properties primarily within the genera comprising the LAB. Still other LAB bacteriocins demonstrate broad-spectrum activity among Gram-positive bacteria other than lactic acid bacteria (Table V): pediocins produced by *Pediococcus acidilactici* (24, 25, 54), sakacin produced by *Lb. sake* (55), and lacticin 481 produced by *Lc. lactis* (56). Although there have been a few reports of inhibition of Gram-negative bacteria by bacteriocin-producing LAB, there has been no follow-up to such findings, leading one to believe that such reports have either been erroneous (i.e., inhibition due to lactic acid) or so weakly inhibitory that the data may be questionable or have little practical application. However, recent work has indicated that LAB bacteriocins may be effective against Gram-negative organisms when used in conjuction with chelating agents as a means of bypassing the outer membrane of Gram-negative organisms (57).

The eukaryotic antimicrobial peptides described earlier, especially the magainins and cecropins, have shown strong inhibitory activity against Gram-negative and Gram-positive bacteria, fungi, parasites, and protozoa (Table V). The extremely large spectrum of organisms affected by eukaryotic antimicrobial peptides has prompted applications as antimicrobial agents in topically applied medical chemotherapeutics. The growing number of antimicrobial peptides found in the intestines of mammals and their inherent activities indicate their natural defensive role in the host organism. The identification of cecropin P4 in pig intestines (58) indicates that these antimicrobials are indeed broadly based defense mechanisms among insects, amphibia, and mammals. Kimbrell (59) suggests that such antibacterial peptides can conceivably be used in insects which are, themselves, vectors for infectious agents and in gene replacement therapy for humans who suffer deficiencies in granule defense mechanisms (i.e., human defensins). Lactic acid bacteria that are capable of colonizing the human intestine have been suggested as suitable hosts for cloning broad-spectrum eukaryotic peptides for subsequent introduction into the human intestinal tract where they may serve as a constant source of these probiotics (60). These compounds may ultimately find use as biopreservatives in food and/or animal feeds where there is strong concern for reduction/elimination of spoilage and pathogenic organisms (42, 60). For instance, honeybee royalisin is secreted into royal jelly from the honeybee pharyngeal gland and is implicated to act as a "natural preservative" in royal jelly against bacterial contamination (46). Disclosures of other such natural examples of applications of antimicrobial peptides may provide greater incentive for their use in foods.

Table V. Broad-Spectrum Activity Among LAB and Eukaryotic Peptides

Peptide	Organism	Peptide	Organism
Nisin	lactic acid bacteria *Bacillus* spp. *Clostridium* spp. *Listeria* spp. *Micrococcus* spp. *Staphylococcus* spp. *Streptococcus* spp.	Pediocins	lactic acid bacteria *Bacillus* spp. *Brochothrix thermosphacta* *Clostridium* spp. *Listeria* spp. *Staphylococcus* spp.
Magainins	*Acinetobacter caloaceticus* *Citrobacter freundii* *Enterobacter cloacae* *Escherichia coli* *Klebsiella pneumoniae* *Proteus vulgaris* *Pseudomonas aeruginosa* *Salmonella typhimurium* *Serratia marcescens* *Shigella* spp. *Staphylococcus* spp. *Streptococcus* spp. *Candida albicans* *Saccharomyces cerevisiae* *Acanthamoeba castellani* *Paramecium caudatum* *Tetrahymena pyriformis*	Cecropins	*Acinetobacter calcoaceticus* *Bacillus* spp. *Escherichia coli* *Micrococcus luteus* *Pseudomonas aeruginosa* *Serratia marcescens* *Staphylococcus aureus* *Streptococcus* spp. *Xenorhabdus nematophilus* *Plasmodium falciparum*

Mode of Action and Structural Characteristics of Antimicrobial Peptides

Studies examining the mode of action of LAB bacteriocins have predominantly been performed with nisin which has existed as a commercially available compound since 1960 (Nisaplin, Aplin & Barrett Ltd, London, England). In addition, many LAB bacteriocins have only recently been identified (Table II) and emphasis is often first directed towards characterizing the inhibitory agent. The mode of action of nisin on prokaryotic, eukaryotic, and artificial membranes has been examined and demonstrated to result from membrane interaction (61, 62, 63). The ability of nisin to promote membrane permeability depends on the presence of a sufficiently high membrane potential (62) whereas lactococcin A has been shown to induce membrane permeability in a voltage-independent manner (36). Other LAB bacteriocins have also been implicated in membrane interaction and leakage of cytosolic constituents. Lactostrepcin 5 has been shown to cause leakage of ATP and K^+ ions from *Lc. lactis* (64) while pediocin AcH was shown to release UV absorbing material and K^+ ions and increase permeability to ONPG from *Lb. plantarum* (65).   It is well known that certain structural motifs such as amphiphilic α-helices and β-sheets, or uniformly hydrophobic α-helices are commonly involved with membrane interactions of proteins (66).

Membrane interaction by the magainins and cecropins has also been demonstrated and has been complemented with molecular analyses such that working models for membrane-channel formation have been developed.   The magainins have been shown to bind negatively-charged lipid vesicles and liberate trapped flourescent markers, and form anion-permeable channels in lipid bilayers and cell membranes, causing disruption of membrane potential (67, 68, 69, 70). The cecropins also show a propensity for forming ion channels in lipid vesicles and planar lipid membranes (71, 72).   The amino acid sequences of the magainins and cecropins were identified by protein sequencing and translation of cloned cDNA nucleic acid sequences (39, 40, 49, 50). Knowledge of the peptide sequences allowed the chemical synthesis of synthetic analogues which were indistinguishable from their natural counterparts (73, 74, 75).   Studies with synthetic substitution and omission analogues have demonstrated the importance of specific residues and α-helicity towards the inhibitory activity exhibited by these peptides (72, 76, 77, 78, 79, 80). Together, these data have allowed predictive modelling for seconday structures and provided further insight into their mode of action.   Predictive molecular modelling indicated that the magainins could adopt a rodlike α-helical conformation (Figure 1).

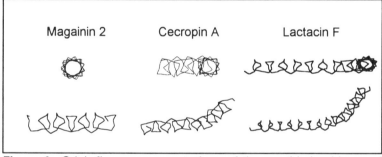

Figure 1. Stick-figure representations of the peptide backbone of magainin 2, cecropin A, and lactacin F. Top, end view of α-helix; bottom, side view. Structures were obtained using HyperChem software.

Furthermore, the incorporation of helix-promoting residues increased the inhibitory activity of magainin 2 by as much as 100-fold (77). Berkowitz et al. (39) suggested that a cluster of magainin 2 α-helices can form a positively-charged transmembrane pore. Contrary to the rodlike magainin structure, the cecropins were suggested to have separate N- and C-terminal helices separated by a "hinged" structure (-Gly-Pro-)(Figure 1; 72). Elimination of the hinged region among synthetic analogues reduced their effectiveness in forming voltage-dependent ion channels (72).   Christensen et al. (72) therefore proposed the cecropin-membrane interaction model whereby cecropin molecules are first attracted to the membrane via electrostatic interactions whereby the amphiphilic helix then inserts horizontally onto the membrane; when a sufficient number of horizontally-embedded helices have accumulated, they "flip" vertically to form a positively-charged ion channel by virtue of their hydrophilic surfaces.

Molecular analysis of LAB bacteriocins has not yet extended to the level of determining structure-function relationships among this important group of peptides. However, certain natural variants of nisin have indicated the importance of certain amino acid residues to inhibitory activity. Chan et al. (81) identified two degradation products of nisin, nisin$^{1-32}$ which suffered a deletion of the last two amino acids, and $(des$-$\Delta Ala^5)$-nisin$^{1-32}$ which incurred an additional degradation at Ala^5. Although $(des$-$\Delta Ala^5)$-nisin$^{1-32}$ was relatively inactive, nisin$^{1-32}$ was as active as nisin, indicating that the last two amino acids were not necessary for nisin activity. In a different vein, Mulders et al. (82) demonstrated by nucleic acid sequence analysis and protein NMR studies that a nisin-variant strain of L. lactis NIZO 22186 (nisin Z) contained an amino acid substitution (His^{27} → Asn^{27}) which was encoded at the level of the nisin gene. Although nisin Z was indicated to have inhibitory activity, the authors did not investigate the effect of the substitution on the level of inhibitory activity relative to native nisin. Recently, molecular modelling applied to NMR data on the structure of nisin in aqueous solution suggested that nisin exhibits two domains, an N-terminal amphiphilic region (residues 3-19) and an overwound α-helix (residues 23-28), separated by a flexible hinge region (83). Muriana and Klaenhammer (7) recently suggested that lactacin F secondary structure predictions indicated a propensity for β-structures, however, Muriana (84) noted that lactacin F helical wheel projections segregate very nicely into an amphiphilic helix with an exclusively hydrophobic surface and an opposing hydrophilic surface. Chen et al. (77) have indicated that although the magainins do not demonstrate α-helical structures in an aqueous solvent, these structures are induced when the peptides are placed in a hydrophobic solvent (i.e., such as lipid/cell membranes). Similar findings have also been reported for the cecropins which do not appear to have much structure in water but have been shown to assume a conformation with a high proportion of α-helical structure in trifluoroethanol (74). Using molecular modelling software, lactacin F is presented as having dual helices similar to those suggested for cecropin A (Figure 1). Lactacin F also has the "Gly-Pro" region between the two proposed helices which was described as the "hinge" region in cecropin A. Although the demonstration of helical secondary structure is lacking among most LAB bacteriocins, a combination of physical characteristics and predictive analyses based on sequence implicates the involvement of such structures. The hydrophobicity demonstrated by many LAB bacteriocins could likely result from α-helical involvement, yet the exact model of membrane interaction among the various peptides remains to be demonstrated (i.e., singular α-helix like the magainins or dual-helices like the cecropins). And indeed, if the mode of interaction is similar, as say for cecropin A and lactacin F, it will become important to find what other characteristics lend themselves to the observed differences in inhibitory spectrum among a limited (lactacin F) or broad (cecropin A) range of organisms. Incorporation of helix-promoting amino acids has demonstrated a potential for increasing inhibitory activity through enhanced membrane interaction. It would be interesting to see if incorporation of charged residues on the hydrophilic helical surface would similarly provide greater inhibitory activity through a greater degree of cellular attachment via electrostatic interactions.

Application of Antimicrobial Peptides in Foods and Interaction with Food Components

The characteristic production of lactic acid by LAB is a trait which is exploited in the fermentation and culturing of products to enhance the preservation of foods. Even the presence of lactic acid bacteria as a low-level inoculum has demonstrated sufficient inhibitory capability that the USDA has approved the use of *P. acidilactici* in reduced-nitrite bacon ("Wisconsin Process") to aid in the prevention of botulinum toxin production by outgrowth of *C. botulinum* spores (85). Hutton et al. (86) have demonstrated that the "Wisconsin Process" can be adapted to low-acid refrigerated foods (i.e., chicken salad) as an effective method of excluding botulinum toxigenesis. The widespread use of LAB in food fermentations, their ubiquitous occurrence as indigenous contaminants in non-sterile foods, and their demonstrated capacity to inhibit pathogens when present as a purposely added inoculum is demonstrative of their safe applications in foods. These examples suggest that bacteriocin-producing LAB would render additional antimicrobial characteristics to food products in which they are utilized. Studies intending immediate application of bacteriocin-mediated inhibitory barriers have mainly examined the use of bacteriocin-producing starter cultures as a means of introducing the bacteriocin into a cultured/fermented food product because of the GRAS status of lactic starter cultures. Although other studies have examined the use of bacteriocin preparations, their practical application would require regulatory acceptance as a food additive.

Nisin is the only bacteriocin approved for use in foods as a food additive and is currently used worldwide in various applications including pasteurized, flavored, and long-life milks, aged and processed cheeses, and canned vegetables and soups (for a review on applications of nisin see 84 and 87). In the U.S., however, nisin has only been approved for use in pasteurized/processed cheese spreads to control the outgrowth of *C. botulinum* spores (88). Several recent studies have examined the use of nisin-producing cultures to inhibit foodborne pathogens. Okereke and Montville (89) examined the nisin-producing strain *Lc. lactis* 11454 and other bacteriocin-producing LAB at low cell densities for their ability to inhibit *C. botulinum* spores. These strains were later used to test their efficacy at inhibiting *C. botulinum* at refrigeration (4°C) and abuse (10°, 15°, 35°C) temperatures (90). At all temperatures tested, they found that *Lc. lactis* 11454 (nisin) was more effective than either *P. pentosaceus* 43200 (pediocin A) or *Lb. plantarum* BN (plantaricin BN) on inhibition of *C. botulinum* spores in laboratory media. Maisnier-Patin et al. (91) were able to obtain a 3.3-log reduction in *L. monocytogenes* V7 during the manufacture of Camembert cheese using a nisin-producing strain (*Lc. lactis* CNRZ 150). They observed a 2.4-log difference in the levels of *L. monocytogenes* in Camembert cheese made with Nis⁺ strains compared to Nis⁻ strains, concluding that bacteriocin-mediated inhibition of foodborne pathogens could be a viable strategy in helping to reduce the numbers of listeria when present in modest numbers. However, it was noted that cheese systems are subject to aging, a pH increase, and proteolysis which may reduce the longterm effectiveness of nisin. Furthermore, they cautioned that an aspect which may be more disturbing is the potential development of nisin-resistant (Nisʳ) strains of *L. monocytogenes* when challenged at sublethal levels of nisin or when challenged with high populations of listeria. The possibility of the development of Nisʳ strains of *L. monocytogenes* was examined by Harris et al. (92). They found Nisʳ mutants of 3 strains of *L. monocytogenes* (ATCC 19115, Scott A, UAL500) to arise at frequencies of 10^{-6} to 10^{-8} when nisin was challenged with 10^9 CFU/ml. This indicates that resistant *L. monocytogenes* are likely to arise in foods if nisin is used as the sole inhibitory agent in foods; however, the inhibitory effect of nisin was accentuated when complemented with either NaCl or acidification (92). While Nis^r cells survived, the level of *L. monocytogenes* cells (ca. 10^9 CFU/ml) used in this study were not levels characteristically found in foods.   Although bacteriocin applications have not demonstrated complete inhibiton of target pathogens, the use of bacteriocins as part of a "barrier system" suggests that they would not be used alone as the primary/sole inhibitory agent.

The worldwide application of nisin in foods, its poor solubility and stability near neutral pH, and various recurring problems with foodborne pathogens has generated a strong surge of interest in examining practical applications of other lactic bacteriocins in food. Pucci et al. (93) examined the effect of a lyophilized pediocin PA-1 preparation (produced by *P. acidilactici* PAC 1.0) on *L. monocytogenes* LM01 in cottage cheese, half-and-half cream, and cheese sauce maintained at 4°C. Inhibition of *L. monocytogenes* was observed in all samples by as much as 10^3-10^5 CFU/ml in 14 days relative to controls without bacteriocin, however, resurgence of *L. monocytogenes* occurred after 7 days in half-and-half cream and cheese sauce.   Although pediocin PA-1 may be effective in high-moisture refrigerated foods, the results also suggest that bacteriocin-mediated inhibition may be less effective in high-protein and/or high-fat food systems. Nielsen et al. (94) examined the inhibitory effect of beef pieces saturated with a pediocin PA-1 preparation on subsequent challenge with *L. monocytogenes* and found that the beef pieces retained listericidal activity for 28 days at 5°C.   Although the aspect of beef protein interaction with pediocin PA-1 was not the intended purpose of the study, Nielsen et al. (94) noted that inhibitory activity maintained on the beef pieces gradually declined with time, suggesting that available pediocin PA-1 may be bound by the high protein content of the beef or reduced by native proteases. This phenomenon was confirmed by Degnan and Luchansky (95) in studies which specifically addressed the recovery of pediocin AcH after addition to slurries of beef tallow and beef muscle as a means of assessing bacteriocin availability in high-fat or high-protein meat systems. Significant losses in added bacteriocin activity were incurred in both slurries; typically, recovery of pediocin AcH activity ranged from 60% to 40% from tallow, whereas recovery from beef ranged from 30% to 20% depending on sampling time (95). Heating of either tallow or beef to inactivate native proteases before addition of bacteriocin improved recoverable activity by approximately 5% whereas encapsulation of bacteriocin in liposomes improved recovery of pediocin AcH activity by approximately 28%. Food component interaction with nisin has long been demonstrated and has recently been shown in milk systems. Jung et al. (96) examined the effect of milk fat and emulsifiers on nisin activity against *L. monocytogenes*. Nisin activity in fluid milk was observed to decrease as fat levels increased, reaching a maximum effect at approximately 4% milk fat content (96). This reduction in nisin activity was partially reversed by the addition of emulsifier (0.2% Tween 80). These studies indicate that antimicrobial activity elicited by bacteriocins may be variable depending on the composition of the intended foodstuffs.

Bacteriocin-producing strains have also demonstrated the ability to inhibit foodborne pathogens when food products are either fermented with such strains, or contain them as inhibitory flora.   In sausage fermentation trials using bacteriocin-producing *Pediococcus* strain JD1-23 (97) or *P. acidilactici* PAC 1.0 (98), a greater reduction of *L. monocytogenes* ($> 1 \log_{10}$ CFU/gm) was obtained than with the use of non-bacteriocin-producing *P. acidilactici* starters. Yousef et al. (99) found both *P. acidilactici* JBL1095 (pediocin AcH producer) or pediocin AcH preparations similarly effective as biocontrol agents in wiener exudates co-inoculated with *L. monocytogenes*. When *P. acidilactici* JBL1095 was applied in actual vacuum-packaged wieners surface-inoculated with *L. monocytogenes*, Degnan et al. (100) found that both organisms survived and that JBL1095 did not produce acid or bacteriocin when maintained at 4°C. However, at 25°C (abuse temperature) *L. monocytogenes* in control wiener packages increased $3.2 \log_{10}$ CFU/gm whereas in packages co-inoculated with *P. acidilactici* JBL1095, *L. monocytogenes* demonstrated a $2.7 \log_{10}$ CFU/gm decrease. Identical results were obtained in a similar study with frankfurters (101) in which no inhibition of *L. monocytogenes* was observed when co-inoculated with a pediocin-producing strain (*P. acidilactici* JD1-23) at 4°C while approximately a $0.5\text{-}\log_{10}$ reduction was reported at abused (15°C) temperatures. Berry et al. (101) attribute this reduction to bacteriocin-mediated inhibiton, however, the inhibitory effect on *L. monocytogenes* appears to occur at the time of co-inoculation and not during the course of temperature-abused incubation.

A bacteriocin produced by *Lb. sake* (sakacin A) has also been used to control *L. monocytogenes* in minced meat. Schillinger et al. (102) demonstrated that inoculation of *Lb. sake* Lb706 into pasteurized minced meat or pork sausages ("Mettwurst") co-inoculated with *L. monocytogenes* prevented the growth of listeria during the first few days after manufacture and suggested this as a potential application for comminuted fresh meat products. They also noted that the bacteriocin-mediated inhibitory effect was less effective in meat than was demonstrated in broth media, implicating that food component interaction may be a characteristic drawback experienced by all LAB bacteriocins.

Heat Stability Among LAB Bacteriocins

There are many examples now that have demonstrated that LAB bacteriocins can function as an "additional" barrier against pathogenic organisms in fermented, vacuum-packaged, or comminuted foods. Many LAB bacteriocins also show a tendency towards high heat stability which is likely related to their small size and/or "tight" molecular structure. This characteristic has allowed nisin to be used in numerous canning, retort, and pasteurization applications (Figure 2; Muriana, P.M., Purdue University, unpublished data). Although numerous LAB bacteriocins have been described as "heat-resistant", there have been relatively few examples (other than nisin) which demonstrate their effectiveness in (thermally) processed foods. Bacteriocin-producing LAB are easily isolated from retail foods (55, 103, 104) which suggests that the public is unaffected by the consumption of viable LAB present as ubiquitous flora on non-sterile foods.

Experimental. Retail foods such as vacuum-packaged meats, deli meats, vegetables, fruits, and other retail foods were purchased for isolation of LAB. Individual retail products were aseptically immersed in MRS broth in sterile bags and incubated overnight at 37°C. Bacteriocin-producing LAB were then detected by the method of deferred antagonism as described previously (51, 105). Once detected, bacteriocin-producing colonies were recovered from the sandwiched overlay media by streaking onto fresh MRS

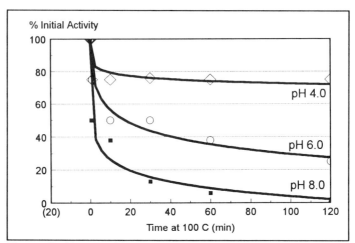

Figure 2. Heat stability of nisin after treatment at 100°C at pH 4.0, 6.0, and 8.0. Nisin was obtained from culture supernatant broth of *Lc. lactis* 11454 and resuspended in citrate-phosphate buffer.

agar plates, patch-plating to new MRS plates, and overlayed with an indicator strain (*Lb. delbrueckii* 4797) for confirmation of bacteriocin production. Bacteriocin-producing isolates were speciated using biochemical assays, carbohydrate fermentation patterns, physiological growth parameters, and SDS-PAGE protein profiles for identification (K. Garver and P.M. Muriana, Purdue University, manuscript in preparation). Bacteriocin-producing strains were distinguished from one another by plasmid analysis whereas bacteriocins were distinguished from each other by differential inhibition of bacteriocin-sensitive strains and by differential proteolytic inactivation (Garver and Muriana, manuscript in preparation).

Bacteriocin-containing supernatant was obtained by incubating 400 ml MRS broth for 12 hr with a 1% inoculum of an overnight culture of a bacteriocin-producing strain. Cells were removed by centrifugation (15 min, 10k RPM, 4°C) and the bacteriocin precipitated by addition of ammonium sulfate (to 75% saturation) as described previously (35). The protein pellet was resuspended with a minimal volume ($\sim$ 3 ml) of deionized water which was later diluted with 200 mM citrate-phosphate buffer at pH 4.0, 6.0, or 8.0. Aliquots of bacteriocin-buffer solutions were dispensed into multiple tubes which were placed in boiling water for either 0.5, 1, 10, 30, 60, or 120 min and then quickly frozen by contact with a cold ethanol bath (-20°C); a control tube (0 min) was frozen directly without being subjected to heating. Bacteriocin activity was determined following a two-fold serial dilution and "spot assay" as described previously (35, 105).

Results and Discussion. The heat stability of the bacteriocins produced by four strains of *Lc. lactis* (FS84, FS90, FS91-1, FS97) isolated from retail foods was examined. The four bacteriocins tested were shown to be different from nisin by their inhibitory spectra and/or protease inactivation profile (Garver and Muriana, Purdue University, manuscript in preparation). All four bacteriocins demonstrated similar trends in thermal stability at both pH 4.0 and pH 8.0 (Figure 3). It is interesting that decreases in initial activity were realized within a short time (~ 10 min) but stabilized to 25-55% of initial activity for nearly two hours thereafter (pH 4.0). Although similar trends were observed among the

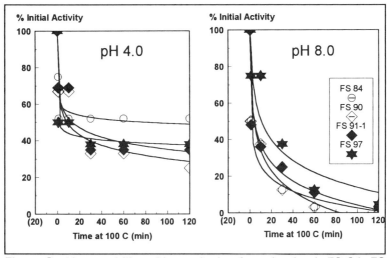

Figure 3. Heat stability of bacteriocins from *Lc. lactis* FS 84, FS 90, FS 91-1, and FS 97 in citrate-phosphate buffer at pH 4.0 and 8.0.

four bacteriocins, the bacteriocin produced by strain FS84 demonstrated the greatest stability at pH 4.0 after 2 hr at 100°C, whereas it demonstrated the least stability at pH 8.0 (Figure 3). Differences in thermal stability of peptides may likely be attributed to differences in degree of secondary structure and specific amino acid composition which may contain side chains which are interactive at various pH optima. Although these bacteriocins may not demonstrate the level of stability of nisin at pH 4.0 (Figure 2), they are comparable in stability at pH 8.0 (Figure 3). Furthermore, nisin has found application in thermally-processed foods even though nisin itself is susceptible to thermal-induced losses in activity (Figure 2).

From the data presented in Figure 3, it is likely that heat-stable LAB bacteriocins (other than nisin) may find suitable application as biologically-produced antimicrobials ("biopreservatives") in thermally-processed foods. Future studies directed towards the examination of a variety of novel bacteriocins in thermal processing applications may identify new LAB bacteriocin(s) which have comparable, or perhaps superior, properties as nisin. Further research is needed to determine the presence of bacteriocins in retail foods which may help to reduce the regulatory hesitancy towards their practical application as safe food ingredients. Data provided by research with the magainins and cecropins indicates that simple peptides, unlike nisin, can be easily synthesized as "nature-identical" and perhaps obtained by chemical synthesis rather than by isolation from culture broths. In an era where applications of biotechnology have far outdistanced our ability to implement them, simple bacteriocin peptides encoded by simplistic genes/operons offers an easy target in producing enhanced bacteriocins through the application of genetic engineering.

Literature Cited

1. Oudega, B.; de Graaf, F.K. *Curr. Top. Microbiol. Immunol.*, **1986**, *125*, p183.
2. Austin-Prather, S.L.; Booth, S.J. *Can. J. Microbiol.*, **1983**, *30*, pp268.
3. Deslauriers, M.; ni Eidhin, D.; Lamonde, L.; Mouton, C. *Oral Microbiol. Immunol.*, **1990**, *5*, pp1.
4. Lyon, W.J.; Glatz, B.A. *Appl. Environ. Microbiol.*, **1991**, *57*, pp701.
5. Grinstead, D.A.; Barefoot, S.F. *Appl. Environ. Microbiol.*, **1992**, *58*, pp215.
6. Worobo, R.W.; Henkel, T.; Roy, K.L.; Vederas, J.C.; Stiles, M.E. *Abst. Gen. Meet. Amer. Soc. Microbiol.*, **1992**, Abstr.#P56, pp333.
7. Muriana, P.M.; Klaenhammer, T.R. *J. Bacteriol.*, **1991**, *173*, pp1779.
7a.Barefoot, S.F.; Klaenhammer, T.R. *Antimicrob. Agents Chemother.*, **1984**, *45*, pp328.
8. Toba, T.; Samant, S.K.; Yoshioka, E.; Itoh, T. *Lett. Appl. Microbiol.*, **1991**, *13*, pp281.
9. Toba, T.; Yoshioka, E.; Itoh, T. *Lett. Appl. Microbiol.*, **1991**, *12*, pp228.
10. Toba, T.; Yoshioka, E.; Itoh, T. *Lett. Appl. Microbiol.*, **1991**, *12*, pp106.
11. Mortvedt, C.I.; Nissen-Meyer, J.; Sletten, K.; Nes, I.F. *Appl. Environ. Microbiol.*, **1991**, *57*, pp1829.
12. Daeschel, M.A.; McKenney, M.C., McDonald, L.C. *Food Microbiol.*, **1990**, *7*, pp7.
13. Rammelsberg, M.; Radler, F. *J. Appl. Bacteriol.*, **1990**, *69*, pp177.
14. Piard, J.C.; Muriana, P.M.; Desmazeaud, M.J.; Klaenhammer, T.R. *Appl. Environ. Microbiol.*, **1992**, *58*, pp279.
15. Holo, H.; Nilssen, O.; Nes, I.F. *J. Bacteriol.*, **1991**, *173*, pp3879.
16. van Belkum, M.J.; Kok, J.; Venema, G. *Appl. Environ. Microbiol.*, **1992**, *58*, pp572.
16a.Nissen-Meyer, J; Holo, H.; Havarstein, L.S.; Sletten, K.; Nes, I.F. *J. Bact.*, **1992**, *174*, pp5686.
17. Kojic, M.; Svircevic, J.; Banina, A.; Topisirovic, L. *Appl. Environ. Microbiol.*, **1991**, *57*, pp1835.
18. Dufour, A.; Thuault, D.; Boulliou, A.; Bourgeois, C.M.; Le Pennec, J.-P. *J. Gen. Microbiol.*, **1991**, *137*, pp2423.
19. Powell, I.B.; Ward, A.C.; Hillier, A.J.; Davidson, B.E. *FEMS Microbiol. Lett.*, **1990**, *72*, pp209.
20. Davey, G.P.; Richardson, B.C. *Appl. Environ. Microbiol.*, **1981**, *41*, pp84.
21. Lewus, C.B.; Sun, S.; Montville, T.J. *Appl. Environ. Microbiol.*, **1992**, *58*, pp143.
22. Hastings, J.W.; Sailer, M.; Johnson, K.; Roy, K.L.; Vederas, J.C.; Stiles, M.E. *J. Bacteriol.*, **1991**, *173*, pp7491.
23. Daba, H.; Pandian, S.; Gosselin, J.F.; Simard, R.E.; Huang, J.; Lacroix, C. *Appl. Environ. Microbiol.*, **1991**, *57*, pp3450.
24. Henderson, J.T.; Chopko, A.L.; van Wassenaar, P.D. *Arch. Biochem. Biophys.*, **1992**, *295*, pp5.
25. Bhunia, A.K.; Johnson, M.C.; Ray, B. *J. Appl. Bacteriol.*, **1988**, *65*, pp261.

26. Schnell, N.; Entian, K.-D.; Schneider, U.; Gotz, F.; Zahner, H.; Kellner, R.; Jung, G. *Nature*, **1988**, *333*, pp276.
27. Gross, E.; Morell, J.L. *J. Am. Chem. Soc.*, **1971**, *93*, pp4634.
28. Ingram, L. *Biochim. Biophys. Acta*, **1970**, *224*, pp263.
29. Buchman, G.W.; Banerjee, S.; Hansen, J.N. *J. Biol. Chem.*, **1988**, *263*, pp16260.
30. Kaletta, C.; Entian, K.-D. *J. Bacteriol.*, **1989**, *171*, pp1597.
31. Dodd, H.M.; Horn, N.; Gasson, M.J. *J. Gen. Microbiol.*, **1990**, *136*, pp555.
32. Horn, N.; Swindell, S.; Dodd, H.; Gasson, M. *Mol. Gen. Genet.*, **1991**, *228*, pp129.
33. Rauch, P.J.G.; de Vos, W.M. *J. Bacteriol.*, **1992**, *174*, pp1280.
34. Joerger, M.; Klaenhammer, T.R. *J. Bacteriol.*, **1990**, *172*, pp6339.
35. Muriana, P.M.; Klaenhammer, T.R. *Appl. Environ. Microbiol.*, **1991**, *57*, pp114.
36. van Belkum, M.J.; Kok, J.; Venema, G.; Holo, H.; Nes, I.F.; Konings, W.N.; Abee, T. *J. Bacteriol.*, **1991**, *173*, pp7934.
37. Harmon, K.S.; McKay, L.L. *Appl. Environ. Microbiol.*, **1987**, *53*, pp1171.
38. Marugg, J.D. *Food Biotech.*, **1991**, *5*, pp305.
38a. Marugg, J.D.; Gonzalez, C.J.; Kunka, B.S.; Ledeboer, A.M.; Pucci, M.; Toonen, M.Y.; Walker, S.A.; Zoetmulder, L.C.M.; Vandenbergh, P.A. *Appl. Environ. Microbiol.*, **1992**, *58*, pp2360.
39. Berkowitz, B.A.; Bevins, C.L.; Zasloff, M.A. *Biochem. Pharmacol.* **1990**, *39*, pp625.
40. Zasloff, M. *Proc. Natl. Acad. Sci. USA.* **1987**, *84*, pp5449.
41. Soravia, E.; Martini, G.; Zasloff, M. *FEBS Lett.*, **1988**, *228*, pp337.
42. Moore, K.S.; Bevins, C.L.; Brasseur, M.M.; Tomassini, N.; Turner, K.; Eck, H.; Zasloff, M. *J. Biol. Chem.* **1991**, *266*, pp19851.
43. Boman, H.G.; Faye, I.; Gudmundsson, G.H.; Lee, J.-L.; Lidholm, D.-A. *Eur. J. Biochem.* **1991**, *201*, pp23.
44. Kreil, G. *FEBS Lett.* **1973**, *33*, pp241.
45. Casteels, P.; Ampe, C.; Jacobs, F.; Vaeck, M.; Tempst, P. *EMBO J.*, **1989**, *8*, pp2387.
46. Fujiwara, S.; Imai, J.; Fujiwara, M.; Yaeshima, T.; Kawashima, T.; Kobayashi, K. *J. Biol. Chem.*, **1990**, *265*, pp11333.
47. Ganz, T.; Selsted, M.E.; Lehrer, R.I. *Eur. J. Haematol.*, **1990**, *44*, pp1.
48. Hultmark, D.; Steiner, H.; Rasmuson, T.; Boman, H.G. *Eur J. Biochem.* **1980**, *106*, pp7.
49. Steiner, H.; Hultmark, D.; Engstrom, A.; Bennich, H.; Boman, H.G. *Nature*, **1981**, *292*, pp246.
50. Hultmark, D.; Engstrom, A.; Bennich, H.; Kapur, R.; Boman, H.G. *Eur. J. Biochem.* **1982**, *127*, pp207.
51. Barefoot, S.F.; Klaenhammer, T.R. *Appl. Environ. Microbiol.* **1983**, *45*, pp1808.
52. West, C.; Warner, P.J. *FEMS Microbiol. Lett.*, **1988**, *49*, pp163.
53. Mortvedt, C.I.; Nes, I. *Can. J. Microbiol.*, **1990**, *136*, pp1601.
54. Hoover, D.G.; Walsh, P.M.; Kolaetis, K.M.; Daly, M.M. *J. Food Prot.*, **1988**, *51*, pp29.
55. Schillinger, U.; Lucke, F.K. *Appl. Environ. Microbiol.*, **1989**, *55*, pp1901.
56. Piard, J.C.; Desmazeaud, M. *Lait*, **1992**, *72*, pp113.
57. Stevens, K.A.; Sheldon, B.W.; Klapes, A.; Klaenhammer, T.R. *Appl. Environ. Microbiol.*, **1991**, *57*, pp3613.
58. Lee, J.Y.; Boman, A.; Sun, C.; Andersson, M.; Jornvall, H.; Mutt, V.; Boman, H.G. *Proc. Natl. Acad. Sci. USA*, **1989**, *86*, pp9159.
59. Kimbrell, D.A. *BioEssays*, **1991**, *13*, pp657.

60. Casteels, P. *Res. Immunol.*, **1990**, *141*, pp940.
61. Ruhr, E.; Sahl, H.-G. *Antimicrob. Agents Chemother.*, **1985**, *27*, pp841.
62. Kordel, M.; Schuller, F.; Sahl, H.-G. *FEBS Lett.*, **1989**, *244*, pp99.
63. Kordel, M.; Sahl, H.-G. *Microbiol. Lett.*, **1986**, *34*, pp139.
64. Zadjel, J.L.; Ceglowski, P.; Dobrzanski, W.T. *Appl. Environ. Microbiol.*, **1985**, *49*, pp969.
65. Bhunia, A.K.; Johnson, M.C.; Ray, B.; Kalchayanand, N. *J. Appl. Bacteriol.*, **1991**, *70*, pp25.
66. von Heijne, G. *Biochim. Biophys. Acta*, **1988**, *947*, pp307.
67. Westerhoff, H.V.; Juretic, D.; Hendler, R.W.; Zasloff, M. *Proc. Natl. Acad. Sci.*, **1989**, *86*, pp6597.
68. Duclohier, H.; Molle, G.; Spach, G. *Biophys. J.*, **1989**, *56*, pp1017.
69. Cruciani, R.A.; Barker, J.L.; Zasloff, M.; Chen, H.-C.; Colamonici, O. *Proc. Natl. Acad. Sci.*, **1991**, *88*, pp3792.
70. Matsuzaki, K.; Harada, M.; Funakoshi, S.; Fujii, N.; Mihahima, K. *Biochim. Biophys. Acta*, **1991**, *1063*, pp162.
71. Steiner, H.; Andreu, D.; Merrifield, R.B. *Biochim. Biophys. Acta*, **1988**, *939*, pp260.
72. Christensen, B.; Fink, B.; Merrifield, R.B.; Mauzerall, D. *Proc. Natl. Acad. Sci.*, **1988**, *85*, pp5072.
73. Zasloff, M.; Martin, B.; Chen, H.-C. *Proc. Natl. Acad. Sci USA*, **1988**, *85*, pp910.
74. Merrifield, R.B.; Vizioli, L.D.; Boman, H.G. *Biochem.*, **1982**, *21*, pp5020.
75. van Hofsten, P.; Faye, I.; Kockum, K.; Lee, J.-Y.; Xanthopoulos, K.G.; Boman, I.A.; Boman, H.G.; Engstrom, A.; Andreu, D.; Merrifield, R.B. *Proc. Natl. Acad. Sci. USA*, **1985**, *82*, pp2240.
76. Cuervo, J.H.; Rodriguez, B.; Houghten, R.A. *Pept. Res.*, **1988**, *1*, pp81.
77. Chen, H.-C.; Brown, J.H.; Morell, J.L.; Huang, C.M. *FEBS Lett.*, **1988**, *236*, pp462.
78. Juretic, D.; Chen, H.-C.; Brown, J.H.; Morell, J.L.; Hendler, R.W.; Westerhoff, H.V. *FEBS Lett.*, **1989**, *249*, pp219.
79. Andreu, D.; Merrifield, R.B.; Steiner, H.; Boman, H.G. *Proc. Natl. Acad. Sci. USA*, **1983**, *80*, pp6475.
80. Andreu, D.; Merrifield, R.B.; Steiner, H.; Boman, H.G. *Biochem.*, **1985**, *24*, pp1683.
81. Chan, W.C.; Bycroft, B.W.; Lian, L.-Y.; Roberts, G.C.K. *FEBS Lett.*, **1989**, *252*, pp29.
82. Mulders, J.W.M.; Boerrigter, I.J.; Rollema, H.S.; Siezen, R.J.; de Vos, W.M. *Eur. J. Biochem.*, **1991**, *201*, pp581.
83. van de Ven, F.J.M.; van den Hooven, H.W.; Konings, R.N.H.; Hilbers, C.W. *Eur. J. Biochem.*, **1991**, *202*, pp1181.
84. Muriana, P.M. In *Advances in Aseptic Processing Technologies*; Singh, R.K.; Nelson, P.E., Eds.; Elsevier Science Publishers Ltd.: New York, NY, 1992; pp 217-244.
85. FDA. *Code Fed. Reg.*, **1990**, *Title 9*, part 318.7, section b3(ii), pp199.
86. Hutton, M.T.; Chehak, P.A.; Hanlin, J.H. *J. Food Safety*, **1991**, *11*, pp255.
87. Delves-Broughton, J. *Food Technol.*, **1990**, *44*, pp100.
88. FDA. *Fed. Reg.*, **1988**, *53*, pp11247.
89. Okereke, A.; Montville, T.J. *J. Food Prot.*, **1991a**, *54*, pp349.
90. Okereke, A.; Montville, T.J. *Appl. Environ. Microbiol.*, **1991b**, *57*, pp3423.
91. Maisnier-Patin, S.; Deschamps, N.; Tatini, S.R.; Richard, J. *Lait*, **1992**, *72*, pp249.
92. Harris, L.J.; Fleming, H.P.; Klaenhammer, T.R. *J. Food Prot.*, **1991**, *54*, pp836.

93. Pucci, M.J.; Vedamuthu, E.R.; Kunka, B.S.; Vandenbergh, P.A. *Appl. Environ. Microbiol.*, **1988**, *54*, pp2349.
94. Nielsen, J.W.; Dickson, J.S.; Crouse, J.D. *Appl. Environ. Microbiol.*, **1990**, *56*, pp2142.
95. Degnan, A.J.; Luchansky, J.B. *J. Food Prot.*, **1992**, *55*, pp552.
96. Jung, D.-S.; Bodyfelt, F.W.; Daeschel, M.A. *J. Dairy Sci.*, **1992**, *75*, pp387.
97. Berry, E.D.; Liewen, M.B.; Mandigo, R.W.; Hutkins, R.W. *J. Food Prot.*, **1990**, *53*, pp194.
98. Foegeding, P.M.; Thomas, A.B.; Pilkington, D.H.; Klaenhammer, T.R. *Appl. Environ. Microbiol.*, **1992**, *58*, pp884.
99. Yousef, A.E.; Luchansky, J.B.; Degnan, A.J.; Doyle, M.P. *Appl. Environ. Microbiol.*, **1991**, *57*, pp1461.
100. Degnan, A.J.; Yousef, A.E.; Luchansky, J.B. *J. Food Prot.*, **1992**, *55*, pp98.
101. Berry, E.D.; Hutkins, R.W.; Mandigo, R.W. *J. Food Prot.*, **1991**, *54*, pp681.
102. Schillinger, U.; Kaya, M.; Lucke, F.-K. *J. Appl. Bacteriol.*, **1991**, *70*, pp473.
103. Sobrino, O.J.; Rodriguez, J.M.; Moreira, W.L.; Fernandez, M.F.; Sanz, B.; Hernandez, P.E. *Int. J. Food Microbiol.*, **1991**, *13*, pp1.
104. Lewus, C.B.; Kaiser, A.; Montville, T.J. *Appl. Environ. Microbiol.*, **1991**, *57*, pp1683.
105. Muriana, P.M.; Klaenhammer, T.R. *Appl. Environ. Microbiol.*, **1987**, *53*, pp553.

RECEIVED February 9, 1993

Chapter 25

Physiological Strategies To Control Geosmin Synthesis in Channel Catfish Aquaculture Systems

Christopher P. Dionigi

Food Flavor Quality Research, Southern Regional Research Center, Agricultural Research Service, U.S. Department of Agriculture, 1100 Robert E. Lee Boulevard, New Orleans, LA 70124

Several microbial taxa produce $1\alpha,10\beta$-dimethyl-9α-decalol (geosmin) from farnesyl precursors resulting in an intense musty "off-flavor" that negatively impacts the flavor quality of channel catfish (*Ictalurus punctatus*) and other food and water resources. Methods to selectively control geosmin production are not available, and the sporadic occurrence of geosmin hinders identification of potential microbial sources of geosmin. Physiological investigations were conducted to provide information necessary to develop molecular controls of geosmin production. Growth of the geosmin-producing filamentous bacterium *Streptomyces tendae* was inhibited by exogenous farnesol, whereas geosmin did not inhibit growth. Media were identified that both promote and repress geosmin production, and cultures of *S. tendae* grown on geosmin-promoting media were more resistant to farnesol than cultures grown on media that repress geosmin production, suggesting that geosmin production may provide a mechanism to remove farnesyl moieties. This information is being used to search for non-geosmin producing isolates to competitively exclude problematic strains and to provide sequence information required for gene probes to identify potential sources of geosmin.

Impact of Off-flavors

Geosmin ($1\alpha,10\beta$-dimethyl-9α-decalol) and 2-methylisoborneol ((1-R-exo)-1,2,7,7-tetramethyl-bicyclo-[2,2,1]-heptan-2-ol) (MIB) are isoterpenoid (**Figure 1**) derivatives that can impart intense earthy/muddy "off-flavors" to a wide variety of food and water resources (*1 - 4*). As little as 0.01 µg/L of geosmin in water can be detected by olfaction (*4*), yet surface waters have been reported to contain 2 µg/L

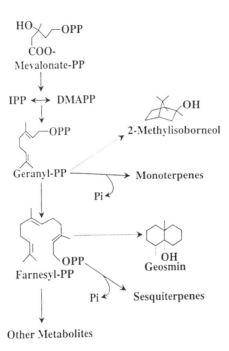

Figure 1. A schematic representation of the terpene pathway showing the possible derivation of the musty/earthy "off-flavor" microbial metabolites 2-methylisoborneol (M*B) and geosmin. Abbreviations: PP, pyrophosphate; IPP, isopentyl pyrophosphate; DMAPP, dimethylally pyrophosphate; Pi, inorganic phosphate (Adapted from ref. *42*).

of geosmin (5). Geosmin may also round-off the flavor of whiskey, red beetroot, and perfumes by adding earthy notes (6). Several types of phototrophic bacteria (algae), heterotrophic bacteria, fungi and other organisms have been reported to produce geosmin in culture (**Table I**). In addition to geosmin and MIB, the dehydration products of MIB, 2-methyleneborane and 2-methyl-2-bornene, have been implicated as potential off-flavor metabolites (46, 47). However, recent reports indicate that 2-methyleneborane and 2-methy-2-bornene obtained from channel catfish (*Ictalurus punctatus*) filets did not exhibit an earthy or musty odor after purification (48).

Economic impacts associated with "off-flavors" have encouraged research into the bioregulation of geosmin and MIB synthesis. Although this chapter focuses primarily upon controlling geosmin production in the context of the channel catfish aquaculture industry, the approaches discussed may have relevance in the control of undesirable metabolites in other systems.

Despite the widespread perception that channel catfish production is a small-scale "cottage" industry, commercial catfish production in the U.S. is a large-scale, technologically sophisticated, capital intensive, and expanding industry. To be economically competitive, producers must maintain highly efficient, relatively large scale operations. Catfish impoundments contain dense populations of fish and receive large amounts of feed (75 kg/ha/day; 49). Channel catfish are fed by broadcasting a pelleted mixture of feed grains, oils, fish meal, vitamins, minerals, and other components onto the water. Most of the commercial channel catfish production in the United States is located in the Southeastern States. This area often experiences warm temperatures and abundant sunlight. These climatic factors coupled with nutrients from feed contribute to the establishment of dense algal and bacterial populations (blooms) on the surface and near the surface of the water (50).

Unfortunately, a wide variety of soil and water microorganisms produce off-flavor metabolites (**Table I**) that negatively impact fish production, processing, marketing, and consumption. Fish can rapidly absorb off-flavor metabolites through their gills and other tissues (51). To avoid consumer rejection and market share losses, producers and processors screen channel catfish for off-flavors and do not harvest and process fish found to be "off-flavor." This rejection, in turn, delays harvesting, restocking, and pond management, and disrupts cash flow and processing schedules. Fish that are harvested but not accepted for processing must be returned to the pond which requires additional handling and transportation.

Fish mortality may be extensive during handling. Fish not harvested continue to require feed, aeration, and other inputs while continuing to grow. This results in fish that are larger than the optimal size for marketing and a further lowering of profits. Additionally, it is not feasible to check every fish for flavor quality prior to processing. Because only a small proportion (one or two fish) of a consignment of fish can be evaluated for flavor quality prior to processing, some off-flavor fish may escape detection. Processors that allow off-flavor catfish to be sold risk the alienation of customers. Conversely, producers may have an entire consignment of catfish rejected because a single or small number of off-flavor fish. In 1987, channel catfish producers in Mississippi reported that about 64 million kg of their fish were off-flavor (52).

Table I. Organisms that have been reported to contain geosmin in culture. Data presented are an update of those presented in reference (7). Organism (*Ref.*)

BLUE-GREEN ALGAE:
Anabaena macrospora (8, 9, 10)
A. cirinalis (11, 12, 13)
A. schieremetiev (7)
A. viguiera (14)

Aphanizomenon sp. (15)

Fischerella muscicola (16)

Lygbya aestuarii (17)

Oscillatoria sp. (18, 19)
O. agardii (17, 20)
O. amphibia (20)
O. amoena (15, 19, 21)
O. brevis (20, 22, 29)
O. cortiana (17)
O. flos-aquae (19, 21)
O. limosa (20)
O. macrospora (19, 29)
O. prolifica (17)
O. simplicissma (9)
O. splendida (15, 19, 20, 21)
O. tenuis (22, 23, 24)
O. variablis (10)

Phomidium autumnale (20, 25)

Pseudoanabaena catenata (20)

Scchizothrix muelleri (26)

Symploca muscorum (23, 27, 28)

FUNGI:
Basidiobolus ranarum (30)
Chaetomium globosum (26, 31)
Pennicilium expasum (32)
P. citrium (33)
P. farenosum (33)

MYXOBACTERIA:
Nannocystis exedens (34)

AMOEBA:
Vannella sp. (35)

ACTINOMYCETES:
Actinomyces biwako (10)

Microbispora resea (28)

Nocardia sp. (28)

Streptomyces spp. (36)
S. albidoflavus (17)
S. alboniger (28)
S. antibioticus (28)
S. chibaensis (19)
S. fradiae (28)
S. fargilis (19)
S. griseoflavus (19)
S. griseoluteus (37, 38)
S. griseus (38, 39, 40)
S. lavendulae (28)
S. odorifer (38, 41)
S. neyegawaensis (19)
S. phaeofaciens (19)
S. prunicolr (19)
S. tendae (42, 43, 44)
S. versipellis (19)
S. virdochromogenes (29)
S. werraensis (19)

LIVERWORT:
Symphogyna brongniartii (45)

Dissipation and Degradation of Off-flavor Metabolites

It has been suggested that microbial cata ɔolism of off-flavor metabolites may be enhanced to provide an effective method ɔ control off-flavors. Indeed, microbial cultures have been reported to degradɐ relatively high levels of off-flavor metabolites (53, 54, 55), but whether microbial activity can be used practically to improve the flavor quality of aquaculture systems remains to be determined.

In addition to biological processes, off-flavor metabolites may be adsorbed onto substrates and lost by volatilization (56). These physical factors must be considered in the evaluation of the effectiveness of control strategies, and may account for some of the reductions in metabolite concentration attributed to bacterial action (56).

Fish can rapidly accumulate off-flavors from water containing even very low concentrations of off-flavor metabolites (46). Therefore, to be totally effective, removal systems must continually biodegrade or dissipate even trace amounts of metabolites from very large volumes of water which reduces the potential effectiveness of this approach. An increased understanding of the flux rate of metabolites within a system will enhance channel catfish production, but research efforts must also focus upon preventing the occurrence of off-flavor metabolite synthesis to be fully effective.

Nonselective Control Agents

If microbes cause off-flavors, it might seem logical that the destruction of microbial populations with chemical control agents (biocides) would prevent the occurrence of off-flavor metabolites. Unfortunately, nonselective biocides can initiate a series of events that may be ultimately more problematic than off-flavors (49, 58, 59).

Although nonselective biocides may reduce off-flavor-producing populations along with other populations (49), widespread destruction of microbial and/or macrophyte biomass can render large amounts of nutrients available for decomposition and a complete disruption of the biological community (59). Increased biological oxygen demand associated with biomass decomposition coupled with a loss of phototrophic populations could rapidly deplete oxygen reserves, resulting in extensive fish mortality. Algal populations are beneficial in that they reduce the penetration of light into the water and thereby reduce macrophyte weed growth. Macrophyte weeds hinder pond maintenance, harvesting, and may harbor geosmin-producing bacteria (54). Microbial populations are also critical to nutrient cycling and contribute to fish vigor by removing nitrogenous wastes products from the water.

Even if fish populations could be maintained by aeration, biocides may not be effective in controlling off-flavors. Most of the geosmin present (ca. 90-99%) is associated with cells rather than the media (16, 43, 60, 12). Biocide induced cell lysis may release off-flavor metabolites from cells (61) and render them available for adsorption by fish (8, 9, 13). Off-flavor metabolites require long periods of time to clear from fish (51). Off-flavor-producing populations may become reestablished before off-flavor metabolite clearance is complete. In addition, the biocides used may be carried off-site or taken up by fish.

Although chemical agents may control certain macrophyte weed populations, their effectiveness for the control of off-flavors is limited. For example, application of a saponified castor oil derivative (Solricin-135), reported to inhibit algal growth, did not reduce off-flavor incidence (*62*). Application of dyes to the water to reduce algal populations by shading may actually foster other problematic algal taxa (*63*). Copper sulfate and other copper containing compounds are widely used algicides (*64*). However, copper sulfate is not completely effective (*59*). Additionally, the use of copper compounds has resulted in the occurrence of copper-tolerant populations of problematic algae (*25*), water quality problems (*59*), and adverse affects on fish quality and health (*65*). While peroxide evolving compounds have been reported to affect off-flavor metabolite-producing populations in channel catfish ponds, their practical effectiveness may be limited (*66*).

Although many terrestrial herbicides may also control phototrophic bacteria (algae; *67 - 70*), several species of non-phototrophic bacteria, such as *Streptomyces*, are prolific producers of off-flavor metabolites (**Table I**). The growth of non-phototrophic populations may be fostered by nutrients released from phototrophic populations exposed to herbicides (*71*), thereby limiting the potential effectiveness of adapting terrestrial herbicides for the control of off-flavor metabolite synthesis.

Selective Chemical Control Agents

The agricultural chemical research community has developed and marketed many agents to selectively control aquatic and terrestrial weeds. However, there are no agents currently available to selectively inhibit off-flavor-producing populations. Typically, control agents are discovered and developed through large scale screening programs preformed on crops and pests that represent major markets. Agents that exhibit phytotoxic properties of interest receive further research attention. The selective properties of an agent are often identified secondarily rather than by a direct search for a specific set of properties.

The process of registration and marketing is extremely costly and can only be justified for markets that represent a significant profit potential. The current aquaculture market is not sufficient to justify an effort to synthesize and develop specific agents to selectively inhibit off-flavor metabolite production.

Recovery of Flavor Quality

One approach to the problem of off-flavor fish may be to allow fish to purge these compounds from their tissue prior to processing (*72*). Purging fish offers producers and processors the advantages of being able to harvest and process fish on schedule while avoiding the marketing of off-flavor fish. However, several technical concerns must be addressed before purging systems can be made fully effective and practical (*72*).

Purging requires either a continuous flow of water through the system or the recirculation of water through filters. The large amounts of "clean" water required by flow-through systems is expensive and may not be widely available (*73*). Recirculating systems use much less water than flow-through systems. However, the filters used to purify the water may become a source of off-flavors (*74*).

Purging systems offer a more limited and manageable environment to attempt the selective control of off-flavors than an open impoundment (*72*). Methods similar to those used to treat off-flavors in potable water sources may be applied to recirculating systems (*39*), and might include comparatively uncomplicated approaches, such as limiting the influx of light to reduce phototrophic populations and flushing the system with fresh water. However, to be fully effective, off-flavor metabolite synthesis must be controlled in purging facilities (*72*).

Geosmin Biosynthesis

The episodic and species specific occurrence of MIB and geosmin suggest that they are secondary metabolites, and that their biosynthesis may be induced by environmental stimuli (for reviews see *75-88*). Unlike the primary metabolites, secondary metabolites are not essential for growth and tend to be strain specific (*89*). Secondary metabolites may exhibit a wide range of chemical structures and biological activities, and are derived by unique biosynthetic pathways. These pathways are often long and complex, and are under the control of specific regulatory genes that control the synthesis of these metabolites so that they are well integrated into the physiology of the organism (*89*). For example, in strains genetically capable of producing geosmin, expression of geosmin synthesis may be associated with specific developmental stages and environmental stimuli (*83, 85*).

The biosynthesis of geosmin is of interest because an enhanced understanding of the pathways and enzymes involved may support the development of effective off-flavor control strategies. Increased biochemical information could also enhance the use of biochemical systems to produce large amounts of geosmin for the flavor and fragrance industries, which may be interested in this compound to provide a desirable earthy note to certain products (*88*). Currently, difficulties associated with the synthesis of the three chiral carbons renders the chemical synthesis of geosmin extremely difficult (*88*).

Although the many details of the biosynthetic pathway of geosmin synthesis are not known (*88*), research with blue-green algae (*90*) and *Streptomyces spp.* (*91*) indicate that geosmin and MIB are derived from the terpene pathway which produces a wide variety of secondary metabolites (**Figure 1**). Research with *Streptomyces spp.* indicates that geosmin may be derived from a sesquiterpene precursor, such as farnesyl pyrophosphate (*91*), and appears to be coregulated with aerial mycelium synthesis genes carried on giant linear plasmids (*92*). Mixed-function oxidase activity may account for hydroxylation (*44*), L-methionine, and folic acid may be involved in the methylation of geosmin precursors (*93*).

The adaptive advantage of geosmin production (if any) is not known. However, secondary metabolite; i.e. geosmin, production requires carbon and energy indicating that geosmin synthesis may confer an adaptive advantage to the organism or geosmin synthesis would not be widely expressed in natural populations (*90*). For example, geosmin synthesis has been associated with nitrogen fixation in blue-green algae (*72*), dissipation of excess carbon associated with photosynthetic pigment production (*22*), and with the coordination of cytodifferentiation among populations (i.e. aerial mycelium and spore formation) (*86*). Cells separated from the media contained much greater concentrations (ca.>99%) of geosmin than the medium (*43*) which agrees with reports indicating that less than 1% of the total geosmin produced by the blue-green algae *Fischerella muscicola* (*16*) and *Oscillatoria tenuis* (*22*) was released from the cell prior to culture senescence. The large proportion of geosmin retained within these cells suggests that geosmin may have a role within the organism in addition to any possible exogenous roles.

Detection of Geosmin-Producing Organisms

Off-flavor episodes may correspond to changes in species composition within a particular aquatic system and/or the induction of metabolite synthesis within existing populations (*93, 94*). Culturability is a significant problem in investigation of the microbial component of environmental samples (*95*), since estimates indicate that only 0.01 -12.5% of the viable bacterial population from marine samples can be cultured (*96*). Thus it is very difficult to determine effectively which taxa or isolates are actually responsible for a particular off-flavor episode. Even if an organism can be successfully cultured, an individual isolate may not produce off-flavors, but other isolates, at other times or under different culture conditions may exhibit off-flavors. Therefore, the taxon the culture represents cannot be definitively ruled out as a potential producer. If the culture is found to contain off-flavors, all that can be determined definitively is that the individual culture has the ability to produce off-flavors in the laboratory. However, the taxon the culture represents may or may not be a significant producer of off-flavors in the environment of interest. An additional problem lies in the length of time required before a particular culture can be grown to a sufficient density to allow an accurate quantification of geosmin or MIB. Because fish take up lipophilic off-flavor metabolites very rapidly, the identification of problematic taxa after a long incubation period would not be timely enough to provide useful information to catfish producers.

This lack of information concerning which organisms are actually responsible for off-flavor episodes, and the need to culture organisms greatly hinders efforts to monitor problematic populations. However, if specific genes and/or gene products can be identified that are co-regulated or directly involved with MIB or geosmin production, samples could be screened rapidly for the presence of the "genetic machinery" associated with off-flavor metabolite synthesis. This would provide a rapid indication of the potential for off-flavors without culturing.

The use of molecular probes to track specific microbes in the environment, specifically those not easily cultured, has been recently reviewed (*95, 97, 98*). The sensitivity of these probes may be further enhanced by using amplification strategies (e.g., polymerase chain reaction or PCR), to amplify segments of DNA from samples obtained from production systems (*95, 99*). However, gene probes for geosmin or MIB synthesis are not currently available.

Induction of Geosmin Biosynthesis

The search for probes for off-flavor metabolite-producing taxa is facilitated by the development of model systems to induce metabolite synthesis. *Streptomyces spp.* have been reported to produce odorous compounds under several environmental conditions (*57*). *Streptomyces tendae* and other actinomycetes are a widely occurring filamentous bacteria that are useful as sources of antibiotics (*30, 100*). The relatively simple culture requirements and high concentrations of geosmin produced by some of these bacteria make them useful model systems for investigations concerning induction off-flavor metabolite synthesis. Sivonen (*101*) qualitatively examined odor production and sugar utilization by several taxa of filamentous bacteria and Weete et al. (*102*) and Yagi et al. (*36*) investigated the effects of chemical and physical parameters on geosmin synthesis by cultures of *Streptomyces* grown in broth. Dionigi et al. (*43*) reported that cultures of *S. tendae* [ATCC 31160] cultures could be grown either Actinomyces (ACT; *103*) or Hickey-Tresner (HT; *104*) media covered with a polycarbonate membrane (0.05 µm pore size). Growing cultures on a membrane allowed quantitative separation of the cells from the medium. Cells and media samples were subjected to gas chromatography with flame ion detection (see *42, 105*) and factorial analysis according to McIntosh (*106*). Although cultures produced biomass on both the HT and ACT media (**Figure 2**), the point of onset of geosmin biosynthesis differed (*43*). Geosmin was not detected by olfaction or gas chromatography in ACT medium-grown cultures before 55 hr after inoculation. However after 55 hours, geosmin was detected in these cultures (**Figure 2**). In contrast, geosmin was detected in HT medium-grown cultures throughout the time course indicating that growth on HT media promoted geosmin synthesis to a greater extent than did growth on ACT medium (**Figure 2**).

The observation that cultures from the same inoculum grown on differing media exhibit a very different level of geosmin synthesis provides an opportunity to search for gene products; e.g., proteins, that are co-regulated with geosmin synthesis or regulate geosmin synthesis. Preliminary results indicate that at least one protein band appeared to be associated with geosmin production (*Dionigi, C. P.; Spanier, A. M. personal communication, 1992*). Further research is required to characterize this protein. Purified protein can be used to develop immunological markers to localize geosmin production at a subcellular level. Sequence information obtained from the protein could be used to construct probes to identify potential geosmin producers without culturing.

Selection of Geosmin Synthesis Mutants

Mutant strains of bacteria that have lost the ability to produce geosmin could be used to exclude problematic populations from environments, such as purging systems, by competition. Additionally, genetic analysis of strains in which the insertion of a transposon (*107*) prevented geosmin synthesis could provide DNA sequence information associated with geosmin synthesis. This information can be used to construct probes to rapidly identify sources of geosmin production from environmental samples without culturing (see *95, 97, 98*). However, a large number

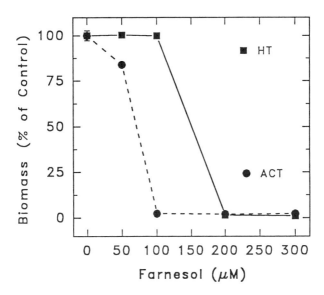

Figure 2. Geosmin production and fresh biomass accumulation by *Streptomyces tendae* cultures on Hickey Tresner and Actinomyces Media. Data are expressed ± 2 SE of a mean (Adapted from ref. *43*).

of transposon mutants would have to be generated (*107*). Gas chromatographic analysis would have to be preformed on each mutant rendering this approach prohibitively time consuming. A prescreening strategy is required to reduce the number of mutants requiring gas chromatographic analysis.

Terpenes, such as geraniol and farnesol, can inhibit the growth of microorganisms (*108-110*), and Dionigi et al. (*42*) hypothesized that geosmin synthesis may provide a mechanism to convert potentially inhibitory farnesyl precursors into a less inhibitory compound, geosmin. Indeed, cultures of *S. tendae* grown on HT medium containing 300 µM farnesol produced 97% less (Pr≥0.0001) biomass than untreated controls (*42*). In contrast, biomass accumulation in cultures exposed to 300 µM geosmin was not affected (Pr≥0.85) compared to untreated controls, indicating that, geosmin was less inhibitory than farnesol (*42*). If geosmin production provides a detoxification mechanism that allows cultures to avoid the accumulation of farnesyl moieties, then cultures that produce relatively greater amounts of geosmin should be more tolerant of farnesol than cultures that produce less geosmin. Cultures grown on ACT medium (a non-geosmin promoting medium) exhibit an increased (Pr≥0.0001) sensitivity to farnesol compared to those grown on HT medium (a geosmin-promoting medium) (**Figure 3**). For example, cultures

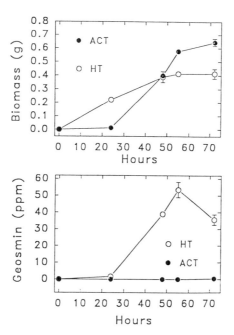

Figure 3. Effects of farnesol on fresh matter accumulation by *Streptomyces tendae* cultures grown on either Hickey-Tresner (HT) or Actinomyces (ACT) medium. Data are expressed ± 1 SE of the mean (Adapted from ref. *42*).

grown on HT medium containing 100 µM farnesol accumulated as much biomass as untreated controls, whereas 100 µM farnesol coincided with a nearly complete inhibition of biomass accumulation on AB medium. However, 200 and 300 µM farnesol produced a nearly complete inhibition of biomass production on both AB and HT medium.

Prescreening a mutant library to indicate which mutants exhibit an increased sensitivity to sublethal doses of farnesol, would greatly reduce the number of samples requiring gas chromatographic analysis. A proportion of the farnesol sensitive mutants identified by prescreening may exhibit increased farnesol sensitivity because they have lost the ability to produce geosmin. Non-geosmin producing strains (e.g., deletion mutatants) inoculated into environments, such as those used to purge off-flavors from fish, may competitively exclude populations that produce off-flavors. If the loss of geosmin production in selected strains is non-adaptive relative to "wild types" in these environments, the establishment of selected strains may be facilitated by a selective substrate readily utilizable by the desired strain (*111*). Genes that preclude off-flavor metabolite production may be contained on plasmids and transferred to other populations, thereby further inhibiting off-flavor metabolite production. More far reaching approaches may include specific inhibition of gene expression (*112*) which may provide the selective inhibition of geosmin or MIB synthesis needed for the continued expansion of the aquaculture industry.

Conclusions

Although the earthy/muddy off-flavors caused by the microbial metabolites geosmin and MIB hinder the production, processing, and marketing of channel catfish, other microbial metabolic activities are critical to the overall stability and productivity of the aquatic ecosystem. Therefore, nonselective control agents that cause a massive destruction of aquatic biomass in an effort to control off-flavors do not represent a viable control option. Until effective and safe agents are available to selectively control off-flavors, the channel catfish industry will have to continue to avoid the production and marketing off-flavor fish. One approach may include the use of systems that allow fish to recover flavor quality prior to processing. However, these systems will have to be kept free of off-flavors. This will require a sensitive means to detect the occurrence of problematic populations, and effective management strategies to deal with these populations once they occur. Genetic probes that are specific for geosmin and/or MIB producers would provide a rapid detection of problematic taxa and could be used to direct the efforts of production and processing facility managers. An increased understanding of the physiology and genetics of geosmin and MIB production will contribute to the development of genetic probes for geosmin and MIB production and strategies to effectively control off-flavor metabolite production.

Acknowledgments

I thank Arthur Spanier, Paul Engel, Peter Johnsen, Beverly Montalbano, Mary Clancy, Karen Bett, Judy Bradow, and Jennifer Celino for their help with portions of this manuscript.

Literature Cited

1. Henatsch, J. J.; Juttner, F. *Fed. Eur. Microbiol. Soc. Microbiol. Lett.* **1986**, *35*, 135-139.
2. Lovell, R. T.; Lelana, I. Y.; Boyd, C. E.; Armstrong, M. S. *Trans. Am. Fish. Soc.* **1986**, *115*, 485-489.
3. Persson, P. E. *Finn. Fish Res.* **1980**, *4*, 1-13.
4. Maga, J. A. *Food Rev. Int.* **1987**, *3*, 269-284.
5. Juttner, F. *Appl. Environ. Microbiol.* **1984**, *47*, 814-820.
6. Berger, R. G.; Drawert, F.; Tiefel, P. *Bioformation of Flavors*; Patterson, R. L. S., Charlwood, B. V.; MacLeod, G.; Williams, A. A., Eds.; Royal Soc. Chem.: Cambridge, UK, 1992; pp. 21-32.
7. Wood, S.; Williams, S. T.; White, W. R. *Int. Biodeterior. Bull.* **1983**, *19*, 83-97.
8. Aoyma, K.; Tomita, B.; Chaya, K.; Saito, M. *J. Hygen. Chem.* **1991**, *37*, 132-136.
9. Izaguirre, G.; Hwang, C. J.; Kranser, S. W.; McGuire, M. J. *Appl. Environ. Microbiol.* **1982**, *43*, 708-714.
10. Miwa, M.; Morizane, K. *Wat. Sci. Tech.* **1988**, *20*, 197-203.
11. Henley, D.E.; Glaze, W.H.; Silvey, J.K.G. *Enivron. Sci. Technol.* **1969**, *3*, 268-271.
12. Rosen, B. H.; Maclead, B. M.; Simpson, M. R. *Wat. Sci. Tech.* **1992**, *25*, 185-190.
13. Bowmer, K. H.; Padovan, A.; Oliver, R. L.; Korth, W.; Gant, G. S. *Wat. Sci. Tech.* **1992**, *25*, 259.
14. Wu, J. T.; Ma, P. I.; Chou, T. L. *Arch. Microbiol.* **1991**, *151*, 66-69.
15. Tsuchiya, Y.; Matsumoto, A.; Shudo, K.; Okamoto, T. *Yakugaku Zashi*, **1980**, *100*, 468-471.
16. Wu, J. T.; Juttner, F. *Arch. Microbiol.* **1998**, *150*, 580-583.
17. Tabecheck, J. L.; Yurkowski, M. *J. Fish. Res. Board Canada.* **1976**, *33*, 25-35.
18. Hayes, K. P.; Burch, M. D. *Wat. Res.* **1989**, *23*, 115-121.
19. Tsuchiya, Y.; Matsumoto, A. *Wat. Sci. Tech.* **1988**, *20*, 149-155.
20. Van Breeman, V. W. C. A.; Dits, J. S.; Ketelaars, H. A. M. *Wat. Sci. Tech.* **1992**, *25*, 233.
21. Matsumto, A.; Tsuchiya, Y. *Wat. Sci. Tech.* **1988**, *20*, 170-183.
22. Naes, H.; Utkilen, H. C.; Post, A. F. *Wat. Sci. Tech.* **1988**, *20*, 125-131.
23. Medsker, L. L; Jenkins D.; Thomas, J. F. *Environ. Sci. Technol.* **1968**, *2*, 461-464.
24. Wu, J. T.; Hsu, Y. M. *Bot. Bull. Academia Sinica*, **1988**, *29*, 183-188.
25. Izaguirre, G. *Wat. Sci. Tech.* **1992**, *25*, 217.
26. Kikuchi, T.; Kadota, S.; Suehara, H.; Nishi, A.; Tsubaki, K.; Yano, H.; Harimaya, K. *Chem. Pharm. Bull.* **1973**, *21*, 2342-2343.
27. Safferman, R. S.; Rosen, A. A.; Mashni, C. I.; Morris, R. E. *Environ. Sci. Technol.* **1967**, *1*, 429-430.
28. Gerber, N. N. *Biotechn. Bioeng.* **1967**, *9*, 321-327.
29. Nakashima, S.; Yagi, M. *Wat. Sci. Tech.* **1992**, *151*, 207-216.

30. Lechevalier, H. A. *Bull. Instuit. Pasteur.* **1974**, *72*, 159-175.
31. Kikuchi, T.; Kadota, S.; Suehara, H.; Nishi, A.; Tsubaki, K. *Chem. Pharm. Bull.* **1981**, *29*, 1782-1784.
32. Mattheis, J. P.; Roberts, R. G. *Appl. Environ. Microbiol.* **1992**, *58*, 3170-3172.
33. Pisarnitsky, A. F.; Egorov, I. A. *Prikl. Biokhim. Microbiol.* **1988**, *24*, 760-764.
34. Trowitzch, W.; Witte, L.; Reichenbeck, H. *FEMS Mircobiol. Lett.* **1981**, *12*, 257-260.
35. Hayes, S. J.; Hayes, K. P.; Robinson. *J. Protozool.* **1991**, *38*, 44-47.
36. Yagi, O.; Sugiura, N.; Sudo, R. *Verh. Int. Verein. Limnol.* **1981**, *21*, 641-645.
37. Rosen, A.; Safferman, R. S.; Mashni, C. I.; Romano, A. H. *Appl. Microbiol.* **1968**, *16*, 178-179.
38. Gerber, N. N. *Wat. Sci. Tech.* **1983**, *15*, 115-125.
39. Whitmore, T. N.; Denny, S. J. *J. Appl. Bacteriol.* **1992**, *72*, 160.
40. Gerber, N. N.; Lechevalier, H. A. *Appl. Environ. Micro.* **1977**, *34*, 857-858.
41. Collins, R. P.; Knaak, L. E.; Soboslai, J. W. *Lloyida*, **1987**, *33*, 199-200.
42. Dionigi, C. P.; Millie, D. F.; Johnsen, P. B. *Appl. Environ. Microbiol.* **1991**, *57*, 3429-3432.
43. Dionigi, C. P.; Millie, D. F.; Spainer, A. M.; Johnsen, P. B. *J. Agric. Food Chem.* **1992**, *40*, 122-125.
44. Dionigi, C. P.; Greene, D. A.; Millie, D. F.; Johnsen, P. B. *Pestic. Biochem. Physiol.* **1990**, *38*, 76-80.
45. Spöle, J.; Becker, H.; Allen, N. S.; Gupta, M. P. *Z. Naturforsch.* **1991**, *46c*, 183-1888.
46. Martin, J. F.; Bennet, L.W.; Graham, W. H. *Wat. Sci. Tech.* **1988**, *20*, 99.
47. Martin, J. F.; Fisher, T. H.; Bennett, L. M. *Agric. Food Chem.* **1988**, *36*, 1257.
48. Mills, O. E.; Chung, S. Y.: Johnsen, P. B. *J. Agric. Food Chem. In Press*
49. Tucker, C. S.; Boyd, C. E. *Channel Catfish Culture*; Elsevier Science: Amsterdam, 1985; pp 135-227.
50. Millie, D. F.; Baker, M. C.; Tucker, C. S.; Vinyard, B. T.; Dionigi, C. P. *J. Phycol.* **1992**, *28*, 281-290.
51. Johnsen, P. B. *J. Ag. Food Chem.* (in press).
52. Harvey, D. *Aquaculture sitation outlook report.* **1991**. U. S. Dept. Ag. AQUA-7.
53. Izaguirre, G; Wolfe, R. L.; Means, E. G. *Appl. Environ. Microbiol.* **1989**, *55* 2424-2431.
54. Narayan, L. V.; Nunnez, W. J. *J. Am. Wat. Works Assoc.* **1974**, *66*, 532-536.
55. Mac Donald, J. C.; Bock, C. A.; Slater, G. P. *Appl. Microbiol. Biotechnol.* **1987**, *25*, 392-395.
56. Pirbazari, M.; Borow, H. S.; Craig, S.; Ravindran, V.; Mc Guire, M. *J. Wat. Sci. Tech.* **1992**, *25*, 81.
57. Cross, T. *J. Appl. Bacteriol.* **1981**, *50*, 397-423.
58. Tucker, C. S.; Van der Ploeg, M. *For Fish Farmers.* **1991**, 2-5.
59. Safferman, R. S.; Morris, M. E. *J. Am. Wat. Works Assoc.* **1964**, *56*, 1217-1224.

60. Utkilen, H. C.; Frøshaug, M. *Wat. Sci. Tech.* **1992**, *25*, 199.
61. Ando, A.; Miwa, M.; Kajino, M.; Tatsumi, S. *Wat. Sci. Tech*, **1992**, *25*, 299-306.
62. Tucker, C. S.; Lloyd, S. W. *Aquaculture.* **1987**, *25*, 217.
63. Martin, J. F. *Wat. Sci. Tech.* **1992**, *25*, 315.
64. Ranman, R. K. *J. Am. Wat. Works Assoc.* **1985**, *77*, 41-43.
65. Ansari, I. A. *Geobios*, **1984**, *11*, 188-190.
66. Martin, J. F. *Wat. Sci. Tech.* **1992**, *25*, 315-321.
67. Hawxby, K.; Tubea, B.; Ownby, J.; Basler, E. *Pestic. Biochem. Physiol.***1977**, *7*, 203-209.
68. Maule, A.; Wright, S. J. L. *J. Appl. Bacteriol.*, **1984**, *57*, 369-379.
69. Richardson, J. T.; Frans, R. E.; Talbert, R. E. *Weed Sci.* **1979**, *27*, 619-624.
70. Tucker, C. S.; Busch, R.L.; Lloyd, S. W. *J. Aquat. Plant Manage.* **1983**, *21*, 7-11.
71. Blevins, W. T. *Introduction to Environmental Toxicology*; Shapiro, F. E.; Perry, J. J., Eds.; Elsevier: New York, 1980; pp 350-357.
72. Jonhsen, P. B.; Dionigi, C. P. *J. Appl. Aquacult.* (in press).
73. Pote, J. W.; Wax, C. L.; Tucker, C. S. *Mississippi Ag. Forest Exp. Sta. Special Bulletin.* **1988**, 88-3.
74. Burman, N. P. *Soc. Appl. Bacteriol.* Symp. series 2, **1973**, 219.
75. Bach, T. J. *Plant Physiol. Biochem.* **1987**, *25*, 163-178.
76. Gershenzon, J.; Croteau, R. In *Recent Advances in Phytochemistry*, Biochemistry of the Mevalonic Acid Pathway to Terpenoids; Towers, G. H. N.; Stafford, H. A., Eds.; Plenum Press: New York, 1990. pp.99-159.
77. Barz W.; Koster, J. In *The Biochemistry of Plants*; Academic Press: New York, 1981; pp 35-83.
78. Gray, J. C. In *Advances in Botanical Research.* Callow, J. A., Ed.; Academic Press, New York, 1987, 14; pp. 25-91.
79. Banthorpe, D. V.; Charlwood, B. V.; In *Secondary Plant Products*; Bell, A.; Charlwood, B. V., Eds.; Encyclopedia of Plant Physiology; Spranger-Verlag: Berlin, 1980, Vol. 8; pp 185-220.
80. Wink, M. In *Secondary Products from Plant Tissue Culture*; Charlwood, B. V.; Rhodes, M.J.C., Eds.; Clarendon Press: Oxford, 1990; pp 23-41.
81. Luckner, M.; Nover, L. Expression of secondary metabolism. In *Secondary Metabolism and Cell Differentiation*; Luckner, M.; Nover, M. L.; Bohm H. Eds.; Springer-Verlag: Berlin. 1977; pp 1.
82. Grafe, U. In *Regulation of Secondary Metabolism in Actinomycete*; Shapiro, S., Ed.; CRC Press: Boca Raton, FL, 1989; pp 75-135.
83. Skulberg, O. M., *Wat. Sci. Tech.* **1988**, *20*, 167-178.
84. Schreier, P. In *Bioformations of Flavors*; Patterson R. L. S.; Charlwood, B. V.; Macleod, G.; Williams, A. A. Eds.; Royal Soc. Chem. Press: Oxford, UK, 1992; pp 1-20.
85. Berger, R. G., Drawwert, F.; Tiefel, P. *In Bioformations of Flavors*; Patterson R. L. S.; Charlwood, B. V.; Macleod, G.; Williams, A. A., Eds.; Royal Soc. Chem. Press: Oxford, UK, 1992; pp 21-32.

86. Croteau, R. B. Chem. Rev. **1987**, *87*, 929-954.
87. Croteau, R. B.; Cane, D. E. In *Steroids and Isoprenoids*; Law, J. H.; Rilling, H. C., Eds.; Methods of Enzymology; Academic Press: New York, NY, 1985, Vol. 110, pp. 383-405.
88. Croteau, R. B. In *Flavor Precursors*; Teranishi, R.; Takeoka, G. R.; Guntert, M., Eds.; Monoterpene Biosynthesis: cyclization of geranyl pyrophosphate to (+)-sabinene; Amercian Chemical Society Symposium Seriers 490: Washington, DC, 1991, pp. 8-20.
89. Vining, L. C. *Gene.* **1992**, *115*, 135-140.
90. Naes, H.; Utliken, H. C.; Post, A. F. *Arch Micro.* **1989**, *151*, 407-410.
91. Bentley, R.; Meganathan, R. *FEBS Lett.* **1982**, *125*, 220-222.
92. Ishibashi, Y. *Wat. Sci. Tech.* **1992**, *25*, 171-176.
93. Aoyama, K. *J. Appl. Bacteriol.* **1990**, *68*, 405-410.
94. Wu, J. T; Juttner, F. *Wat. Sci. Technol.* **1988**, *20*, 143-148.
95. Puckup, R. W. *J. Gen.Microbiol.* **1991**, *137*, 1009-1019.
96. Ferguson, R. L.; Buckley, E. N.; Palumbo, A. V. *Appl. Environ, Microbiol.* **1984**, *47*, 49-55.
97. Sayler, G. S.; Layton, A. C. *Annu. Rev. Microbiol.* **1990**, *44*, 626-648.
98. Atlas, R. M.; Sayler, G. S.; Burlage, R. S.; Asim, K. B. *Biotechniques.* **1992**, *12*, 708-717.
99. Steffan, R. J.; Atlas, R. M. *Annu. Rev. Microbiol.* **1991**, *25*, 137-1661.
100. Silvey, J. K. G.; Roach, A. W. In *Reviews in Enviromental Control.* CRC Press, Boca Raton, Fla. 1975, pp 233-273.
101. Sivonen, K. *Hydrobiologia.* **1982**, *86*, 165-170.
102. Weete, J. D.; Blevins, W. T.; Wilt, G. R.; Durham, D. *Ag. Exper. Sta. Bull.* Auburn, Al. **1977**, 490, 46pp.
103. Ajello, L.; Georg, L. K.; Kaplan, W.; Kaufman, L. *Laboratory manual for medical mycology*; U. S. Dep. of Health, Education and Welfare, Public Health Service: Atlanta GA, 1963. Publication No. 994.
104. Hickey, R. T.; Tresner, H. D. *J. Bacteriol.* **1952**, *64*, 981-983.
105. Johnsen, P. B.; Kuan, J. W. *J. Chromatography.* **1987**, *409*, 337-342.
106. McIntosh, M. S. *Agron. J.* **1983**, *75*, 153-155.
107. Engle, P. *Appl. Environ. Microbiol.* **1987**, *53*, 1-3.
108. Bard, M.; Albrecht, M. R.; Gupta, N.; Guynn, C. J; Stillwell, W. *Lipids.* **1988**, *23*, 534-538.
109. Knoblock, K.; Pauli, A.; Iberl, B.; Weis, N.; Weigand, H. In *Bioflavor' 87*; Schreier, P., Ed.; Walter de Gruyter: Berlin. 1988, pp. 288-299.
110. Knoblock, K.; Pauli, A.; Iberl, B. *J. Ess. Oil Res.* **1989**, *1*, 119-128.
111. Lajoie, C. A.; Chen, S. -Y.; Oh, K. -C.; Strom, P. F. *Appl. Environ. Microbiol.* **1992**, *58*, 655-663.
112. Goodchild, J. In *Oligodeoxynucleotides Antisense Inhibitors of Gene Expression*; Cohen, J. S., Ed.; Topics in Molecular and Structural Biology; CRC Press: Boca Raton, Florida, 1989; pp 53-78.

RECEIVED October 28, 1992

INDEXES

Author Index

Affiliation Index

Subject Index

Production: Meg Marshall
Indexing: Deborah H. Steiner
Acquisition: Barbara C. Tansill
Cover design: A. M. Spanier

Printed and bound by Maple Press, York, PA

Highlights from ACS Books

Good Laboratory Practice Standards: Applications for Field and Laboratory Studies
Edited by Willa Y. Garner, Maureen S. Barge, and James P. Ussary
ACS Professional Reference Book; 572 pp; clothbound ISBN 0–8412–2192–8

Silent Spring Revisited
Edited by Gino J. Marco, Robert M. Hollingworth, and William Durham
214 pp; clothbound ISBN 0–8412–0980–4; paperback ISBN 0–8412–0981–2

The Microkinetics of Heterogeneous Catalysis
By James A. Dumesic, Dale F. Rudd, Luis M. Aparicio, James E. Rekoske,
and Andrés A. Treviño
ACS Professional Reference Book; 316 pp; clothbound ISBN 0–8412–2214–2

Helping Your Child Learn Science
By Nancy Paulu with Margery Martin; Illustrated by Margaret Scott
58 pp; paperback ISBN 0–8412–2626–1

Handbook of Chemical Property Estimation Methods
By Warren J. Lyman, William F. Reehl, and David H. Rosenblatt
960 pp; clothbound ISBN 0–8412–1761–0

Understanding Chemical Patents: A Guide for the Inventor
By John T. Maynard and Howard M. Peters
184 pp; clothbound ISBN 0–8412–1997–4; paperback ISBN 0–8412–1998–2

Spectroscopy of Polymers
By Jack L. Koenig
ACS Professional Reference Book; 328 pp;
clothbound ISBN 0–8412–1904–4; paperback ISBN 0–8412–1924–9

Harnessing Biotechnology for the 21st Century
Edited by Michael R. Ladisch and Arindam Bose
Conference Proceedings Series; 612 pp;
clothbound ISBN 0–8412–2477–3

From Caveman to Chemist: Circumstances and Achievements
By Hugh W. Salzberg
300 pp; clothbound ISBN 0–8412–1786–6; paperback ISBN 0–8412–1787–4

The Green Flame: Surviving Government Secrecy
By Andrew Dequasie
300 pp; clothbound ISBN 0–8412–1857–9

For further information and a free catalog of ACS books, contact:
American Chemical Society
Distribution Office, Department 225
1155 16th Street, NW, Washington, DC 20036
Telephone 800–227–5558